Animal Cell Biology

Animal Cell Biology

Edited by **Ralph Becker**

SYRAWOOD
PUBLISHING HOUSE

New York

Published by Syrawood Publishing House,
750 Third Avenue, 9ᵗʰ Floor,
New York, NY 10017, USA
www.syrawoodpublishinghouse.com

Animal Cell Biology
Edited by Ralph Becker

International Standard Book Number: 978-1-68286-144-8 (Hardback)

Printed in the United States of America.

Contents

Preface

Over the recent decade, advancements and applications have progressed exponentially. This has led to the increased interest in this field and projects are being conducted to enhance knowledge. The main objective of this book is to present some of the critical challenges and provide insights into possible solutions. This book will answer the varied questions that arise in the field and also provide an increased scope for furthering studies.

The researches that fall under animal cell biology integrate diverse biological fields such as molecular biology, development biology, etc. This book strives to provide a fair idea about this discipline and to help develop a better understanding of the recent advances within this field with the help of concepts related to molecular genetics, immunology, cellular metabolism, etc. A number of latest researches by experts from around the globe have been included in this book. It aims to further the scope of research in this field and contribute to its progress.

I hope that this book, with its visionary approach, will be a valuable addition and will promote interest among readers. Each of the authors has provided their extraordinary competence in their specific fields by providing different perspectives as they come from diverse nations and regions. I thank them for their contributions.

Editor

Effect of nutrient supplementation on biodegradation and metal uptake by three bacteria in crude oil impacted fresh and brackish waters of the Niger Delta

L. O. Odokuma[1]* and E. Akponah[2]

[1]University of Port-Harcourt, Port-Harcourt, Rivers State, Nigeria.
[2]Delta State University, Abraka, Delta State, Nigeria.

The effect of nutrients supplements (NPK inorganic fertilizer and poultry litter) on the biodegradation of crude oil and metal uptake by three bacterial isolates (*Bacillus sp, Pseudomonas sp* and *Aeromonas sp*) in crude oil impacted fresh and brackish water aquatic systems of the Niger Delta were investigated. The abilities of these single cultures to reduce the contaminating oil levels as well as to bioconcentrate-associated heavy metals (Fe, Pb, Cd, Zn, Ni and Cu) were carried out for a period of 120 days. The effect of the nutrient supplements were monitored and enhanced by periodic re-addition to desired levels. Habitat water samples without nutrient supplements served as controls. Nutrient supplementation resulted in both increased pH as well as increase in biomass. In microcosms containing *Bacillus sp* and *Pseudomonas sp*, total viable cell counts increased while hydrocarbon, pH and the inorganic nutrients decreased. Significant reductions (64 and 82%) in oil and grease levels were obtained in fresh and brackish water respectively that received NPK. for *Bacillus* and *Pseudomonas*. Application of poultry litter resulted in 59.7 and 78.7% reductions in fresh and brackish water options respectively for both organisms. However, no significant changes occurred in microcosms containing *Aeromonas sp* except that counts declined considerably with time. *Bacillus* and *Pseudomonas* proved suitable for the uptake of the heavy metals present *Aeromonas* was unsuitable though it was resistant to the heavy metals. Bioconcentration of heavy metals followed the pattern: *Pseudomonas* > *Bacillus*. The pattern of heavy metal uptake by the three bacterial isolates was different in the various treatment options. The uptake of these metals was enhanced by increases in microbial biomass of the isolates. Peak uptake of metals occurred in the exponential phase for *Bacillus* while for *Pseudomonas* peak uptake of metals corresponded with the stationary phase. No significant difference in the amounts of heavy metals bioconcentrated when NPK fertilizer was used as supplement than when poultry litter was employed. The initial concentration of metals, pH of the medium as well as the cultural status of the isolates influenced metal uptake. Results indicated that addition of NPK inorganic fertilizer or poultry litter promoted both biodegradation of crude oil by *Bacillus* and *Pseudomonas* and heavy metal uptake by *Bacillus* and *Pseudomonas* in fresh and brackish water aquatic systems of the Niger Delta.

Key words: Nutrient supplements, bioconcentration, microcosms, heavy metals.

INTRODUCTION

Rehabilitation of polluted environments in the Niger Delta has been a major challenge. This has led to the development of a wide range of clean-up techniques including physical, chemical and biological techniques (Atlas, 1981; Odokuma and Ibor, 2002; Odokuma and Dickson, 2003a, b). Of the three methods biological methods are regarded as the best because it is environmentally friendly and less expensive though it is time consuming and high concentrations of the pollutant reduces the population of degrading and microorganisms and slows down the process. Biological methods (Bioremediation) exploit the diverse degradation abilities

*Corresponding author. E-mail: luckyodokuma@yahoo.co.in.

of microorganisms to convert the complex chemical components of crude oil to harmless products by mineralization (Alexander, 1981; Atlas, 1981). Bioremediation makes use of three processes, biodegradation, bioaccumulation and biosorption. Crude oil biodegradation is a slow but natural process limited mainly by scarcity of nitrogen and phosphorus in the environment (Ladousse and Tramier, 1991; Odokuma and Ibor, 1992). Studies have shown that crude oil biodegradation can be accelerated by the addition of nitrogen and phosphorus- containing fertilizers (whether organic or inorganic) in aqueous, terrestrial or sediment environments (Stevens, 1991; Ladousse and Tramier, 1991; Odokuma and Ibor, 1992).

There is a continuous influx of heavy metals into the biosphere from both natural and anthropogenic sources (Perelomov and Prinsky, 2003). Crude oil is also a source of these metals. In aquatic terrestrial ecosystems their bioavailability depends on their chemical forms. These heavy metals are known to cause severe damage to aquatic and terrestrial life (Odiete, 1999; Odokuma and Emedolu, 2005). A specific problem associated with metals in the environment is their accumulation in the food chain and persistence in the environment (Malekzadeh et al., 1996).

Microorganism's uptake metal either actively (bioaccumulation) and/or passively, (biosorption) (Foureset and Roux, 1992; Shummate and Strandberg, 1995). Thus the use of microorganisms for the recovery of metals from waste streams has received growing attention.

In this study the effect of nutrient (NPK fertilizer and organic poultry litter) supplementation on bioaccumulation of heavy metals and biodegradation of Bonny light crude oil in a brackish water aquatic system of the New Calabar River water of the Niger Delta was examined. The objectives were to examine; the effect of both organic and inorganic fertilizers on both metal uptake and biodegradation. Two nutrient supplements- NPK (Nitrogen, Phosphorus and Potassium) (15, 15, 15) fertilizer and organic poultry litter were used in the study. The test conditions were used to assess:

1.) The effect of NPK fertilizer in enhancing biodegradation of crude oil and bioconcentration of it's associated heavy metals.
2.) The effect of the organic poultry litter in enhancing biodegradation of crude oil and bioconcentration of its associated heavy metals.
3.) The effect of periodic replenishment of the above nutrients in enhancing biodegradation of crude oil as well as the bioconcentration of it's associated heavy metals.

Three bacterial (Bacillus, Pseudomonas and Aeromonas) isolates from the river water were obtained for metal uptake (bioaccumulation) studies. The test organisms, Bacillus, Pseudomonas and Aeromonas were chosen after screening for their resistance to heavy metals found in Bonny light crude oil from a population of nine bacterial isolates which were the most predominant indigenous species in the New Calabar River water. Also supporting their use as test organisms for bioaccumulation tests were the results from related studies confirming their resistance to these heavy metals (Odokuma and Abah, 2003; Odokuma and Ijeomah, 2004; Odokuma and Emelodu, 2005).

MATERIALS AND METHODS

Preparation of stock solution of heavy metal salts

The heavy metal salts employed in this study include: Nickel tetraoxosulphate (vi) salt ($NiSO_4$), copper (ii) tetraoxosulphate (vi) salt ($CuSO_4$), lead trioxonitrate (v) salt ($PbNO_3)_2$, iron (ii) tetraoxosulphate (vi) salt ($FeSO_4$), cadmium tetraoxosulphate (vi) salt ($CdSO_4$) and zinc tetraoxosulphate (vi) $ZnSO_4$ salt. A weight of each of these heavy metal salts that gave a corresponding 1g of each of the respective heavy metal was weighed and dissolved in 1000 ml of deionised water. These were left to stand for 30mins to obtain complete dissolution. This was followed by sterilization by membrane filtration (0.2 μm pore size Aerodisc).

Isolation of heavy metal resistant bacteria from the river water

The test organisms, Bacillus sp, Pseudomonas sp and Aeromonas sp were isolated from the brackish water sample collected from the New Calabar River, Port-Harcourt, Rivers State. The spread plate technique (APHA, 1998) using nutrient agar (Oxoid) was employed for their isolation. The plates were incubated at 37°C for 18 - 24 h. Pure bacterial isolates were characterized and identified using various criteria as described by Krieg and Holt (1994). Pure isolates were transferred into nutrient agar slants stored at 4°C and served as the stock cultures for subsequent tests.

Nine predominant bacterial genera; Achromobacter, Alcaligenes, Aeromonas, Bacillus, Chromobacterium, Corynebacterium, Micrococcus, Pseudomonas and Serratia were identified.

Preparation of standard inoculum of isolates

A loopful of cells from the respective stock cultures were incubated into 100ml sterile nutrient broth contained in 250 ml Erlenmeyer flasks. The flasks were incubated at 37°C for 24 h with intermittent shaking. At the end of the incubation period, cells were harvested by centrifugation at 4000 rpm for 30 min and re-suspended in 100ml sterile physiological saline. The total viable counts were carried out to estimate the number of viable organisms. During this process, the cultures were subjected to serial dilutions up to 10^6 dilutions. An amount (0.1ml) from each dilution was inoculated by spread plate technique into freshly prepared nutrient agar plates, which were incubated at 37°C for 24 h. The dilutions that produced between 30 - 300 colonies were chosen and served as inoculum for preliminary screening experiments.

Preliminary screening test

This was carried out to determine the isolates that possess resistance to some of the heavy metals associated with the crude oil. One hundred millilitres of 1mg/l of the respective heavy metal solutions were prepared as earlier described. Nine millilitres were dispensed into test tubes and sterilized. Controls contained 9 mls of physiological saline. One millilitre of respective standardized isolates' inoculum was then added and incubation followed immediately at a temperature of 25°C±2 for duration of 24 h. At the end of the incubation period, 0.1ml were withdrawn and plated onto the surface of freshly prepared nutrient agar plates using the spread plate technique as described by APHA (1998). Incubation followed immediately at 25°C±2 for 18-24 h. Colonies formed were counted and percent log survival were calculated according to Williamson and Johnson (1981).

$$\% \log survival = \frac{\log of\ count\ in\ toxicant\ concentraton}{Log of\ count\ in\ control} \times 100$$

Based on the results, *Pseudomonas, Bacillus* and *Aeromonas* were chosen for further studies.

Production of inoculum for biodegradation experiments

A loopful of each stock culture of *Bacillus sp, Pseudomonas sp* and *Aeromonas sp* were respectively inoculated into 100 ml of freshly prepared sterile nutrient broth contained in 250 ml Erlenmeyer flask. Incubation followed at room temperature for 18 - 24 h. This was then transferred aseptically into 1000 ml of sterile nutrient broth amended with 0.5% crude oil and incubated at room temperature for 7 days. The cultures were centrifuged at 4000 rpm for 30 min using 800 Dmodel centrifuge. The cells were washed thrice in sterile physiological saline. The washed cells were suspended in 750 ml of physiological saline contained in 1L flask. The total viable counts of the cell suspension were 1.81×10^5, 7.9×10^4 and 1.51×10^5 for *Bacillus sp, Pseudomonas sp* and *Aeromonas sp* respectively. This gave the inocula introduced into the various treatment options.

Water sample source and collection

Composite surface water (0 - 15 cm) samples were collected and pooled together in a clean pre-sterilized 20 L container from Abonnema Wharf (brackish water) and Omuhuechi River (fresh water) both in Port Harcourt, Rivers State, Nigeria. Abonnema Wharf is associated with anthropogenic/xenobiotic pollution while Omuhuechi River is subjected to anthropogenic discharges, soil erosion, surface run-off and other human activities. These two types of aquatic habitats are characteristic of crude oil producing areas of the Niger Delta. Physiochemical analyses of both types of water samples were determined prior to crude oil contamination as shown in Table 1.

Crude oil

Bonny light crude oil (45° API) obtained from Nigerian Agip Oil Company (NAOC) Port Harcourt was used in this experiment.

Nutrient supplements

Two nutrient supplements- NPK (Nitrogen, Phosphorus and Potassium) (15, 15, 15) fertilizer and organic poultry litter were used in the study. The NPK 15, 15, 15 fertilizer was obtained from the Rivers State Ministry of Agriculture Port Harcourt Rivers State, Nigeria.

The organic nutrient supplement (poultry litter) was obtained from a poultry farm cited in University of Port-Harcourt, Port Harcourt Rivers State Nigeria.

The bedding material in the poultry farm was composed of shavings and saw dust. Litter was collected from top 5 cm following the standard stratified sampling procedure (Peterson and Levine, 1986). The litter was sieved using a mesh of approximately 1mm size and 10 g of filtrate were added to 500 ml-deionized water. This was mixed using a blender and stirred at 150 rpm for 1 h. The poultry litter mixture was centrifuged at 4000 rpm for 10 min and the supernatant withdrawn using a pipette.

Amounts (5 ml) of the supernatant were added to appropriate assay flask at day 0. At day 35, 5 ml of the poultry litter was re-introduced. On analyses the poultry litter contained 30.1 mg/g and 2.02 mg/g of Nitrogen and Phosphorus respectively.

The fertilizer (0.1 g) was added to the appropriate flasks. The concentration was arrived at after performing a preliminary toxicity test using NPK as toxicant. Also at day 35, 0.1 g of the fertilizer was re-introduced into the appropriate flasks.

Biodegradation / bioconcentration tests

Erlenmeyer flasks (250 ml) were employed in this experiment. The assay vessels comprised 198 flasks and on each analysis day, the entire content of the desired flask was completely utilized for the analysis. The respective water samples were dispensed in 89 ml amounts into appropriately labelled flasks and sterilized by autoclaving at 121°C for 10 - 15 min. On cooling, 1ml of crude oil and 10 ml of appropriate test organism (for *Pseudomonas sp* it contained 7.9×10^4 cfu/ml, 1.81×10^5 cfu/ml of *Bacillus sp* and 1.51×10^5 cfu/ml of *Aeromonas sp*) were introduced. However, there were some exceptions in that microcosms that received poultry litter supplement consisted of 84 ml of appropriate water sample and 5 ml poultry litter. Detailed descriptions of the treatment option are presented in Table 1.

The flasks were attached to a shaker operated at a speed of 150rpm and a temperature of 25°C ± 2. Changes in bacterial growth, pH, phosphate, nitrate, and residual hydrocarbon were monitored periodically on Days 0, 7, 14, 21, 28, 35, 42, 49, 63, 90 and 120. On the above day 1 ml of sample was withdrawn aseptically from desired flask for enumeration of total viable bacterial count while the remaining content were centrifuged at a speed of 4000 rpm for 30 min. The sediment was washed thrice in phosphate buffered saline and digested for heavy metal analysis.

The test tubes used for centrifugation were flushed thoroughly with xylene to remove all traces of crude oil that adhered during centrifugation. This was then poured back into the supernatant. The various physiochemical analyses were done using the supernatant

Physicochemical analyses

The pH was determined using the Jenway pH meter (3015 model). The ascorbic acid method as described in APHA (1998) was employed in the determination of available phosphorus while the nitrate content of samples was determined using the Brucine method (APHA, 1998). Total hydrocarbons (oil and grease) levels were determined by employing the photometric method (APHA, 1998).

Bacteriological analyses

Viable bacteria present in each flask were enumerated on nutrient agar (oxoid) using the spread plate technique (APHA, 1998).

Digestion and bacterial biomass

The wet oxidation method (APHA, 1998) was adopted. Harvested

Table 1. Treatment options of crude oil impacted freshwater and marine water samples.

S/N	Treatment option	Description
1	CFW/PS	Crude oil + freshwater sample + *Pseudomonas*
2	CFW/PL/PS	Crude oil + freshwater + poultry litter + *Pseudomonas*
3	CFW/NPK/PS	oil + freshwater + NPK + *Pseudomonas*
4	CFW/Ba	Crude oil + freshwater + *Bacillus*
5	CFW/PL/Ba	Crude oil + freshwater + Poultry litter + *Bacillus*
6	CFW/NPK/Ba	Crude oil + freshwater + NPK + *Bacillus*
7	CFW/Ae	Crude oil + freshwater + *Aeromonas*
8	CFW/PL/Ae	Crude oil + freshwater + Poultry litter + *Aeromonas*
9	CFW/NPK/Ae	Crude oil + freshwater + NPK + *Aeromonas*
10	CSW/Ps	Crude oil + brackish water + *Pseudomonas*
11	CSW/PL/PS	Crude oil + brackish water + Poultry litter + *Pseudomonas*
12	CSW/NPK/Ps	Crude oil + brackish water + *Pseudomonas*
13	CSW/Ba	Crude oil + brackish water + *Bacillus*
14	CSW/PL/Ba	Crude oil + brackish water + Poultry + *Bacillus*
15	CSW/NPK/Ba	Crude oil + brackish water + NPK + *Bacillus*
16	CSW/Ae	Crude oil + brackish water + *Aeromonas*
17	CSW/PL/Ae	Crude oil + brackish water + Poultry litter + *Aeromonas*
18	CSW/NPK/Ae	Crude oil + brackish water + NPK + *Aeromonas*
19	Control (Ci	Crude oil + Freshwater
20	Control (Cii)	Crude oil + brackish water

Table 2. Heavy metals present in the crude oil impacted water samples.

Heavy metal	Concentration (mgl^{-1}) in fresh water	Concentration (mgl^{-1}) in brackish water
Iron	1.981	2.006
Lead	1.269	1.68
Cadmium	0.603	1.405
Zinc	0.692	1.724
Nickel	0.38	1.45
Copper	0.502	0.912

and washed bacterial biomass was mixed with 1 ml of trioxonitrate (v) acid, perchloric acid and tetraoxosulphate (VI) acid. The mixture was heated for 10 - 15 min. This caused lysis of the cells releasing the cytoplasmic content.

Heavy metal analyses

The crude oil samples obtained were digested using the wet oxidation method and the heavy metals present were determined using the atomic absorption spectrophotometer (UNICAM 929 AA Spectrometer). Also on each analysis day (0, 7, 14, 21, 28, 35, 42, 49, 63, 90 and 120), the harvested, washed and digested bacterial biomass were analysed for the accumulation of the heavy metals present in the crude oil sample.

Statistical analyses of data

The two-way analysis of variance (ANOVA) and correlation analysis were employed (Finney, 1978).

RESULTS

Results of the analysis of heavy metals present in the crude oil impacted water samples are given in Table 2. The isolate obtained from the river water sample were *Achromobacter*, *Alkaligens*, *Aeromonas*, *Bacillus*, *Corynebacterium*, *Chromobacterium*, *Micrococus*, *Pseudomonas* and *Serratia*. The results of the preliminary screening for the resistance of isolates to the toxicity of the various heavy metals are presented in Table 3. Three of the test isolates (*Aeromonas*, *Bacillus* and *Pseudomonas*) showed resistance to the six heavy metal salts.

Figure 1a shows the growth profiles of all three organisms in fresh and brackish-water treatment options. Lower counts were obtained in flasks without nutrient supplements. In fertilized options, there was a general increase in the counts from Day 0 to Day 14 followed by

Table 3. Response of Isolates to the toxicity of the various heavy metals.

Isolates	Fe	Zn	Cd	Cu	Ni	Pb	Control
Alcaligenes	++	++	-	-	-	-	+++
Aeromonas	+++	+++	++	+++	++	++	+++
Bacillus	+++	+++	+++	+++	+++	+++	+++
Achromobacter	+++	++	-	-	+	-	+++
Chromobacterium	++	+	-	+	-	-	+++
Corynebacterium	++	++	-	+	+	-	+++
Micrococcus	+++	++	+	++	+	-	+++
Pseudomonas	+++	+++	+++	+++	++	++	+++
Serratia	+++	+	-	+	+	-	+++

Key:
+++ = > 70% log survival
++ = 50-69% log survival
+ = 30-49% log survival
- = < 29% log survival

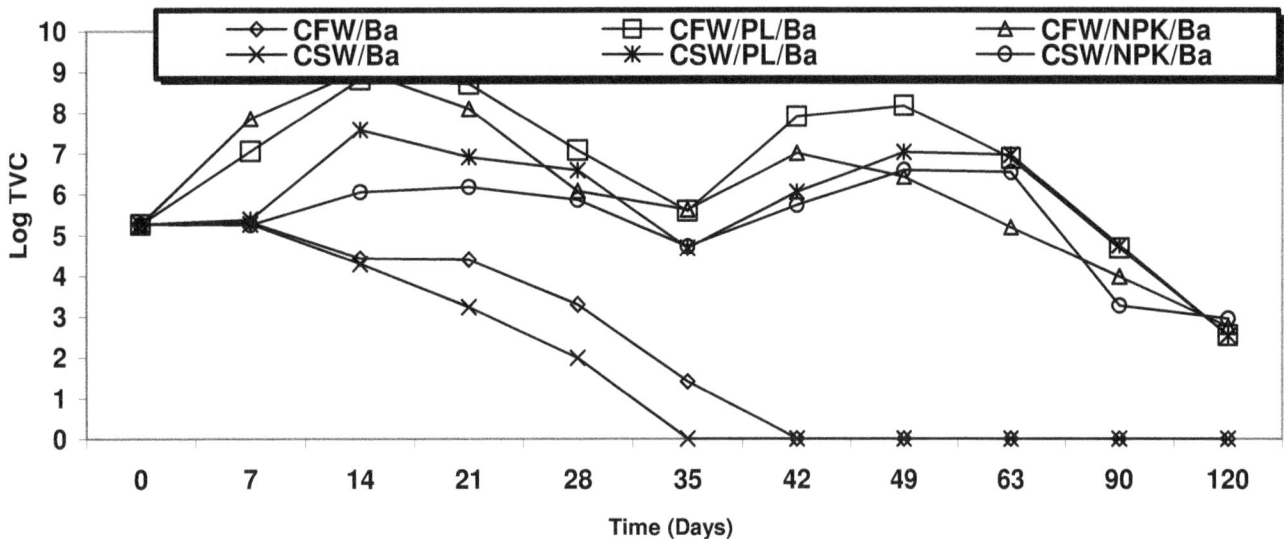

Figure 1a. Log total viable *Bacillus* count in various set-ups.

a reduction from Day 25 to Day 35. The re-application of nutrients at Day 35 resulted in another increase in growth of the organisms. This is evident in Figs 1a (*Bacillus)* and 1b (*Pseudomonas*). The application of nutrients in Figure 1c (*Aeromonas* did not stimulate any further increase in growth in the options containing nutrient supplements.

In Figures 2a and b there was a decrease in hydrocarbon levels with time. The decrease was however greater in fresh water systems supplemented with nutrients than in brackish water systems supplemented with nutrients. Hydrocarbon losses were greater in *Bacillus* and *Pseudomonas* containing test systems. Hydrocarbon losses in *Aeromonas* containing test systems were insignificant. Test systems without nutrient supplementation showed insignificant decreases in hydrocarbon levels.

The changes in nitrate and phosphate levels presented in Figures 3a - b and 3c - d is a reflection of reapplication of nutrients by Day 35 in nutrient supplemented test systems. From Day 0 to day 28 and Day 35 there was a decrease in concentrations of these nutrients. However following reapplication of fresh nutrients by Day 35 there was an increase to Day 42, which was followed by a decrease till Day 120. Test systems without fertilizer/poultry litter application did not show this sequence.

The pH of nutrient supplemented test systems (4a - b) showed a decrease with time. This decrease was evident in both fresh and brackish water test systems. The pH of systems without nutrient supplementation showed insignificant decreases with time.

The abilities of the three bacteria *Bacillus, Pseudomonas*

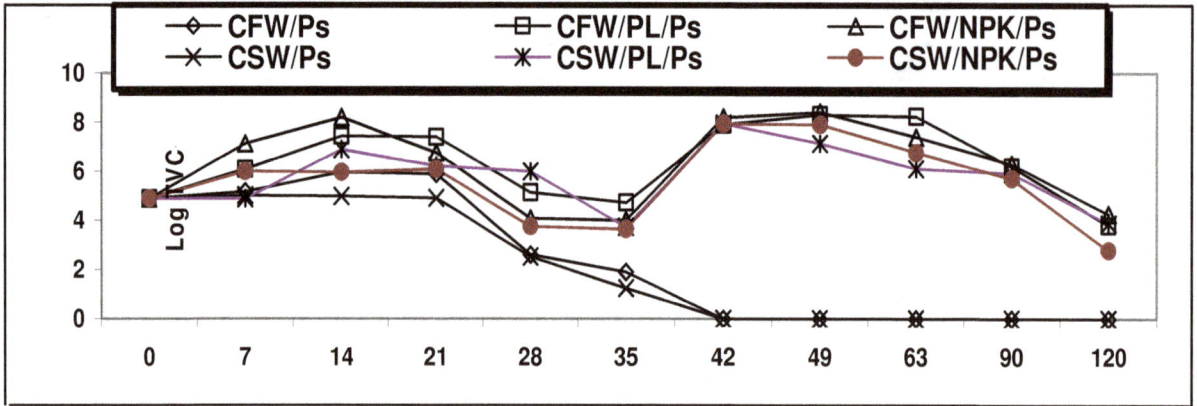

Figure 1b. Total viable *Pseudomonas* count in various set-ups.

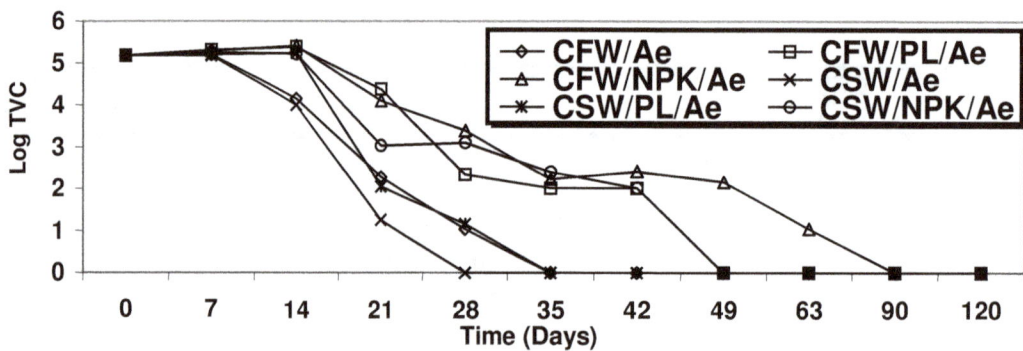

Figure 1c. Total viable Aeromonas count in various set-ups.

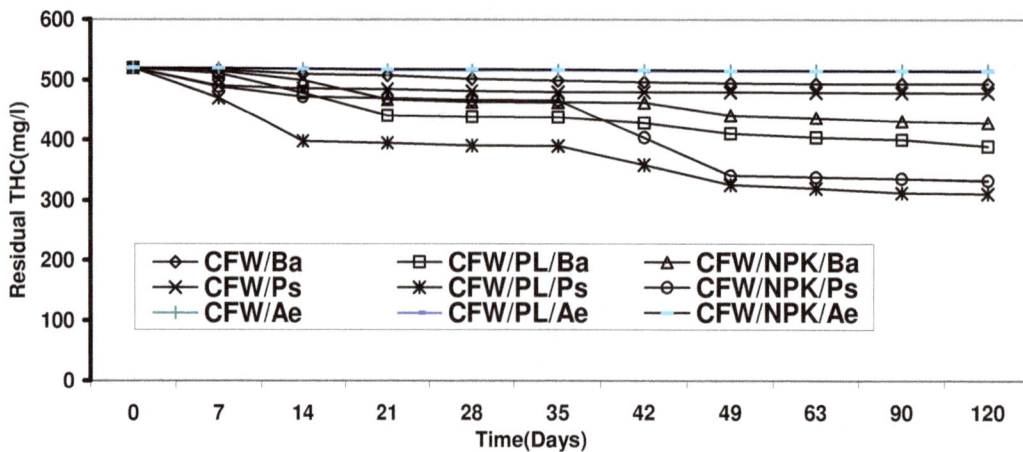

Figure 2a. Changes in total hydrocarbon content (THC) in various freshwater treatment options.

and *Aeromonas* to bioconcentrate the six test heavy metals are presented in Figures 5a - d, 6a - d and 7a - d respectively. The relationship between hydrocarbon loss pH and microbial counts is also shown in these figures. The results showed that in test systems containing

nutrient supplements bioconcentration of heavy metals increased with time. The pH, the total hydrocarbon levels and microbial counts also decreased with time. This was evident in systems containing *Bacillus* and *Pseudomonas.* Bioconcentration increased with increase

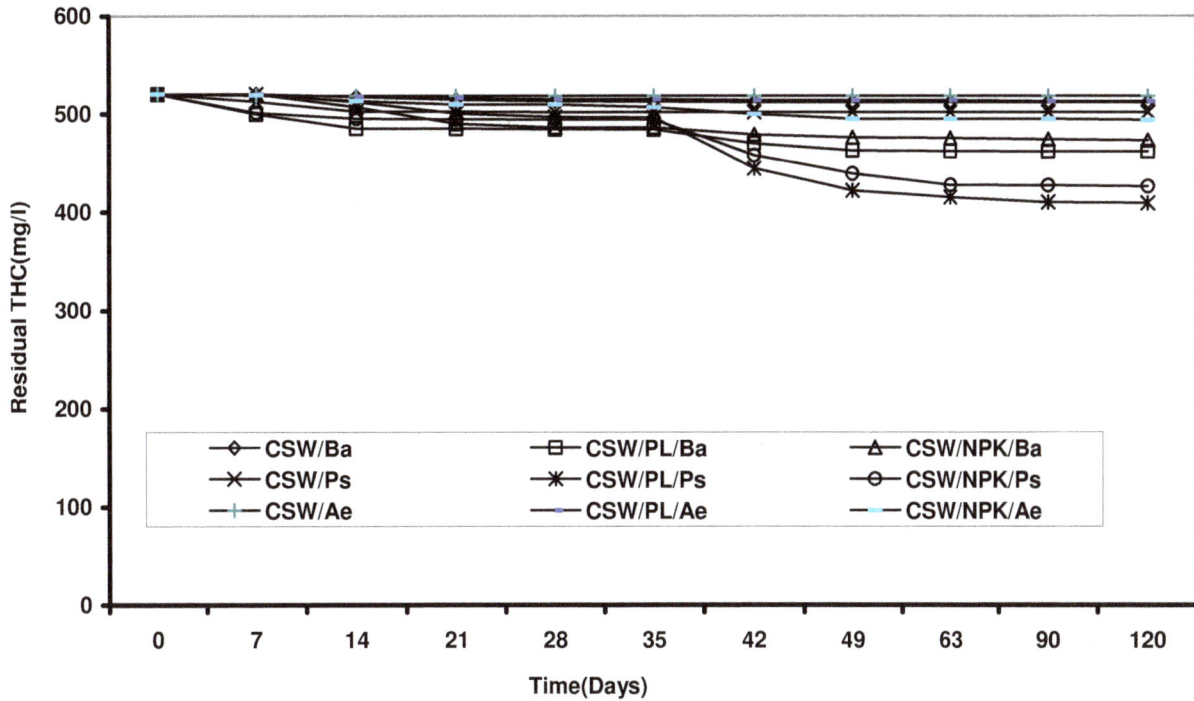

Figure 2b. Changes in total hydrocarbon content in various saltwater treatment options.

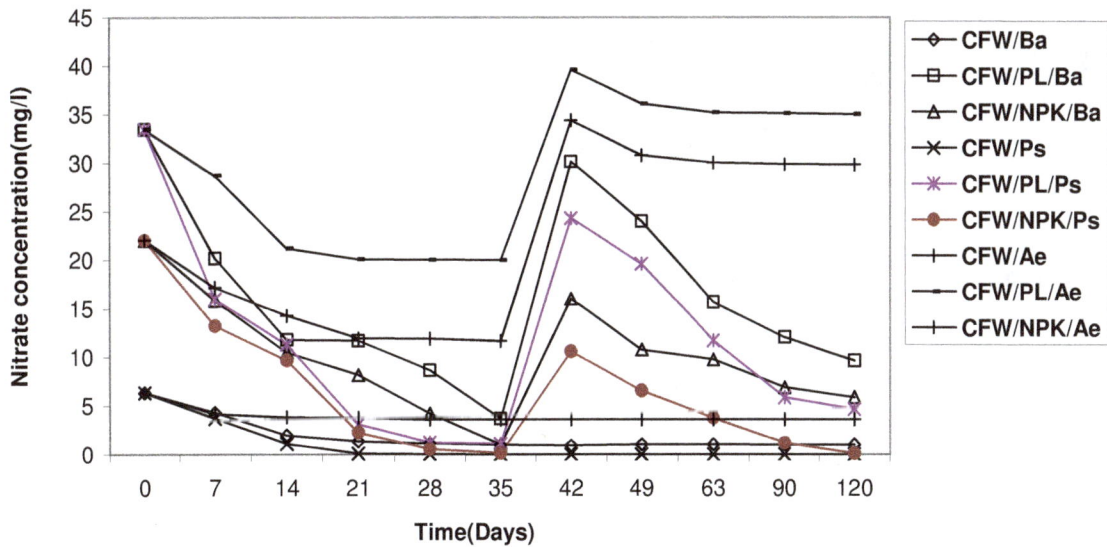

Figure 3a. Changes in nitrate concentration in various fresh water treatment options.

in microbial biomass (microbial counts). Bioconcentration occurred within pH maxima of 8b (optimum pH was within the acidic range). In test systems without nutrient supplements a decrease in bioconcentration of heavy metals was observed with time. In these systems insignificant decreases in pH and total hydrocarbons also occurred with time. This pattern was also evident in systems containing *Aeromonas*.

DISCUSSION

The results of this study showed that the inoculation of crude oil contaminated fresh water and brackish water with two of the three isolates (*Bacillus* and *Pseudomonas*) may be employed as a bioremediation (biodegradation of hydrocarbons and bioconcentration of heavy metals) strategy. These isolates have been

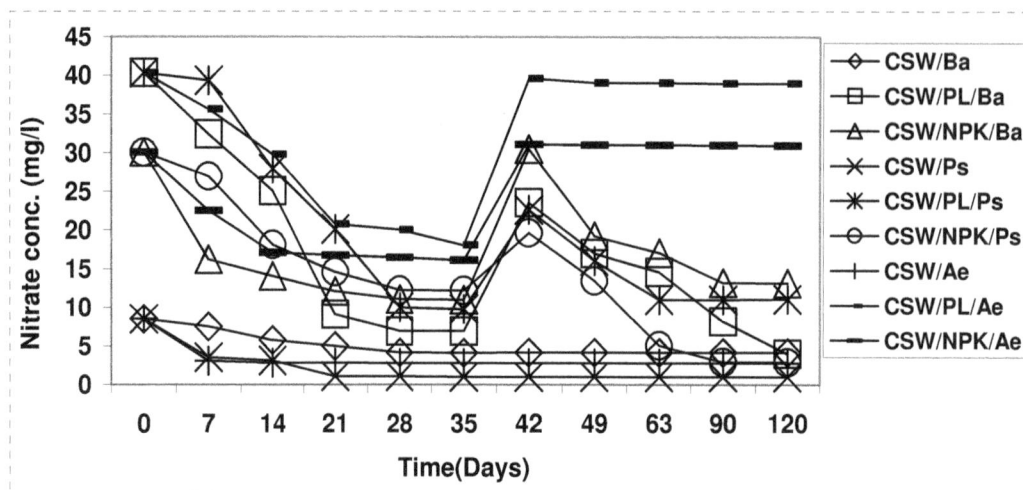

Figure 3b. Changes in nitrate concentrations in various saltwater treatment options.

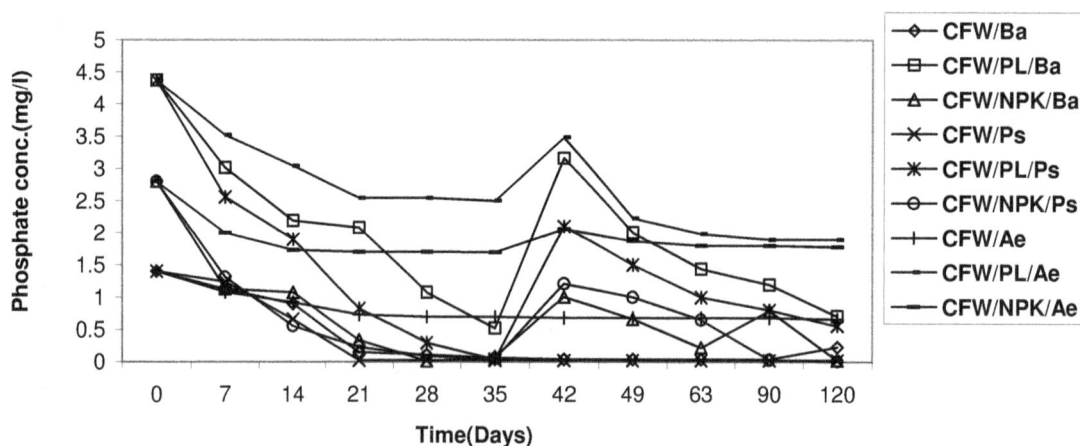

Figure 3c. Changes in phosphate concentration in various freshwater treatment options.

associated with petroleum product degradation (Nwachukwu and Gurney, 2000; Odokuma and Ibor, 2003; Okerentugba and Ezeronye, 2003) in the Niger Delta.

The increase in counts with addition of fresh nutrients (corresponding to Day 0 - 14 and Day 35 - 49) indicates the importance of additional nitrogen and phosphorus to microbial growth in a batch system. The results of inorganic nutrient analysis indicated that phosphate and nitrate became limiting at day 28 - 35. The activities of the isolates in the experimental samples were slowed down probably because of the depletion of these inorganic nutrients. This may have resulted in the reduction in both hydrocarbon levels as well as heavy metal accumulation. This is further supported by the fact that the re-addition of nutrients at day 35 resulted in the resumption of activities by the isolates. This process corrected the nutritional imbalance thereby leading to a

further decrease in hydrocarbon levels and an increase in heavy metal accumulation. Microorganisms require phosphorus as phospholipids in synthesizing cell membranes, as components of nucleic acids and for sugar phosphorylation (Andrew and Jackson, 1996). They also exploit nitrate sources to meet their protein and nucleic acid requirement. NPK fertilizer and poultry litter contain these inorganic nutrients and when they were added to the crude oil contaminated water samples active growth of *Bacillus sp* and *Pseudomonas sp* were obtained with proportional decrease in hydrocarbon levels. The insignificant changes in options containing *Aeromonas sp* may be due to the fact that the isolate lacked required enzymes (genetic constitution) and hence incapable of degrading the complex crude oil components and not as a result of limiting nutrient supply. These observations emphasizes the importance, in all bioremediation programmes for monitoring routinely the

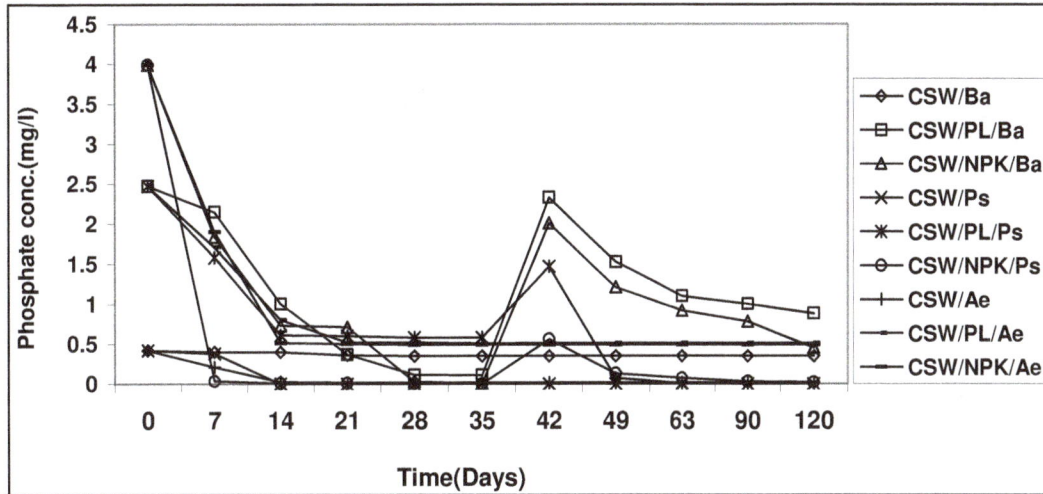

Figure 3d. Changes in phosphate concentrations in various saltwater treatment options.

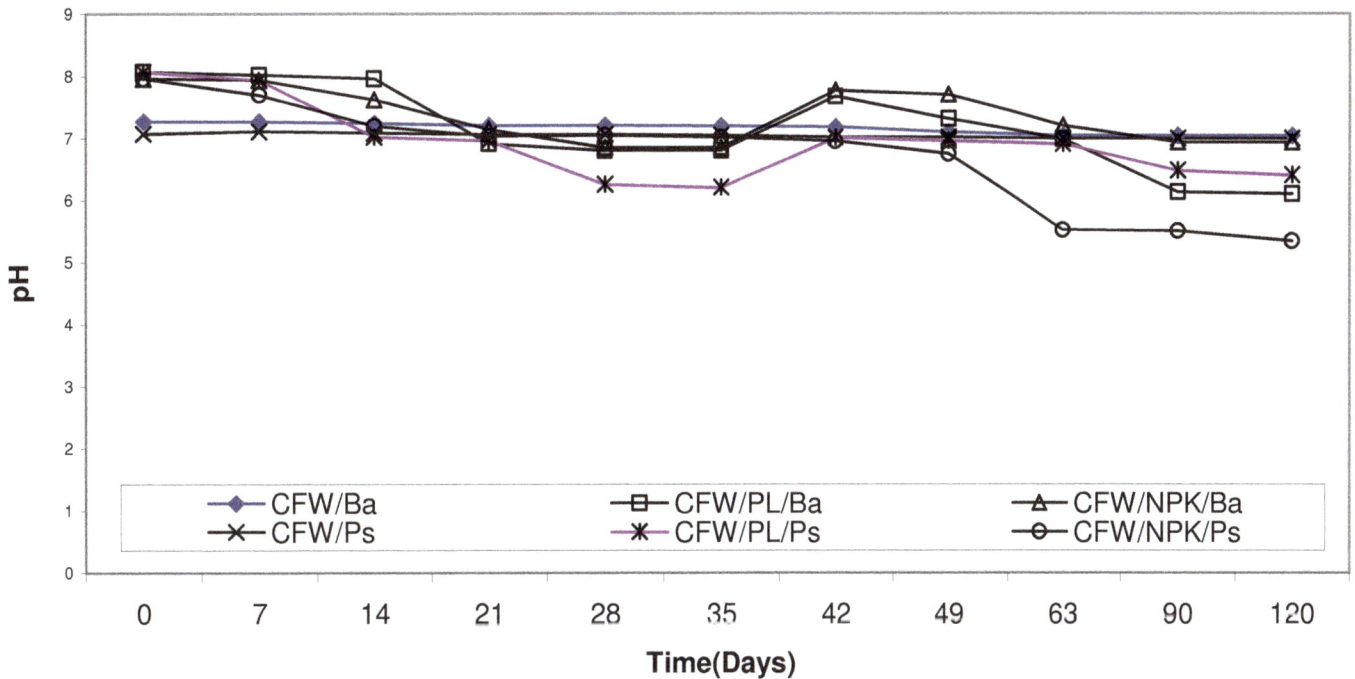

Figure 4a. Changes in pH values in the various freshwater options.

nutritional variables known to influence the biodegradation of pollutants and to correct any imbalances and hence accelerate the recovery rate of the hydrocarbon contaminated environments.

Bioremediation strategies for crude oil impacted water bodies in oil producing regions of developing countries like Nigeria need to be cost- effective and also, able to apply indigenous technology. Thus the appropriate technology may include appropriate strain development in addition to the application of readily available and inexpensive nutrient supplements. *Pseudomonas* and

Bacillus that showed effectiveness in oil degradation are predominant in aquatic habitats in the Niger Delta. *Pseudomonas* has been used extensively in the efficient biodegradation of crude oil and petroleum products (Amund and Igiri, 1990; Rocha et al., 1992; Nwachukwu, 2000; Nwachukwu and Gurney, 2000). However, in chronic oil pollution incidents, there is usually the occurrence of loss of microbial diversity and reduced microbial population leading to the persistence of the contaminant (Atlas, 1991; Wang et al., 1994). Therefore, prospecting for and the inoculation of organisms having

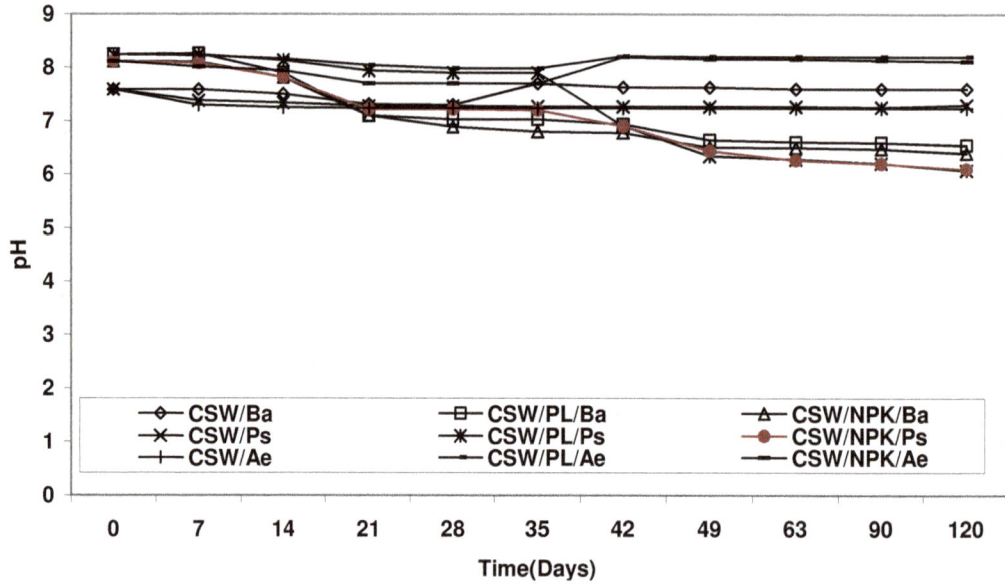

Figure 4b. pH changes in various saltwater treatment options.

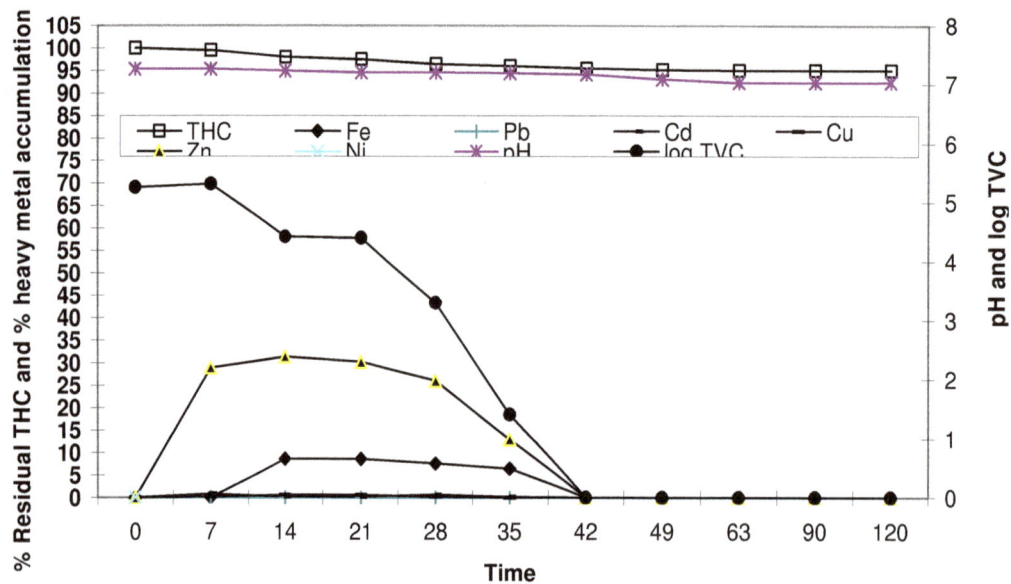

Figure 5a. Activities of *Bacillus* in CFW/Ba.

genes with crude oil degrading potentials as well as employing methods that increase their biomass have been known to promote oil biodegradation in contaminated regions.

Bacillus and *Pseudomonas* revealed great potentials in accumulating the heavy metals associated with crude oil. Similar observations had been made by other investigators demonstrating the capabilities of several bacteria in removing heavy metals such as lead, cadmium, copper, nickel and other heavy metals from polluted effluents (Higham et al., 1985 and Malekzadeh et al., 1995). In this

study an increasing uptake pattern was observed for all metals in *Pseudomonas* and *Bacillus* test systems which must have resulted from the genetic make up of both organisms and increases in biomass generated due to the periodic addition of appropriate nutrient supplements. Saturation of biomass with heavy metals was not observed, indicating that available sites for heavy metal adhesion probably exist. The decreases in values of bioaccumulation obtained corresponded to periods when the population of organisms was reducing. However, determination of saturation levels was not the purpose

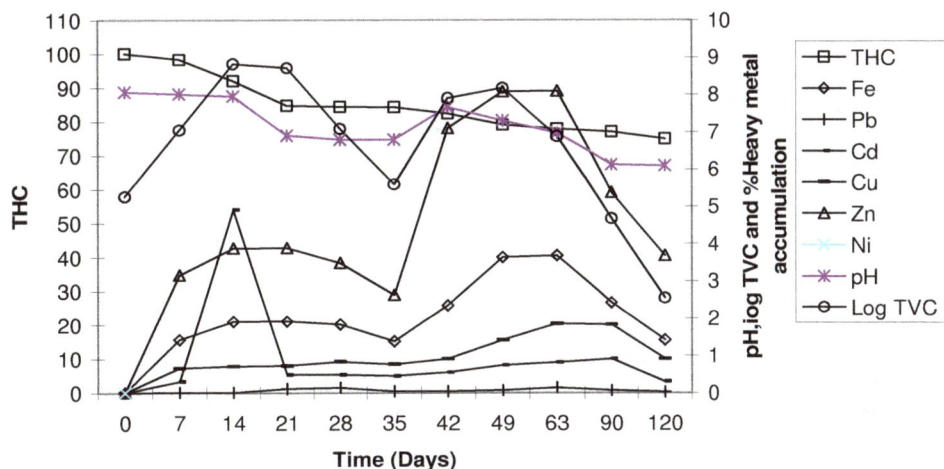

Figure 5b. Activities of *Bacillus* in CFW/PL/Ba.

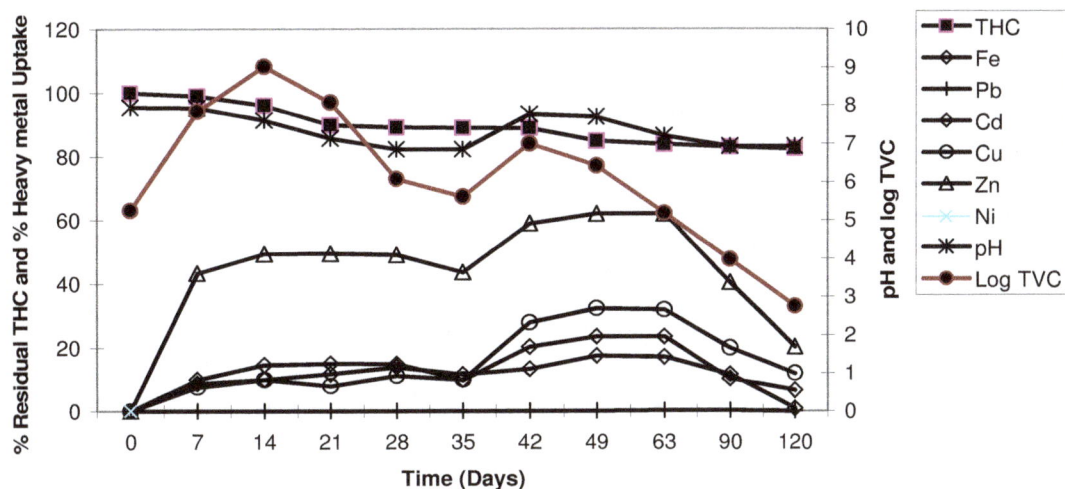

Figure 5c. Activities of *Bacillus* in CFW/NPK/Ba.

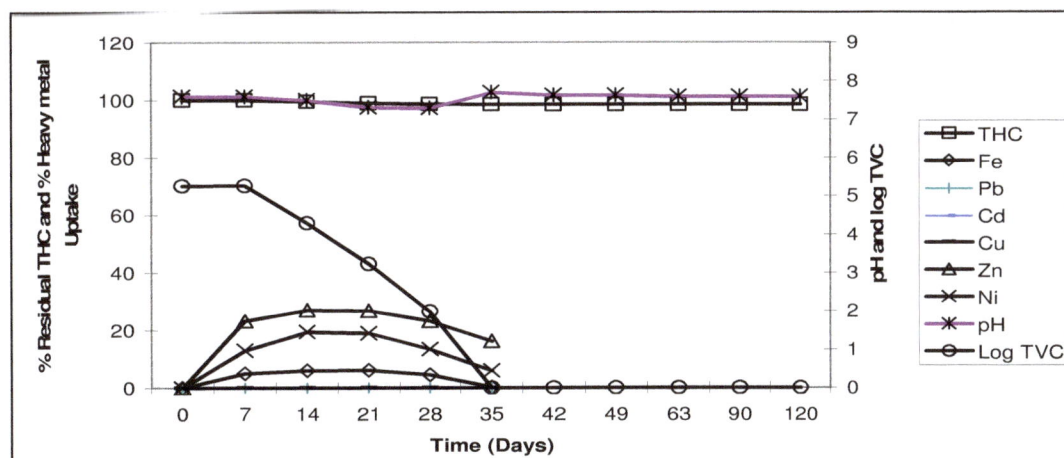

Figure 5d. Activities of *Bacillus* in CSW/Ba.

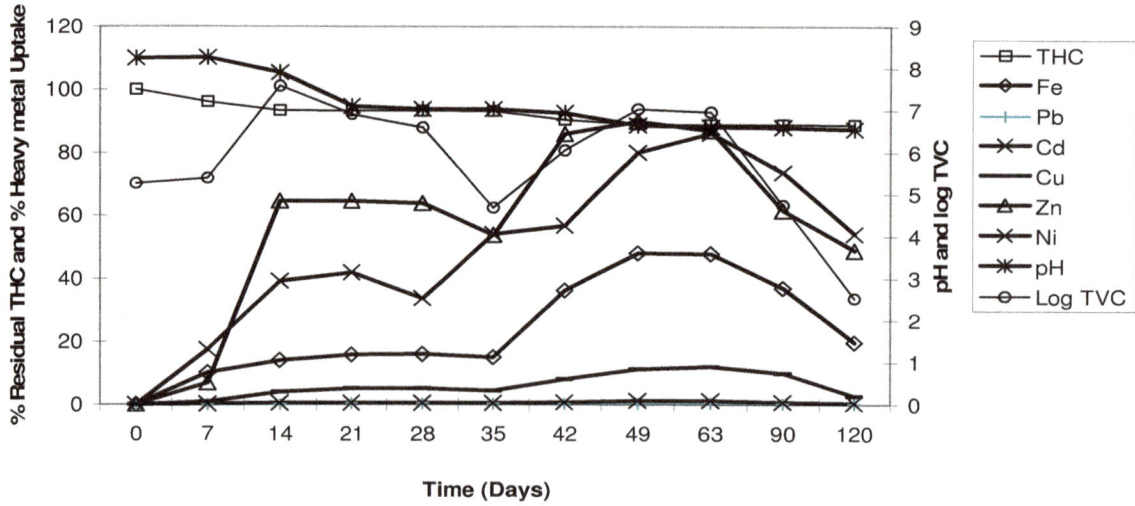

Figure 5e. Activities of *Bacillus* in CSW/PL/Ba.

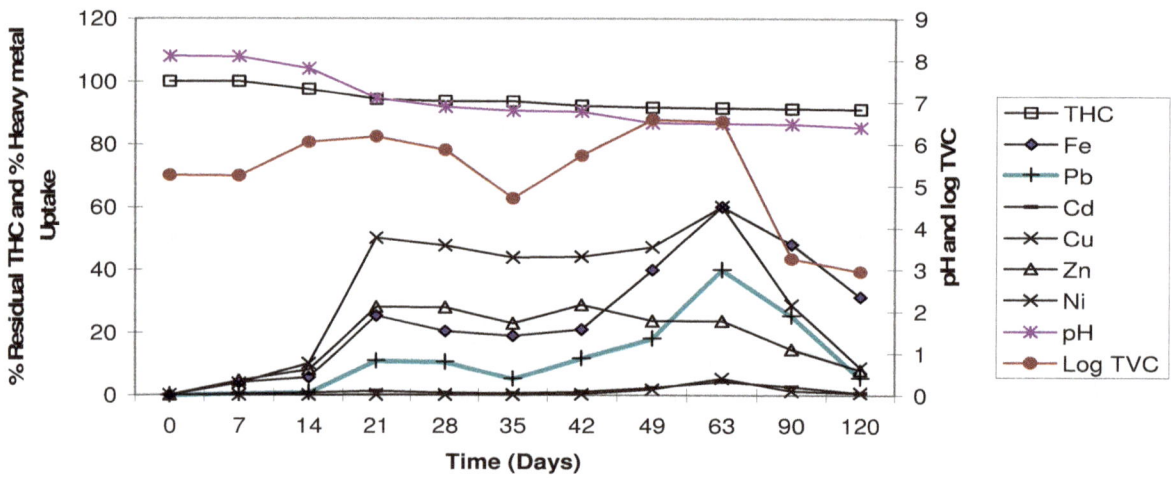

Figure 5f. Activities of *Bacillus* in CSW/PL/Ba.

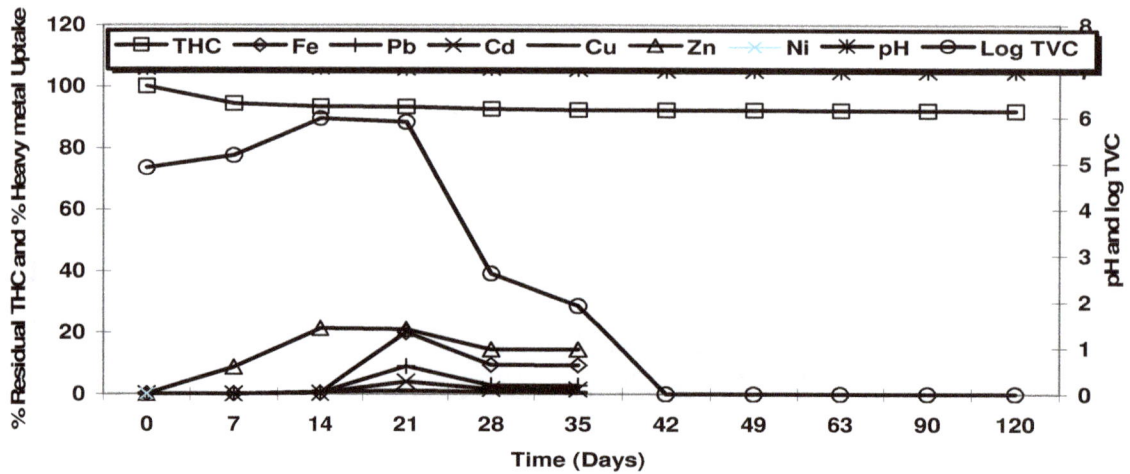

Figure 6a. Activities of *Pseudomonas* in CFW/Ps.

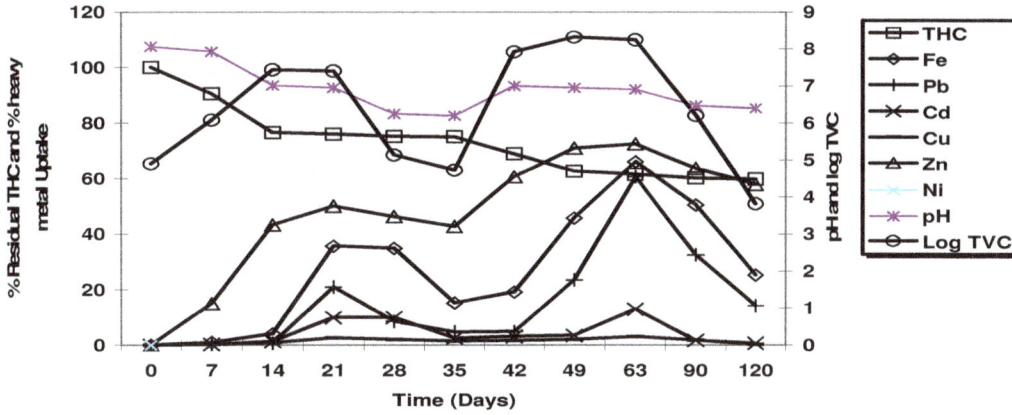

Figure 6b. Activities of *Pseudomonas* in CFW/PL/Ps.

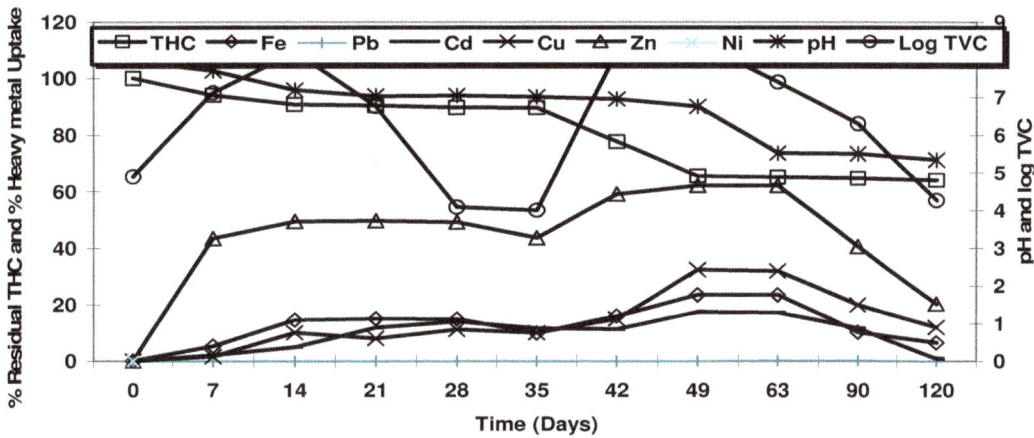

Figure 6c. Activities of *Pseudomonas* in CFW/NPK/Ps.

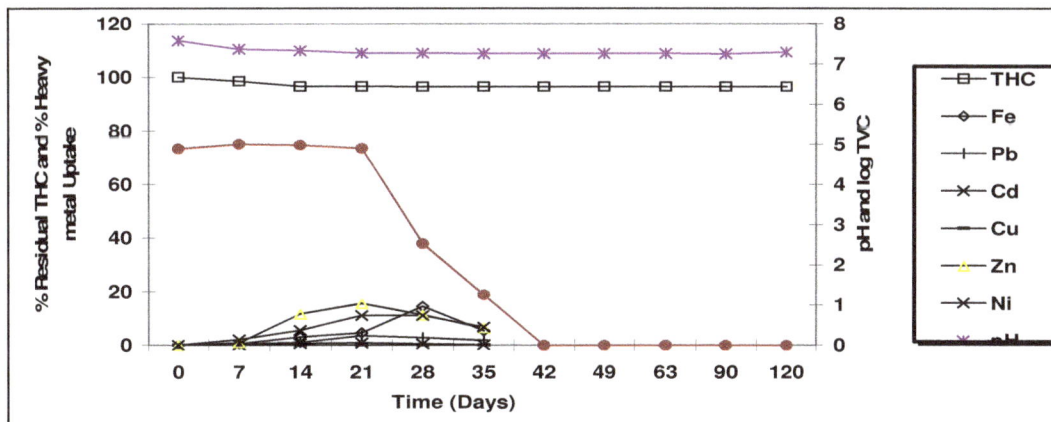

Figure 6d. Activities of *Pseudomonas* in CSW/Ps.

of this study, but the determination of the abilities of the cells to accumulate heavy metals in the presence of other contaminants and hence their use as biosorbents.

Although, the same qualitative results were described for the two test isolates, quantitative bioaccumulation of heavy metals were statistically distinct and followed the

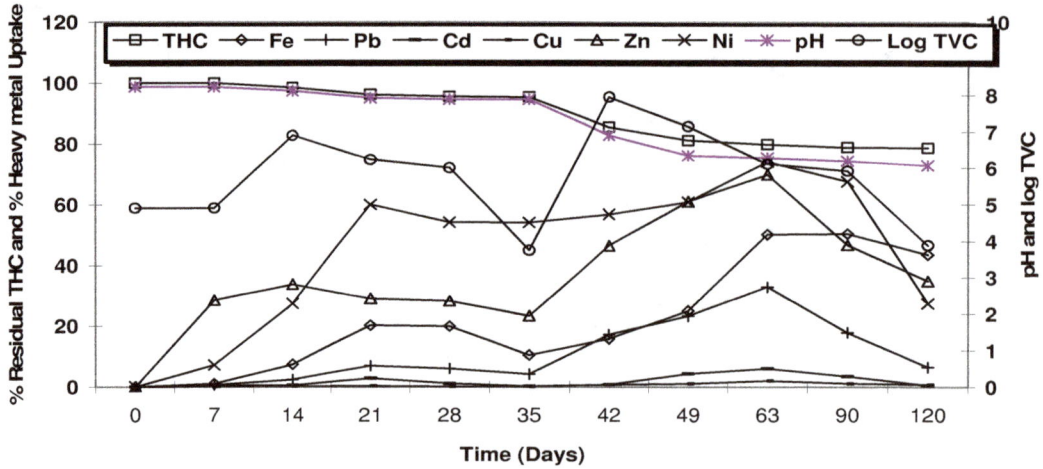

Figure 6e. Activities of *Pseudomonas* in CSW/PL/Ps.

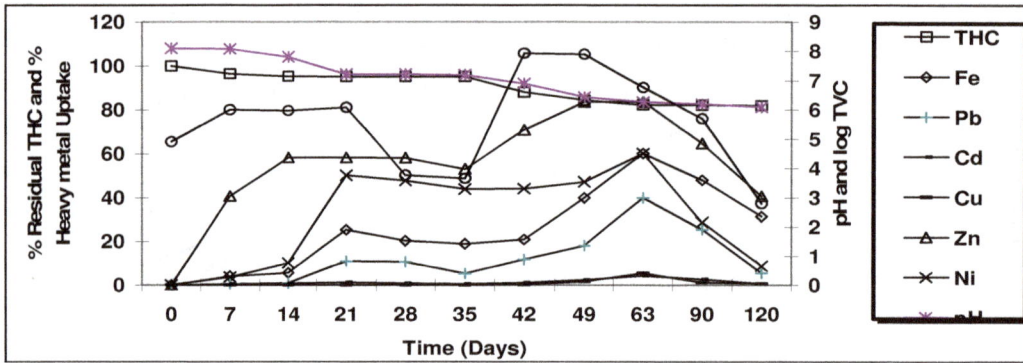

Figure 6f. Activities of *Pseudomonas* in CSW/NPK/Ps.

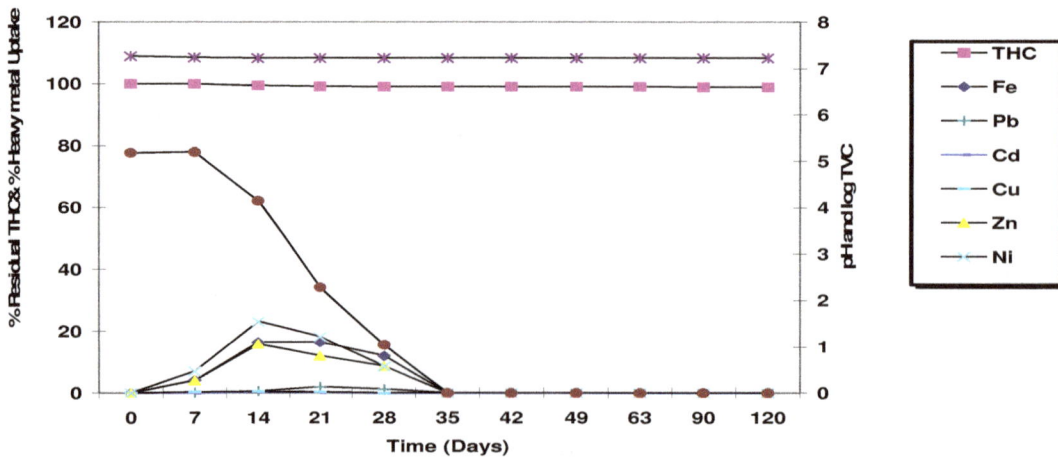

Figure 7a. Activities of *Aeromonas* in CFW/Ae.

pattern: *Pseudomonas* > *Bacillus*. The reduced ability of *Aeromonas* in accumulating heavy metals may be attributable to its low survival rate in the contaminated test systems probably from its genetic make up, which may have resulted from its inability to utilize crude oil as its sole carbon source or it's toxicity to the organism (Not toxicity to heavy metals). The profile for bioaccumulation for the gram negative isolates was such that higher

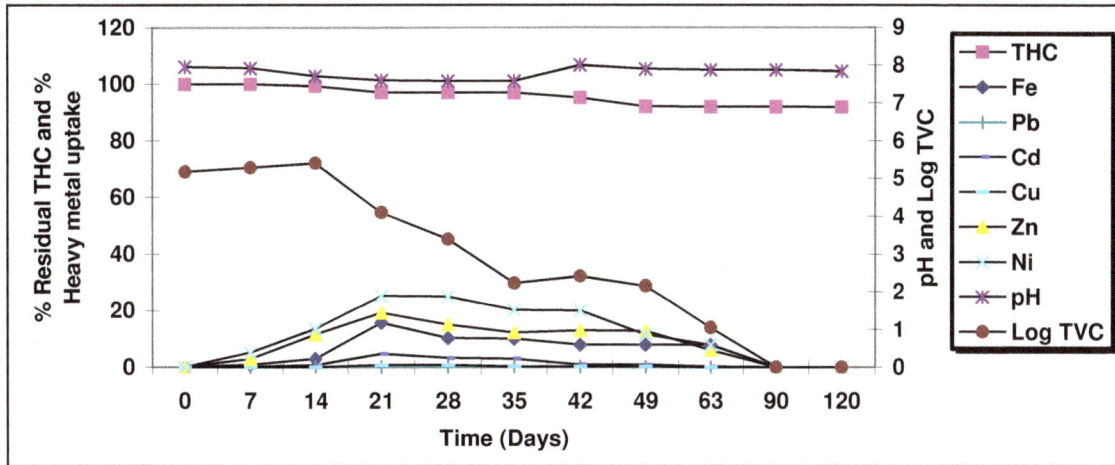

Figure 7b. Activities of *Aeromonas* in CFW/NPK/Ae.

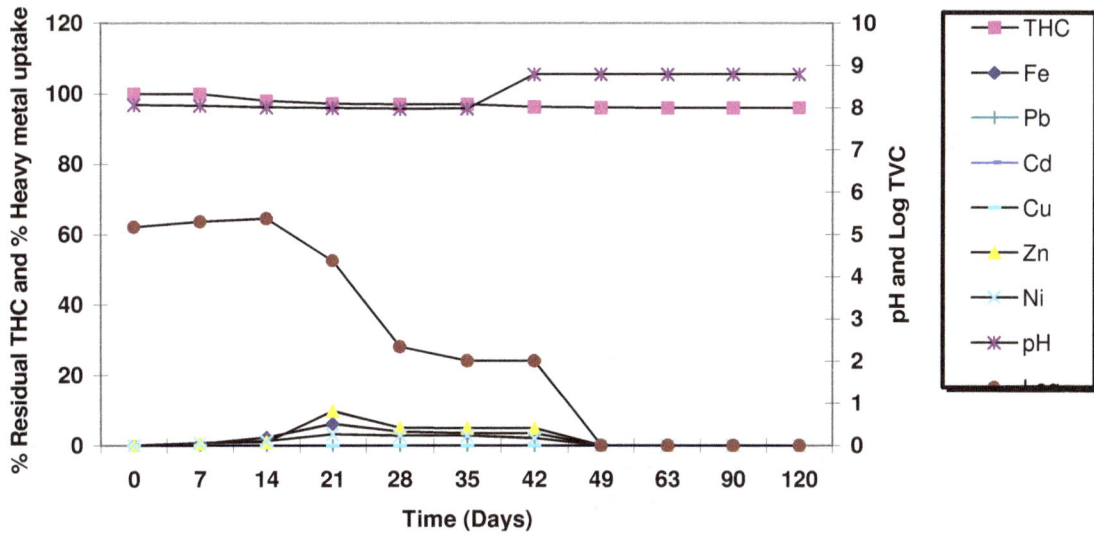

Figure 7c. Activities of *Aeromonas* in CFW/PL/Ae.

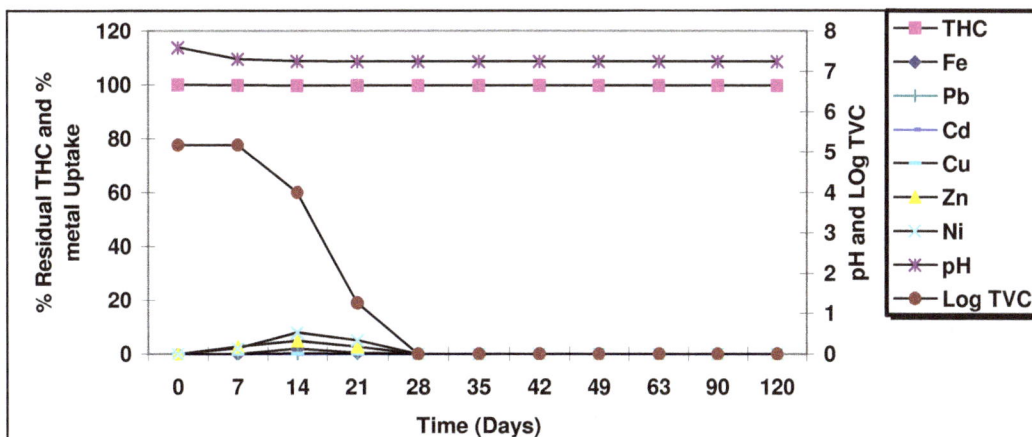

Figure 7d. Activities of *Aeromonas* in CSW/Ae.

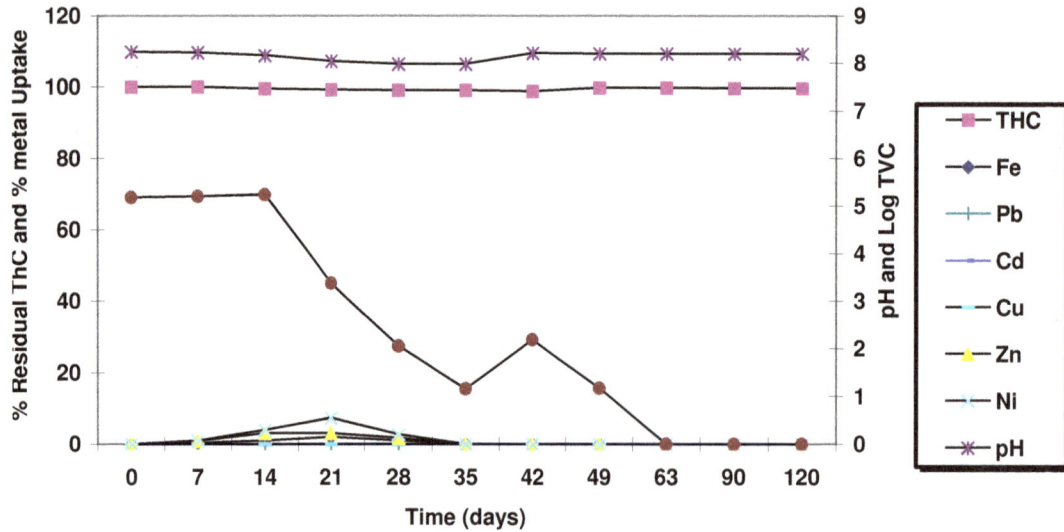

Figure 7e. Activities of *Aeromonas* in CSW/PL/Ae.

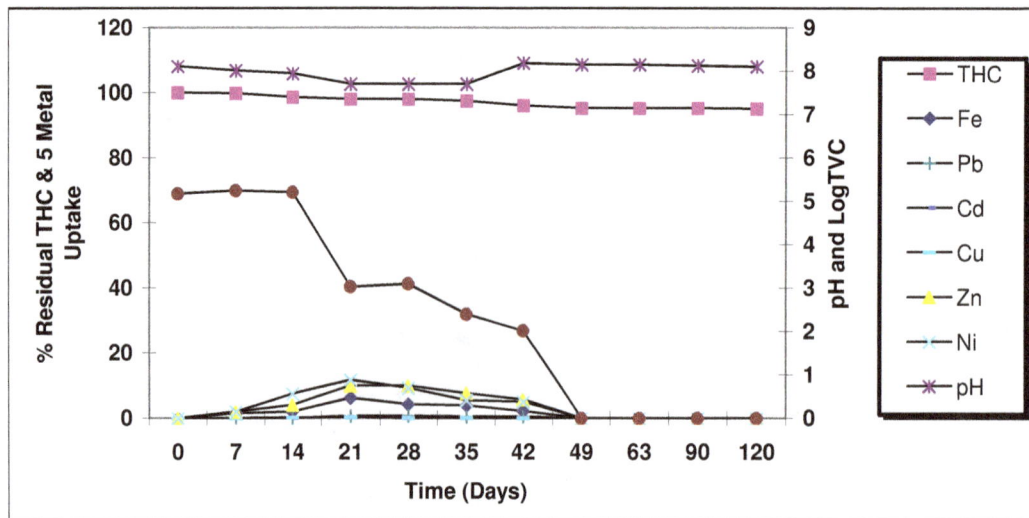

Figure 7f. Activities of Aeromonas CSW/NPK/Ae.

concentrations were accumulated when bacteria growth remained more or less constant corresponding to the stationary growth phase. However, for *Bacillus*, peak accumulation was observed during active growth or exponential growth phase. This findings suggest that metal uptake by *Pseudomonas sp* involves not only diffusion but also surface adsorption which is metabolism independent. In *Bacillus sp*, however, the most likely phenomenon may be diffusion which may result from increased membrane permeability.

Maximum uptake of heavy metals except zinc and nickel by *Pseudomonas* were observed when pH values were in the acidic zones. This may account for the high rates observed in the accumulation of these two heavy metals especially in brackish water environments where

pH values remained in the alkaline range for a longer period of the study. However, lead, copper and cadmium uptake when compared to other metals, were highly repressed in test options containing *Bacillus* and *Pseudomonas*. Their low concentrations in combination to the high pH values in the set-ups might account for this phenomenon. Previous authors have reported that prevailing pH value is one of the main factors in bioaccumulation efficiency by different organisms (Leung et al., 2000; Lopez et al., 2000; Al- Garni, 2005). They reported that low pH affects the network or chemistry of cell wall as well as physiochemistry and hydrolysis of the heavy metal. It was also reported that at low pH values, lead ions, compete with hydrogen ions on the binding sites of the microbial cell. The accumulation of lead, copper and cadmium were in accordance to these findings. However, the inability of *Bacillus sp* to accumulate these three metals could be due to the observation that its accumulation potential was impaired by age of cultures (stationary phase) and it was during this period that pH values reduced to the acidic zone.

The overall result however, showed that higher concentrations of each metal were accumulated beyond Day 35. This may be as a result of the increased biomass yield obtained due to the periodic addition of appropriate nutrient supplements. The study proves that *Bacillus* and *Pseudomonas* are capable of utilizing hydrocarbon contaminants and may be employed in the future for the removal of heavy metals immobilized on waste materials. However, it is important to note that some species of these bacteria are opportunistic pathogens Thus, it is necessary to screen for pathogenicity to confirm their environmental friendliness before field trials.

Conclusion

Periodic nutrient supplementation with NPK fertilizer and poultry litter promoted both biodegradation of crude oil and heavy metal uptake in both crude oil impacted fresh and brackish water test systems containing *Bacillus* and *Pseudomonas*. The hydrocarbon levels of test systems containing *Aeromonas* were fairly constant throughout the duration of the study probably due to toxicity of crude oil components to this organism or the inability of the organism to utilize the oil as sole carbon source (genetic make up). Heavy metal uptake by *Aeromonas* was much reduced when compared with *Bacillus* and *Pseudomonas* probably for the same reason. *Aeromonas* did not serve as a good tool for biodegradation and bioconcentration. Bioconcentration capabilities of organisms revealed the following trend *Pseudomonas* > *Bacillus* > *Aeromonas*. The pattern of heavy metal uptake by the three bacterial isolates was different in the various treatment options. Bioconcentration was also affected by biomass concentrations and pH of the test systems. The study revealed that there was no significant difference in bioconcentration values when comparing NPK fertilizer with poultry litter. The study has shown that a combination of bioaugmentation with indigenous species of *Bacillus* and *Pseudomonas* and biostimulation with inorganic NPK fertilizer or poultry litter may be employed to bioremediate (biodegrade and bioconcentrate) crude oil impacted fresh and brackish water systems of the Niger Delta.

REFERENCES

Alexander M (1981). Biodegradation of Chemicals of Environmental Concern. Sci. 211(4478): 132-138.

Al-Garni SM (2005). Biosorption of lead by Gram-ve capsulated and non capsulated bacteria. Wat. Sci. Technol. 31(3): 345-349.

American Public Health Association (1998). Standard Methods for the Examination of Water and Wastewater. 20th ed Washington, D.C. USA.

Amund OO, Igiri CU (1990). Biodegradation of Petroleum Hydrocarbons Under tropical estuarine conditions. World J. Microbiol. Biotechnol. 6: 225-262.

Andrew RWJ, Jackson JM (1996). Pollution and waste management In: Environ. Sci: The Natural environment and Human Impact, Longman Publishers Ltd pp. 281-297.

Atlas RM (1981). Microbial degradation of petroleum hydrocarbons. An Environmental Perspective. Microbiol. Rev. 45: 180-209.

Fourest E, Roux CJ (1992). Heavy metal biosorption by fungal mycelial by-products: Mechanisms and influence of pH. Appl. Microbiol. Biotechnol. 37(3): 399-403.

Higham DP, Sadler PJ, Scawen MD (1995). Cadmium resistance in *Pseudomonas putida*: Growth Uptake of Cadmium. J. Gen. Microbiol. 131: 2539-2544.

Krieg NR, Holt JJ (1994). Bergey's Manual of Systematic Bacteriology, William and Wilkins Baltimore, Ltd.

Ladousse A, Tramier B (1991). Results of 12 years of research in spilled oil bioremediation. Inipol EA 22. Proceedings of the 1991 oil spill conference. American Pet. Inst. Wasshington, D.C.

Lal R, Saxena DM (1982). Accumulation metabolism and effects of organochlorin insecticides on microorganisms. Microbiol. Rev. 46: 95-127.

Leung WC, Wong MF, Chua H, Lo W, Yu PH, Leung CK (2000). Removal and recovery of heavy metals by bacteria isolated from activated sludge treating industrial effluents and Municipal wastewater. Water Sci. Technol. 44: 233-240.

Lopez A, Lazaro N, Priego I M, Marques AM (2000). Effect of pH on the Biosorption nickel and other heavy metals by Pseudomonas flourescens J. Industrial. Microbiol. Biotechnol. 24: 146-151.

Malekzadeh FA, Farazmaud H, Ghafourian M, Shahamar M, Levin M, Grim C, Colwell RR (1995). Accumulation of heavy metals by a bacterium isolated from electroplating effluent. Proc. Biotechnol. Risk Assess. Symp. Canada pp. 338-398.

Nwanchukwu SCU (2000). Modelling the recovery of a tropical soil ecosystem polluted with crude petroleum using *Pseudomonas spp.* as the test organisms. The Inter. J. Environ. Edu. Info. 19(1): 56-62.

Nwanchukwu SCU, James P, Gurney TR (2000). Inorganic nutrient utilization by adapted *Pseudomonas putida* strain used in the bioremediation of agricultural soil polluted with crude petroleum. J. Environ. Biol. 22(3): 153-162.

Peterson RG, Levine LD (1986). Soil Sampling, In: methods of Soil Analysis, Part1. Physical and Mineralogical Methods. Klute, A.(ed) American Society of Agronomy, Madison pp. 33-51.

Odiete WO (1999). Environmental Physiology of animals and pollution. Diversified Resources Ltd. Surulere, Lagos pp. 225-228.

Odokuma LO, Ibor MN (2002). Nitrogen Fixing Bacteria Enhanced bioremediation of a crude oil polluted soil. Global J. Pure Appl. Sci. 8 (4): 455 – 468.

Odokuma LO, Abah AE (2003). Heavy Metal Biosorption by three Bacteria Isolated from a Tropical River. Global J. Environ. Sci. 2 (2): 98 – 101.

Odokuma LO, Dickson AA (2003). Bioremediation of a Crude Oil Polluted Tropical Rain Forest Soil. Global J. Environ. Sci. 2(1): 29 – 40.

Odokuma LO, Dickson AA (2003). Bioremediation of a Crude Oil Polluted Tropical Mangrove Swamp. Nigerian J. Appl. Sci. Environ. Manage. 7(2): 23 – 29.

Odokuma LO, Ijeomah SO (2003). Seasonal Changes in the Heavy Metal Resistant Bacterial Population of the New Calabar River. Global J. Pure Appl. Sci. 9(4): 425 – 433.

Odokuma LO, Emedolu SN (2005). Bacterial Sorbents of Heavy Metals Associated with Two Nigerian Crude oils. Global J. Pure Appl. Sci. 11(3): 343-351.

Okerentugba PO, Ezenronye OU (2003). Petroleum degrading potentials of single and mixed microbial cultures isolated from rivers and refinery effluent in Nigeria. Afr. J. Biotechnol. 2(9): 288-292.

Okpokwasili GC, Amanchukwu SC (1988). Petroleum hydrocarbon degradation by candida species. Environ. Int. 14: 243-247.

Perelomov LV, Prinsky DI (2003). Manganese, lead and Zinc compunds in Gray forest soils of the central Russian Upland, Eurasian. Soil Sci. 6: 610-618.

Rocha CF, San-Blas G, Vierma L (1992). Biosurfactant Production by two isolates of *Pseudomonas aeruginosa*. World J. Microbiol. Biotechnol. 8: 125-127.

Rhodes AN, Hendricks CW (1990). A continuous flow method for measuring effects of chemicals on soil nitrification. Toxicol. Assess. 5: 77-89.

Shumate ES, Strandberg GW (1985). Accumulation of metals by Microbial cells. Compreh. Biotechnol. 13: 235-247.

Wang Z, Fingas M, Sergy G (1994). Study of 22 years old arrow oil samples using biomarker compounds by GC/MS Environ. Sci. Technol. 28: 1733-1746.

Histomorphometrical study of seminiferous tubule in rats after used *Tribulus terresteris*

Arash Esfandiari* and Reza Dehghani

[1]Deparment of Anatomical Sciences, School of Veterinary Medicine. Islamic Azad University, Kazerun Branch, Kazerun, Iran.
[2]School of Veterinary Medicine, Islamic Azad University, Kazerun Branch, Iran.

The aim of the study was to determine the histological and histomorphometrical change of seminiferous tubule in mature and immature wistar rats after using *Tribulus terresteris* (TT). Twenty male wistar rats were selected and randomly divided into four groups: 1) Mature control group (MCG). 2) Mature experimental group (MEG) (orally received 75 mg/kg TT daily for 14 days). 3) Immature control group (ICG). 4) Immature experimental group (IEG) (orally received 75 mg/kg TT daily for 14 days). The number of leydig cells had increment in experimental group when compared with ICG. Result showed that the thickness of the wall of seminiferous tubule in experimental group significantly increased (P < 0.05). Also, TT treatment groups resulted the accumulation of spermatogenic cells were increased in the seminiferous tubule when compared with control group. In addition, sperm was not observed in ICG but sperm were also observed to increase in the treatment groups. It is concluded that TT may improves the sexual activity, increased testestrone by intensification of leydig cells and may caused early puberty in immature rat.

Key words: Histomorphometrical, rat, seminiferous tubule, *Tribulus terresteris*.

INTRODUCTION

The testis contain of many pyramidal compartments called the testicular lobules. Each lobule is occupied by 1 - 4 seminiferous tubules enmeshed in a web of loose connective tissue that is leydig cells. Seminiferous tubules produce male reproductive cells, the sperma-tozoa, whereas leydig cells secrete testicular androgens. Seminiferous tubules lined with a complex stratified epithelium that consist of a tunic of fibrous connective tissue, a well-defined basal lamina and a complex germinal epithelium (Banks, 1993). TT is a flavnonoid herbal drug that has saponin, essence and steroid.

A lot of various studies were undertaken on increasing testosterone and sexual activity after used TT (Gauthman, 2003, 2008; El-tantawy, 2007; Park, 2006). In addition, the γ interferon reduced sertoli cells and the thickness of seminiferous tubular epithelium (Natwar,

1995). Administration of various concentration of sodium fluoride showed that it did not have effect on the testosterone and spermatogenesis and it can cause diminishing of capsule of testis with 100 ppm dosage (Sprando, 1997). Also, Sprando et al. (2000) indicated that linen seed decreased the volume of the seminiferous tubules. The aim of the present study was to investigate the effect of TT on histological change and histomorphometrical of seminiferous tubule of the testes.

MATERIALS AND METHODS

Ten mature wistar male rats weighing between 200 - 250 g (60 days old) and ten immature wistar male rats weighing between 50 - 75 g (21 days old) were obtained from animal house of Islamic Azad University, Kazerun branch. The animals were divided into four groups:

1) Mature control group.
2) Mature experimental group (orally received 75 mg/kg TT daily for 14 days).
3) Immature control group.
4) Immature experimental group (orally received 75 mg/kg TT daily for 14 days).

---

*Corresponding author. E-mail: esfandiari.arash@gmail.com

Abbreviation: TT, *Tribulus terresteris.*

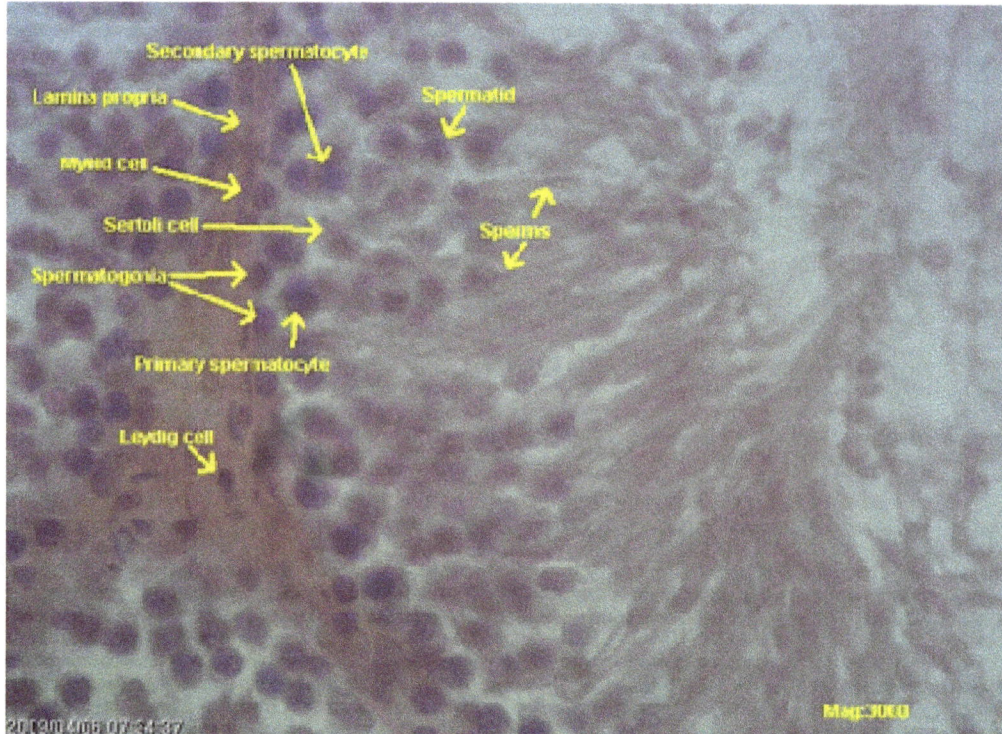

Figure 1. Testicular section from a mature control group shows the spermatogenic cells and histological structure of the seminiferous tubules (H and E ×3000).

All rats fed a standard diet and water. The rats were anesthetized using ether and the peritoneal cavity was opened through a lower transverse abdominal incision. The testes were carefully removed, washed in normal saline solution, blotted and weighed and transferred to 10% formalin buffer for 24 h. Specimens were embedded in paraffin and afterward five-micron thick section were prepared and stained with hematoxylin and eosin (H and E). These sections examined under light microscope and standard micrometry technique. The laboratory care, anesthesia and euthanasia of animals used in this study were performed in accordance with the guide to the care and use of experimental animals.

Statistical analysis

Statistical analysis was performed using SPSS version 16 software. Results were presented as the mean ± SD. Statistical comparisons were made by one way ANOVA. A probability value less than 0.05 was considered statistically significant.

RESULTS

The testes are surrounded by a thick capsule of dense connective tissue, the *Tunica albuginea*. The *Tunica albuginea* is continuous with connective tissue trabeculae. These trabeculae are rather complete septa. The septula testis divides the testicular parenchyma into a varying number of testicular lobules. The intertubular spaces contain loose connective tissue and leydig cells. The leydig cells are large cells with spherical nuclei and acidophilic cytoplasm (Figure 1). The TT is seminiferous tubules lined by the stratified germinal epithelium, surrounded by a lamina propria. Various spermatogenic cells, representing different phases in the development and sertoli cells are located in epithelium of seminiferous tubules (Figure 1). Sperm was not seen in ICG and also, in this group were leydig cells so little (Figure 2). After weighing testes, the mean of weight of testes in the MCG, MEG, ICG and IEG, were 1.56 ± 0.04, 2.14 ± 0.08, 0.8 ± 0.02 and 1.48 ± 0.03 g, respectively, (Table 1). Histomorpho metrical study showed that the mean of thickness of wall of seminiferous tubules in the MCG, MEG, ICG and IEG, were 39.66 ± 4.13, 73.99 ± 17.23, 21.81 ± 2.98 and 55.63 ± 4.39 μm, respectively, (Table 1). In addition, histological study showed that the accumulation of spermatogenic cells and leydig cells in seminiferous tubules were increased in MEG and IEG when compared with MCG and ICG (Figures 2, 3, 4 and 5). Also, sperm was seen in wall of seminiferous tubules in the IEG after used TT (Figure 4).

DISCUSSION

A herbal flavonoid drug and it contains saponin, steroid and different essence. It has a lot of effects such as increasing testosterone and dihydrotestosterone levels (El-tantawy, 2007; Gauthman, 2003, 2008), preventing cell death and destruction of mitochondrial membrane (Liu, 2008), antimicrobial and antifungal effects (Al_bayati,

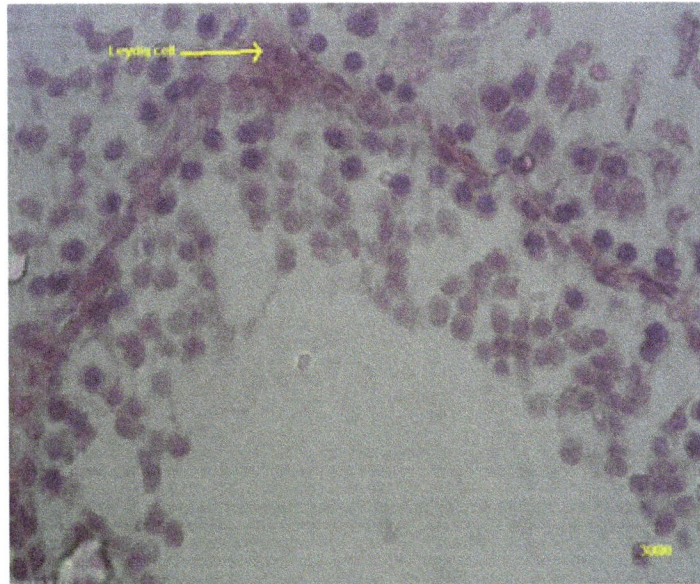

Figure 2. Testicular section from an immature control group shows sperm was not seen and leydig cells were so little (H and E × 3000).

Table 1. Effect of *Tribulus terresteris* on testis weight and thickness of seminiferous tubules (Mean ± SD).

Group		Weight of testis (gram)*	Seminiferous tubules thickness(μm)*
Mature	Control (MCG)	1.56 ± 0.04 a,c	39.66 ± 4.13 a,c
	Experimental (MEG)	2.14 ± 0.08 a,d	73.99 ± 17.23 a,d
	Control (ICG)	0.8 ± 0.02 b,c,d	21.81 ± 2.98 b,c,d
Immature	Experimental (IEG)	1.48 ± 0.03 b,d	55.63 ± 4.39 b,

MCG, Mature control group; MEG, mature experimental group; ICG, immature control group; IEG, immature experimental group. experimental groups received orally 75 mg/kg TT daily for 14 days.
*: The same superscript letter (a, b, c, d) in each column shows significant difference between groups (P < 0.05).

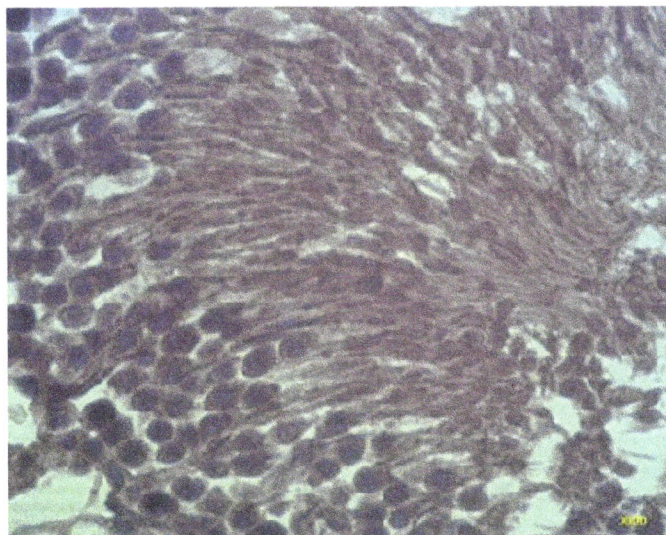

Figure 3. Testicular section from a mature experimental group shows increase the number of spermatogenic cells and leydig cells of seminiferous tubule (H and E × 3000).

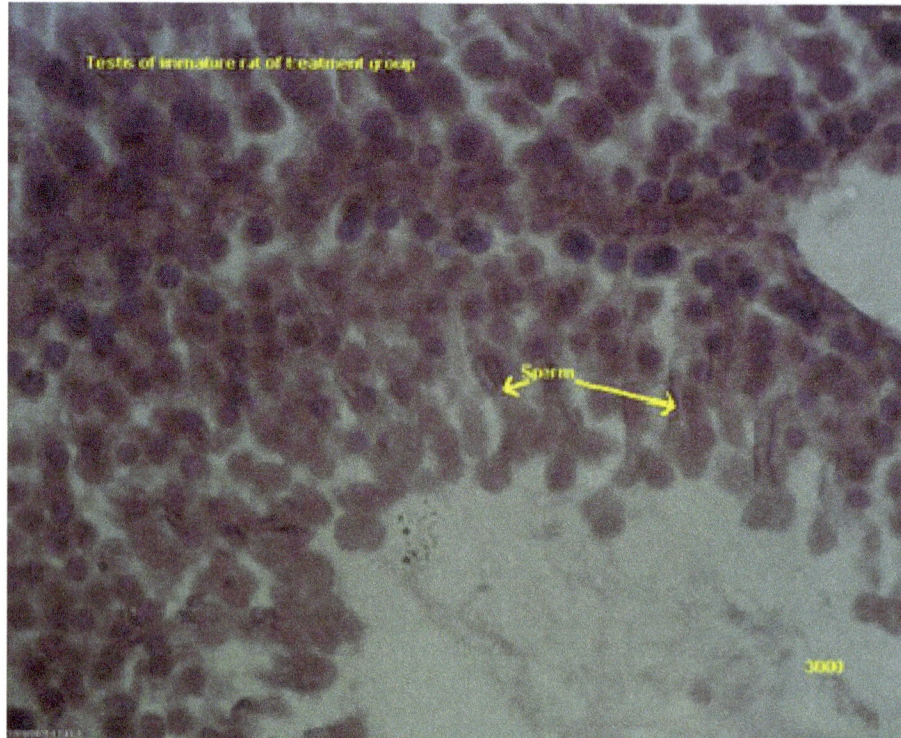

Figure 4. Testicular section from an immature experimental group shows increase the number of spermatogenic cells and leydig cells in seminiferous tubule, Sperms were seen (arrows) (H and E × 3000).

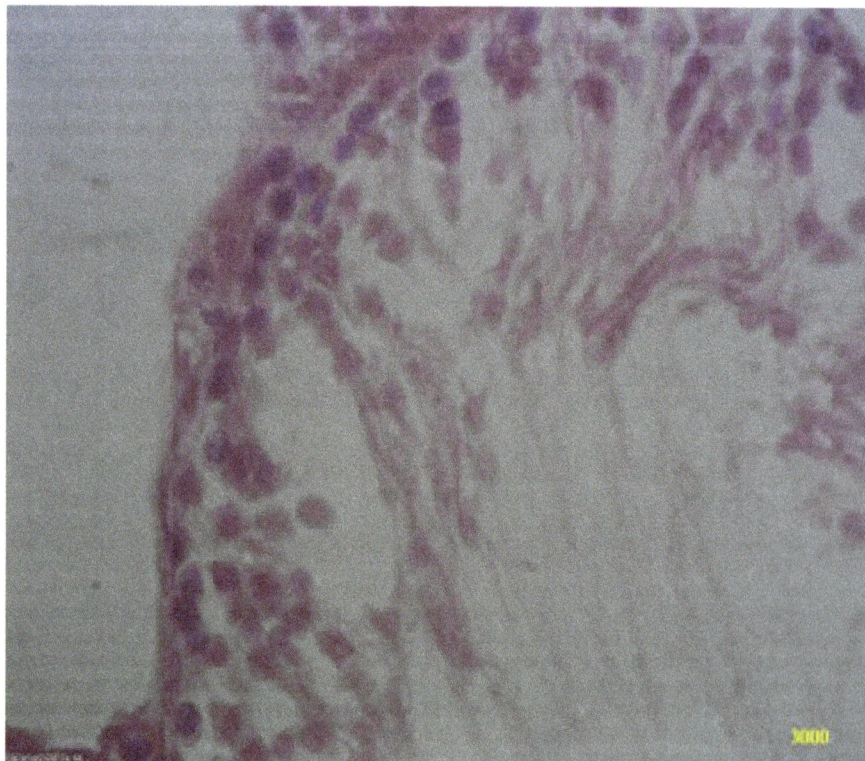

Figure 5. Testicular section from a mature control group shows the accumulation of spermatogenic cells of seminiferous tubule (H and E × 3000).

2008), decreasing blood sugar in diabetic patients, cholesterol and triglyceride (Chu, 2003; Li, 2002; El-tantawy, 2007), increasing muscular and fat mass (Rogerson, 2007), increasing androgen and sexual desires (El-tantawy, 2007), repairing of left ventricle of heart (Guo, 2007), increasing blood pressure of penile artery (Park, 2006) and increasing melanocyte-stimulating hormone (Yang, 2006). The results of our study showed that the thickness of wall of seminiferous tubules were significantly increased in experimental groups as compare to control groups ($P < 0.05$) but, there was no significant difference between MCG and IEG ($P > 0.05$). In addition, sperm was seen and leydig cells were increased after using TT in immature rats.

It seems that TT cause early puberty and it can increase testosterone levels and sexual desires. Our findings are essentially in agreement with El-tantawy et al. (2007) and Gauthman et al. (2003, 2008) findings. Significant increase was observed in weight of testes in experimental groups as compared with control groups ($P < 0.05$), but no significant difference between MCG and IEG was observed ($P > 0.05$). Maybe, these result due to increasing in function and compactness of cells of seminiferous tubules and consequently we have an increased excretion in these tubules. We concluded that TT may cause early puberty and it may increase testosterone levels and it can increase compactness of spermatogenic cells and sperms.

ACKNOWLEDGMENTS

This study was conducted under the sponsorship of the Islamic Azad University, Kazerun Branch.

REFERENCES

Al-Bayati FA, Al-Mola HF (2008). Antibacterial and antifungal activities of different parts of Tribulus Terrestris L. growing in Iraq. J. Zhejiang Univ. Sci. B. pp. 154-159.

Banks WJ (1993). Applied Veterinary Histology. 3rd Ed. Mosby Year Book pp. 429-502.

Chu S, Qu W, Pang X, Sun B, Huang X (2003). Effects of saponin from Tribulus Terrestris on hyperlipidemia. Zhong Yao Cai 26(5): 341-344.

El-Tantawy WH, Hassanin LA (2007). Hypoglycemic and hypolipidemic effects of alcoholic extract of Tribulus alatus in streptozotocin-induced diabetic rats: a comparatire study with T. terrestri (Caltrop). Indian J. Exp. Biol. 45(9): 785-790.

El-Tantawy WH, Temraz A, El-Gindi OD (2007). Free serum testosterone level in male rats treated with Tribulus Alatus extracts. Int. Br. J. Urol. 33(4): 554-559.

Gauthaman K, Ganesan AP, Prasad RN (2003). Sexual effects of productive (Tribulus Terrestris) extract (protodioscin): an evaluation using a rat model. J. Altern. Complement Med. 9(2): 257-265.

Gauthaman K, Ganesan AP (2008). The hormonal effects of Tribulus terrestris and its role in the management of male erectile dysfunction- an evaluation using primates. Phytomedicine 15(7-2): 44-54.

Guo Y, Shi DZ, Yin HJ, Chen KJ (2007). Effects of Tribuli saponins on rentricular remodeling after myocardial infarction in hyperlipidemic rats. Am. J. Chin. Med. 35(2): 309-316.

Li M, Qu W, Wang Y, Wan H, Tian C (2002). Hypoglycemic effect of saponin from Tribulus Terrestris. Zhong Yao Cai 25(6): 420-422.

Liu XM, Huang QF, Zhang YL, Lou JL, Liu HS, Zheng H (2008). Effects of Tribulus Terrestris L. suponion on apoptosis of cortical neurons induced by hypoxia-reoxygenation in rats. Zhong Xi Yi Jie He Xue Bao 6(1): 45-50.

Natwar RK, Mann A, Sharma RK, Aulitzkey W, Frick J (1995). Effects on human gamma interferon on mice testis: a quantitatire analysis of the spermatogenic cells. Aexta Eur Feril, 26(1): 45-49.

Park SW, Lee Ch, Shin DH, Bang NS, Lee SM (2006). Effects of SA1, a herbal formulation, on sexual behavior and penile erection. Biol. Pharm. Bull. 29(7): 1383-1386.

Rogerson S, Riches CJ, Jennings C, Weatherby RP, Meir RA, Gradisnik SM (2007). The effects of five weeks of Tribulus Terrestris supplementation on muscle strength and body composition during preseason training in elite rugby league players. J. Strength Cond. Res. 21(2): 348-353.

Sprando RL, Collins TF, Black TN, Rorie J, Ames MJ, O'donnell M (1997). Testing the potential of sodium fluoride to affect spermatogenesis in the rat. Food Chem. Toxicol. 35(9): 881-890.

Sprando RL, Collins TF, Wiesenteld P, Babu US, Rees C, Black T, Olejnik N, Roric J (2000). Testing the potential of flaxseed to affect spermatogenesis: morphometry. Food Chem. Toxicol. 38(10): 881-892.

Yang L, Lu Jw, An J, Jian X (2006). Effect of Tribulus Terrestris extract on melanocyte-stimulation hormone expression in mouse hair follicles. Nan Fang Yi Ke Du Xue Xue Bao 26(12): 7777-7779.

3

Effects of aqueous extract of *Sorghum bicolor* on hepatic, histological and haematological indices in rats

Akande, I. S.*, Oseni, A. A. and Biobaku, O. A.

Department of Biochemistry, Faculty of Basic Medical Sciences, College of Medicine, University of Lagos, P. M. B. 12003, Idi Araba, Lagos, Nigeria.

Herbal medicine is still the mainstay of about 75 - 80% of the world population, mainly in the developing countries for primary health care. In Nigeria today, there is an upsurge in the acceptance and utilization of these herbal medicine partly because of scientific support for some of their medicinal uses. In recent times, findings from medicinal plants research indicate that extracts from some plants both hepatotoxic and hematotoxic, while some are reported to possess both hepatoprotective and hemopoietic properties. We investigated the effects of aqueous extract of *Sorghum bicolor* leaf sheaths on the biochemical hepatic functions, histological integrity and hematological indices in Sprague-Dawley albino rats. Phytochemical screening of the aqueous extract of *S. bicolor* leaf sheath was carried out. Also, male and female rats (100 – 210 g) and divided into 5 groups were employed for this study. Four groups of 6 rats each were orally administered with 1.0 ml of 200, 400, 800 and 1600 mg/kg body weight daily doses of aqueous extract of *S. bicolor* leaf sheath, respectively for 14 days. The control group consisted of 6 rats treated to a daily dose of 0.5 ml of 0.9% normal saline. At the end of the administration period, the rats were sacrificed; the blood samples were collected through orbital sinus and cardiac puncture. The liver tissues were harvested and used for the hematopoietic and liver functions investigations. Phytochemical analysis of the plant leaf sheath showed the presence of Anthracine glycosides, reducing compounds, saponins, flavonoids, glycosides and polyphenols. Liver function tests revealed that the serum alanine amino transferase (ALT) concentration in the experimental rats showed a significant ($P \leq 0.05$) increase with the increases in dosage concentrations of the extract compared with the control. Aspartate amino transferase (AST) and alkaline phosphatase (ALP) as well as the concentrations of total protein and albumin in male and female experimental rats were not significantly ($P \geq 0.05$) altered compared with the control by the oral administration of the extract. However, red blood cell counts, hemotocrit and haemoglobin concentrations increased significantly ($P \leq 0.05$) on administration of the extract in both male and female experimental rats compared with the control. Histopathological examination did not reveal any lesion or alteration in the morphological features of the liver tissues in all the animals. Data of the present study indicate that aqueous extract of *S. bicolor* leaf sheath is both hepatoprotective and hematopoietic in male and female Sprague-Dawley rats. These findings are therefore of clinical importance given the various reported medicinal potentials of the plant.

Key words: *Sorghum bicolor*, liver, rats, aminotransferases, hemoglobin.

INTRODUCTION

The use of plants for remedies has long been in existence and is among the most attractive sources for developing drugs (Chevellier, 1996). Any part of plant can be considered as herbs including leaves roots, flower, seeds, resins, leaf sheath, bark, inner bark (cambium), berries and sometimes the pericarp or other portion. These ancient indigenous practices were discovered by series of 'trial and error' which then could not be substantiated by proven scientific theories (Holetz et al., 2002).

*Corresponding author. E-mail: akande_idowu@yahoo.com, iakande@unilag.edu.ng.

However, research has shown that these practices have produced results of proven efficacies comparable to conventional modern medicine. In recent times, herbal remedies have become indispensable and forming an integral part of the primary health care system of many nations (Nwogu et al., 2007).

Sorghum bicolor is locally called 'oka pupa' in the Southern part of Nigeria is one of the major grain crops for human food throughout the drier areas of Africa and India and its grain is extensively used for animal feeding, some of its common names are 'Jowar' in India, 'Bachanta' in Ethiopia and here in Africa it has variety of names such as 'Karan dafi' and the Yoruba's of Southern part of Nigeria call it 'Oka pupa' (Dalziel, 1948; Ogwumike, 2002). In West Africa, dye can be extracted from the plant to color leathers, cloths, calabashes and as body pigment (Cobley and Steele, 1976).

Sorghum is used largely for forage in the U.S. It is very important in the world as part of human diet, with over 300 million people dependent on it (Bukantis, 1980). It is grown for grain, forage, syrup and sugar, and industrial uses of stems and fibers. Grain sorghum is a staple cereal in hot dry tropics, the threshed grain ground into wholesome flour. It serve as stalks used as animal feed for silage or green soiling, or for hay when grown irrigated in very dry areas. The grain can be cooked like rice or ground into flour. Sorghum, with large juicy stems containing as much as 10% sucrose, used in manufacture of syrup; sugar can be manufactured from sorghum. Broomcorn used for making brooms. The seed is used as food, in brewing "kiffir beer", the kiffir corn malt and cornmeal is fermented to make *Leting* (a sour mash), the pith is eaten and the sweet culm chewed (Watt and Brayer-Brandwijk, 1962). Arubans make porridge and muffins from sorghum meal. Parched seed are used as coffee substitutes or adulterants (Morton, 1981).

However, the use of the leaf sheath as a remedy against anaemia (reduction of red blood cells or its function) by traditional medicine healers is common in Nigeria as well as within the local people of the Yoruba and Hausa tribes (Ogwumike, 2002). Malted *S. bicolor* grain is higher in protein and lower in fat content than corn and this is partly responsible for its haemopoietic ability (Makokha, 2002). It has been reported that sorghum can be used as antiabortive, cyanogenetic, demulcent, diuretic, emollient, intoxicant and poison. Sorghum is a folk remedy for cancer, epilepsy, flux and stomach ache (Duke and Wain, 1981). The root is used for malaria in Southern Rhodesia; the seed has been employed for the treatment of breast disease and diarrhoea while the stem has been used for tubercular swellings treatment. In India, the plant is considered antihelminthic and insecticidal and in South Africa, in combination with *Erigeron canadense*, it is used for eczema (Watt and breye-Brandwijk, 1962). In China, where the seed are used to make alcohol, the seed husk is braised in brown sugar with a little water and applied to the chest of measles patients. The stomachic seeds are considered beneficial in fluxes (Perry, 1980). According to Morton (Curacao natives drink the leaf decoction for measles, grinding the seeds with those of the calabash tree (*Cresentia*) for lung ailments. Venezuelans toast and pulverize the seeds for diarrhea. Brazilians decoct the seed for bronchitis, cough and other chest ailments, possibly using the ash for goiter. Arubans poultice hot oil packs of the seeds on the back of those suffering pulmonary congestion. According to Grieve (1931), a decoction of a 50 g seed to a litre of water is boiled down to half a litre as a folk medication for kidney and urinary complaints. However it has also been reported that dyes from *S. bicolor* may be carcinogenic (Owolagba et al., 2009 Avwioro et al., 2009). Avwioro et al. (2006) also reported that crude ethanolic extract of *S. bicolor* used as stain on red blood cells, collagen and muscle fibres indicated that the dye may be apigenindin in nature.

Recently focus has been on the leaf sheath of *S. bicolor* being used as herbal remedy for anaemia and having a boosting effect on blood concentration hematinic potentials (Ogwumike, 2002; Friday et al., 2010).

The rising cost of material services and medication including blood tonics are becoming unaffordable for many patients thus preventing these individuals from receiving adequate healthcare and since the importance of blood cannot be over emphasized, then any alternative, easy and locally available means of improving blood concentration is of paramount importance (Little, 2001). According to Okonkwo et al. (2004), accurate laboratory determination of blood parameters remains the only sensitive and reliable foundation for ethical and rational research, diagnosis, treatment and prevention of anaemia.

This study therefore investigated the hematopoietic potential of *S. bicolor* leaf sheath extract in rat model with the aim of showing the effects of aqueous extract of *S. bicolor* leaf sheath on the biochemical indices of liver function and some haematological parameters.

MATERIALS AND METHODS

Identification and preparation of plant Materials

Dry leaf sheaths of *S. bicolor* plants weighing 2.8 g were purchased in May 2008 from herb sellers at Mushin market, Lagos, Nigeria. The sample of the plant specimen was identified and authenticated at the herbarium of the Department of Botany, University of Lagos, Akoka and the specimen was deposited in the herbarium of the same University. The dry leaf sheaths of sorghum plants was air-dried for 3 weeks and subsequently dried to completion in an oven for 48 h at 60°C (after which the lightly coloured part of the leaf sheath were cut off leaving the dark red part of the leaf sheath) and ground into fine powder using electric dry mill. A total of 480 g of the ground powder divided into two (240 g each) was boiled in a 4.5 L of distilled water for 10 min. This was allowed to cool then sifted to remove shafts. The liquid extract was then concentrated using rotary evaporator to produce a gel-like extract which was further evaporated to dryness at 50°C using oven, which weighed 40.2 g.

Subsequently, appropriate concentrations of the extract were made by dilution with water into 200, 400, 800 and 1600 mg/kg body weight and administered to the rats.

Phytochemical screening of the leaf extract of *S. bicolor*

The phytochemical screening of the aqueous extract of the leaves *S. bicolor* was carried out according to the method of Trease and Evans (1983). Qualitative analysis of alkaloids, anthracine glycosides, flavonoids, glucosides, saponins, proteins, tannins and phenolic compounds were studied. Freshly prepared ground samples of *S. bicolor* leaf sheath extract from samples soaked in solvent overnight and filtered was employed for phytochemical screening.

Test for anthracine glycosides

A mixture of 0.5 ml dilute sulphuric acid and 5.0 ml ferric chloride solution was added to 1 ml of the extract. The resulting mixture was boiled for 5 min, cooled and filtered into a separator funnel. The filtrate was shaken with an equal volume of carbon tetrachloride. The lower organic layer was carefully separated into a test tube and 5.0 ml of dilute ammonia solution added with gentle shaking. Pink coloration in ammonia indicated the presence of anthracine glycosides.

Test for alkaloids

1.0 ml of extract of the sample was added and shaken with 5.0 ml of 2% HCl on a steam bath and filtered. Five drops of Meyer's reagent (potassium mercuric iodine solution) was then added to 1.0 ml of the filtrate and observed for cream colored precipitate which is a positive test for alkaloids.

Test for flavonoids

1.0 ml of 10% ferric chloride was added to 1.0 ml of extract. The formation of a greenish brown or black precipitate or color was positive test for phenolic nucleus. To 1.0 ml extract, 1.0 ml of dilute NaOH was added. Addition of 1.0 ml dilute NaOH to 1.0 ml extract gave a precipitate which shows presence of flavonoids.

Test for protein

5.0 ml of distilled water was added to 4.0 ml extract and allowed to stand for 3 h then boiled. 2.0 ml of the boiled mixture was then added to 0.1 ml of mercuric nitrate (Million's reagent) and shaken. A pinkish precipitate indicates presence of protein.

Test for saponins

1.0 ml of extract was boiled with 5.0 ml of distilled water for 5 min and decanted while still hot. The filtrate was used for the following test:

Frothing test

1.0 ml of the filtrate was diluted with 4.0 ml of distilled water shaken vigorously and observed on standing for stable froth which confirms the presence of saponins.

Emulsion test

Two drops of olive oil was added to 1.0 ml of filtrate. The solution was shaken and observed for formation of emulsion which confirms the presence of saponins.

Test for tannins

5.0 ml of extract was added to 2.0 ml of 1% HCl. Deposition of a red precipitate was an evidence for the presence of phlobotannins.

Test for glycoside

5.0 ml of extract was treated with 2.0 ml of glacial acetic acid containing 1 drop of 0.1% ferric chloride, and then mixed with 1.0 ml concentrated sulphuric acid. A brown ring of the interface indicates a deoxysugar characteristic of cardenolides.

Care and treatment of animals

18 males and 12 females Sprague-Dawley albino rats weighing 100 – 210 g were collected from the Laboratory Animal Centre of the Nigerian Institute of Medical Research, Yaba, Lagos, Nigeria. They were housed in wooden cages and allowed to acclimatize for two weeks, fed with rat chow and water *ad libitum* and under standard conditions (ambient temperature, 29.0 ± 2.0 °C and humidity 46%, with a 12 h light/darkness cycle). The rats were divided into five groups of six animals each. The animals were administered various agents as follows: the control group was not given the extract but was received a daily dose of 0.5 ml of 0.9% of normal saline while the second, third, fourth and fifth groups were treated to 1.0 ml 200, 400, 800 and 1600 mg/kg of the extract respectively in daily oral doses for 14 days. All the animal experiments were carried out in accordance with the guidelines of the Institutions Animal Ethical Committee.

Collection of blood and biochemical analysis

The rats were sacrificed after last day of extract administration and blood samples were collected from each animal through ocular bleeding into two sets of plain and EDTA-treated sample bottles, respectively.

The blood in the plain samples bottles were allowed to clot after 3 h. The clotted blood samples were spun in a bench top centrifuge to obtain sera. The serum samples were thereafter separated into another set of plain sample tubes and stored in the refrigerator pending enzyme assay. The whole blood collected into EDTA-treated sample bottles were used for the assay of haematological parameters. All assays were done within 24 h of the sample collection. The assay of alanine amino transferase (ALT), aspartate amino transferase (AST), alkaline phosphatise (ALP), total proteins and albumins assays were carried out according to the procedures described by Roche Laboratories Ltd, USA. Packed cell volume (PCV) was measured by the microhaematocrit technique using a Hawksley microhaematocrit centrifuge and spinning for 5 min at 12,000 xg before reading with the hematocrit reader. Red blood cell counts (RBC counts) were estimated using the hemocytometer method. Haemoglobin levels (Hb) were measured colorimetrically by the oxyhemoglobin methods using Reichert's haemoglobinnometer.

Histology

The rat liver from each group was fixed in 10% formol saline and for

72 h. The organ was dehydrated in graded alcohol, cleaned in xylene and embedded in paraffin. The resulting blocks were exhaustively sectioned. The sections were randomized, while the selected sections were stained in haemotoxylin and eosin. The slides were then examined at magnifications of × 400 under optical microscope.

Statistical analysis

The results from the study were analyzed by one-way analysis of variance (ANOVA) and Students t-test. Statistical significance was tested at $P \leq 0.05$.

RESULTS

The results of this study are presented in Tables 1 - 3, as well as Plates 1 - 4. The phytochemical screening of the leaf sheath indicated the presence of Anthracine glycosides, saponins flavonoids, glycosides and polyphenols (Table 1).

The results obtained in this present study for some serum liver enzymes, total proteins and albumin in normal male and female rats treated with aqueous extract of S. bicolor leaf sheath are presented in Table 2, while the results of hematological analysis are presented in Table 3. It was observed from these results that treatment of rats with aqueous extract of S. bicolor leaf sheath have no significant effect ($P \geq 0.05$) on the activities of the serum liver enzymes as well as the concentrations of serum total proteins and albumins, compared respectively with the control (Table 2).

This observation indicated that the aqueous extract S. bicolor leaf sheath did not show marked hepatotoxic effect in the animal model. Plates 1 and 2 show photomicrograph of the liver tissues of the test rats, while plates 3 and 4 show the photomicrograph sections of the liver tissues of the control rats, respectively. The liver section of the animal in control groups showed a central vein with prominent small-sized nuclei, with the hepatocytes well separated by sinusoids. While the tissue section of the test rats showed a prominent central vein with a relatively large-sized nuclei; also, the sinusoids separating the hepatocytes in the test rats are observed to be relatively more prominent than that of the rats in the control groups. Generally, the liver sections of rats in the control and test groups showed that the cords of hepatocytes well preserved, cytoplasm not vacuolated and the sinusoids well demarcated. Also, no area of infiltration by inflammatory cells and fatty degenerative changes were observed in the tissue sections. These features gave an indication of normal hepatic integrity for rats in both control and test groups.

The results of this study also showed that the red blood cell counts, hematocrit and hemoglobin concentrations increased significantly ($P < 0.05$) on administration of the extract to the rats. Also, no significant effect ($P > 0.05$) on the white blood cell counts was observed to be associated with treatment of the rats with extract (Table 3).

The different dosage administration of the aqueous extract of S. bicolor suggests the hematinic potentials of the plant extract.

Furthermore, very significant increase in the haematological indices took place with the administration of the highest dosage of the extract and it is an indication of the fact that the hematopoietic effects of the extract is dose dependent.

This present study therefore suggests that aqueous extract of S. bicolor leaf sheath is not hepatotoxic, but rather possess hemapoietic property in rats. These findings are therefore, of clinical importance given the various reported medicinal potentials of the plant.

DISCUSSION

The results of this study showed that the administration of the aqueous leaf extracts of S. bicolor has hemopoietic and antioxidant properties in the rat. The present study further confirmed the previous report on the hematinic potentials of S. bicolor (Ogwumike, 2002; Avwioro et al., 2009). The extract contains phenolic compounds and saponins which are powerful antioxidant. The medicinal value of this plant lies in those chemical substances that produce a definite physiological action on the human body such as alkaloids, flavonoids and tannins. This is in agreement with observations made previously (Hill, 1952; Trease and Evans, 1996). These phytochemicals are known to perform many functions in plants and may exhibit different biochemical and pharmacological actions in animal species when ingested (Duke and Wain, 1981; Owolagba et al., 2009). Saponins are known to possess hypocholesterolemic effects (Price et al., 1987) and as such its presence in the leaf sheath extract of S. bicolor may aid in lessening the oxidative stress on the liver. The results of this study also indicate that leaf sheath extract of S. bicolor may function as blood booster in anaemic condition and this possibly could be as a result of its direct effect on the hematopoietic systems (Friday et al., 2010).

Stimulations of hematopoietic growth factors and erythropoietin systems have been reported to enhance rapid synthesis of blood cells (Murray, 2000). It is also safe at the concentrations used in the experiment on the basis of the result of liver function tests and histological studies which did not indicate significant changes between the tests and the control. Avwioro et al. (2006) however, using crude ethanolic extract of S. bicolor to stain red blood cells, collagen and muscle fibres indicated that the dye may be apigenindin. However, result from oral administration of the same extract up to a concentration of 1500 mg/kg on albino rats did not show any adverse reaction while the rats administered 2500 mg/kg died. It is therefore necessary to conduct further works on this plant in order to obtain clearer pictures on its medicinal values since the works done so far are based on short term exposure of the experimented

Table 1. Phytochemical profile of aqueous extract of *Sorghum bicolor* leaf sheath.

Phytochemicals	Occurrence
Anthracine Glycosides	+ve
Alkaloids	-ve
Flavonoids	-ve
Glucosides	+ve
Protein	-ve
Saponins	
Frothing Test	-ve
Emulsion Test	+ve
Tannins	-ve
Phenolic compound	+ve

Note: +ve: present, -ve: absent

Table 2. Effect of *Sorghum bicolor* aqueous leaf extract on liver enzymes, total proteins and albumins levels of Sprague-Dawley albino rats.

Treatment group (mg/kg)	AST	ALT	ALP	Total serum protein	Serum albumin
Control Saline	89.83 ± 17.26	84.50 ± 3.33	154.34 ± 26.12	66.67 ± 15.22	27.88 ± 9.33
Extract 200	102 ± 11.08	86.6 ± 2.7	167.38 ± 18.37	65.80 ± 17.45	26.56 ± 6.21
Extract 400	106.6 ± 19.42	92. ± 4.18	148.24 ± 29.98	68.75 ± 12.22	29.06 ± 8.43
Extract 800	93.25 ± 9.54	88.20 ± 2.95	187.34 ± 20.54	62.63 ± 18.72	28.86 ± 5.11
Extract 1,600	104.8 ± 11.52	88.75 ± 0.5	129.5 ± 15.02	70.21 ± 13.85	30.18 ± 4.25

Table 3. Effect of *Sorghum bicolor* aqueous leaf extract on some hematological parameters of Sprague-Dawley albino rats.

Treatment group (mg/kg)	PCV (%) ± SD	Hb (g/dl) ± SD	RBC × 10^6mm^{-3}	MCV × 10^{-6}	MCH × 10^{-6} (Pg)	MCHC (g/dl)
Control saline	29.00 ± 4.16	9.60 ± 1.51	3.17 ± 0.49	9.16 ± 0.29	3.03 ± 0.10	31.26 ± 0.91
Extract 200	38.67 ± 4.13	12.81 ± 1.37	4.22 ± 0.48	9.18 ± 0.14	3.04 ± 0.05	32.02 ± 0.67
Extract 400	37.80 ± 1.48	12.52 ± 0.49	4.02 ± 0.18	9.41 ± 0.20	3.12 ± 0.07	32.51 ± 1.07
Extract 800	41.00 ± 2.94	13.58 ± 0.98	4.40 ± 0.29	9.32 ± 0.27	3.09 ± 0.09	33.28 ± 1.12
Extract 1,600	39.60 ± 1.95	13.11 ± 0.65	4.26 ± 0.26	9.30 ± 0.21	3.09 ± 0.09	32.02 ± 1.41

PCV = Packed cell volume or hematocrit; Hb = Hemoglobin; RBC = Red blood cell; MCV = Mean corpsclular volume; MCH = Mean corpuscular haemoglobin; MCHC = Mean corpuscular haemoglobin concentration.

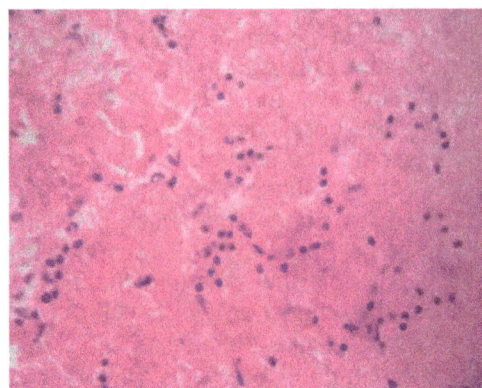

Plate 1. Photomicrograph of hepatocytes of male rats treated with extract of *S. bicolor*.

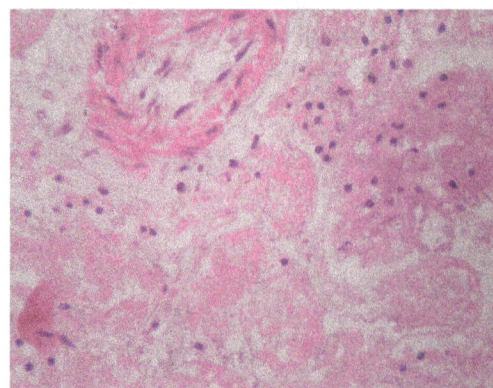

Plate 2. Photomicrograph of hepatocytes of female rats treated with extract of *S. bicolor*.

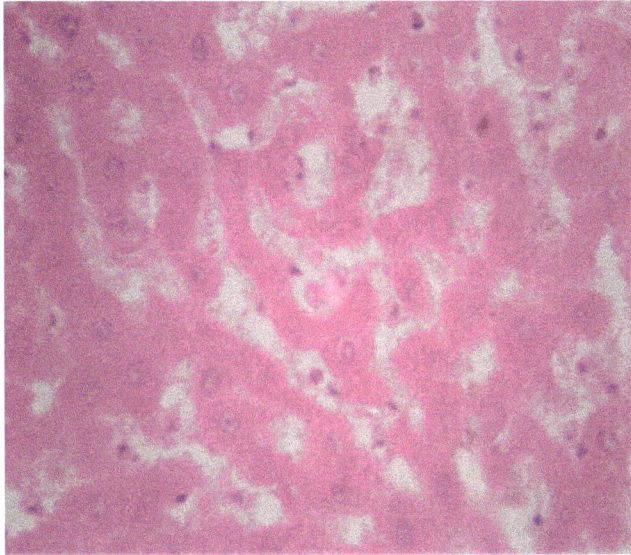

Plate 3. Photomicrograph of hepatocytes of male control rats.

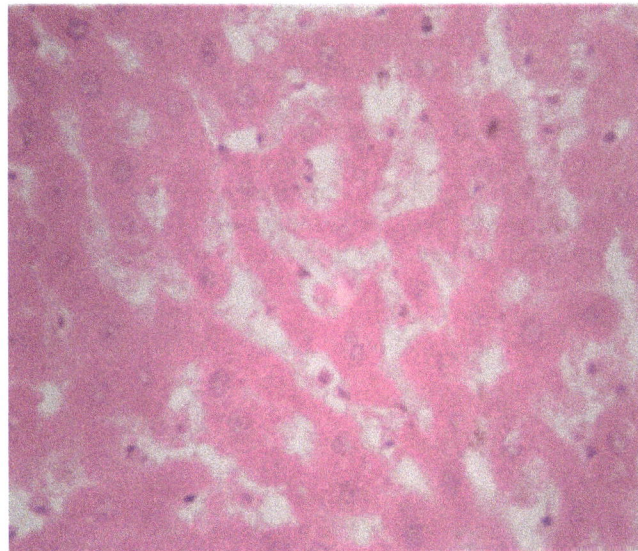

Plate 4. Photomicrograph of hepatocytes of female control rats.

animals.

Conclusion

Data of the present study do suggest that the leaf sheath extract of *S. bicolor* possess hemopoietic potential. There is no indication of the likelihood of it being hepatotoxic or hematotoxic in rats employed in this study. It could therefore be said to possess hepatoprotective and hematopoietic potentials.

REFERENCES

Avwioro OG, Aloamaka CP, Olabampe OB, Oduola T (2006). Collagen and muscle stain obtained from Sorghum bicolor. Scand J. Clin. Lab. Invest. 66: 161-168.

Avwioro OG, Owolagba GK, Anibor E, Bankole JK, Oduola T, Adeosun OG, Aloamaka CP (2009). Biochemical observations in Wistar rats fed with the histological dye extracted from Sorghum bicolor. Int. J. Med. Med. Sci., 1(10): 464-466.

Bukantis R (1980). Energy inputs in sorghum production, p. 103-108. In: Boca Raton, FL. Chevellier A(1996).The Encyclopedia of Medical Plant. London. Dorling Kindersley Ltd. (online), htt://www.chclibrary.org/plant.html.

Cobley LS, Steele WM (1976). 'An introduction to the Botany of Tropical crops'. 2nd edition, pp. 43: 52.

Dalziel JM (1948). 'The useful plants of West Africa'. Published by the Crown Agents for the colonies, London, p. 546.

Duke JA, Wain KK (1981). Medicinal plants of the World. In computer index more than 85,000 entries, 3 vols.

Eisenberg DN, Kessler RC, Foster C, Norlock EE, Calkins DR, Delbano TL (1993). Unconventional medicine in the United States: Prevalence costs and pattern or use. New England Med. J., 328: 246-252.

Friday EU, Iniobong, EO, Moses BE (2010). Effects of aqueous extract of Psidium guajava leaves on liver enzymes, histopathological integrity and haematological indices in rats. Gastroenterol. Res. Elmer Press, 3(1): 32-38.

Grieve M (1931). A modern herbal. Reprinted 1974. Hafner Press, New York.

Hill AF (1952). Economic botany. A textbook of useful plants and plant products. 2nd edition, McGraw-Hill book comp. Inc New York.

Holezt FB, et al. (2002). Screening of some plants used in the Brazilian folk medicine for the treatment of infectious diseases. Mem. Inst. Oswaldo Cruz; 6(7): 1027-1031.

Little M (2001). Health care rationing: Constraints and equity. Med. J. od Australia, 174: 641-642.

Makokha AO, Oniango RK, Njoroge SM, Kinyanjui PK (2002). Effects of malting on protein digestibility of some Sorghum (Sorghum bicolour) varieties grown in Kenya. Afr. J. Foods Nutr. Sci., 2(2): 168-175.

Morton JF (1981). Atlas of medicinal plants of Middle America. Bahamas to Yucatan C.C. Thomas, Springfield, IL.

Murray RK (2000). Red and White Blood Cells. In: Harperss Biochemistry (R.K. Granner, P.A. Mayes and V. W. Rodwell, eds.) McGraw-Hill, USA, pp. 780-786.

Okonkwo JE, Iyadi KC, Effiong CO (2004). Effect of chronic administration of haematological parameters of rats. Nig. J. Physiol. Sci., 19(1-2): 10-13.

Owolagba GK, Avwioro OG, Oduola T, Adeosun OG, Enaowho TN, Wilson JI, Ajumobi KO (2009). Histological observation of 24 h oral administration of the extract of Sorghum bicolor on albino rats. J. Cell Anim. Biol., 3(1): 001-003.

Perry LM (1980). Medicinal plants of east and Southern Acia. MIT Press, Cambridge.

Price KK, Johnson LI, Feriwick D (1987). The Chemical and Biological significance of saponins in food and feeding stuffs. CRC critical Rovigar in Food Sci. Nutri., 26: 127-135.

Trease GE, Evans WC (1996). Phenols and Phenolic glycosides In: Trease and Evans Pharmacognosy, 13th ed. Biliere Tindall, London, p. 832-833.

4

Essential fatty acid content of eggs and performance of Layer Hens fed with different levels of full-fat flaxseed

Huthail Najib* and Yousef M. Al-Yousef

Department of Animal and Fish Production, College of Agricultural Sciences and Food, King Faisal University, Al-Hofuf 31982, Saudi Arabia.

An examination of the earlier studies in the USA and Canada demonstrate that flax seeds are good source of omega 3 fatty acids. This experiment studies the effect of providing laying hens, with various levels of roasted and unroasted locally produced flax seeds. A key components of this experiment involved attentively observing the laying hens performance, and determining the fatty acids content of the eggs produced. Five levels of flax seeds 0, 5, 10, 15 and 20 dry weight %, roasted or unroasted, were fed to 200 pullets in 5 replicates (4 birds/cage). The results indicated that feeding 5 or 10% roasted flax seed supported good egg production. Birds fed higher levels of unroasted flax seed had the lowest feed consumption. Livability, egg weight, yolk color and specific gravity values were not significantly affected by feeding flax seeds. Feeding 15% unroasted flax seeds maintained higher omega-3 levels: that is docosahexaenoic acid (DHA), eicosapentaenoic acid (EPA), docosapentaenoic acid (DPA) and alpha-linolenic acid ($C_{18:3n3}$) levels in egg, whereas feeding 5 or 15 weight % unroasted flax seeds resulted in the highest level of linoleic acid ($C_{18:2n6}$) in the egg. Roasting the seeds did not improve the omega-3 content of the egg. Feeding flax, regardless of heat treatment, marginally increased the amount of cholesterol. The saturate palmitic acid ($C_{16:0}$) was lower at 15% flax inclusion. We conclude that 10% flax seed added to feed supports good egg production. However, 15% inclusion of unroasted flax may relatively lower the egg production rate but would support an excellent profile of omega 3 fatty acids in the egg.

Key words: Docosahexaenoic acid, eicosapentaenoic acid, docosapentaenoic acid, α-linolenic acid, production, flax seeds.

INTRODUCTION

Fish products are high in n-3 fatty acids such as eicosapentaenoic acid (EPA), 20:5, n-3 and docosahexaenoic acid (DHA), 22:6, n-3 but fish may not be readily available everywhere. In addition, a fishy taint in meat and egg products might occur, which is unacceptable. Therefore, other sources for n-3 fatty acids are needed (Novak and Scheideler, 2001).

Brown-seeded flax (*Linum usitatissimum* L.) is one of the most concentrated sources of omega-3 unsaturated fatty acid available in natural feedstuffs for poultry (Caston and Leeson, 1990; Jiang et al., 1991). Common varieties of flaxseeds have very high concentration of polyunsaturated fatty acids, especially α-linolenic acid (Genser, 1994).

Despite the beneficial effect of using flax seed in layer diets, there has been some concern about toxicity associated with flax seed. Aymond and Van Elswyk (1995) reported a decrease in egg production in hens consuming 15% flax seed. Scheideler et al. (1995) showed that egg production was depressed with the addition of 10% flax seed to the diet, which could be the result of anti-nutritional factors contained with full-fat flax seed. Components such as cyanogenic glycosides (linustatin, neolinustatin and linamarin; Oomah et al., 1992) generate prussic acid and thiocyanate that can be detrimental to the animal. Cyanogenic compounds can be removed from flax seeds by boiling the seeds in water, wet autoclaving or by acid treatments, followed by autoclaving (Mazza and Oomah, 1995). The current experiment studied the effect of different levels of roasted or unroasted flax seeds on the performance and on the egg fatty acids components of hens.

*Corresponding authors:E-mail: huthailn@yahoo.com.

Table 1. The proximate analysis of the flax seed.

Description	Unit	Results
Moisture	%	5.68
Crude Protein	%	25.91
Ether Extract	%	34.95
Crude Fiber	%	4.00
Ash	%	2.52
Salt as NaCl	%	0.12

*All the results are expressed as is on dry matter.

Table 2. The fatty acids composition of flax seeds[1].

Fatty acid profile of flax seed used	Unit	Result[2]
palmitic acid (saturate)	%	10.77
Palmitoleic acid (monounsaturate)	%	0.14
stearic acid (saturate)	%	9.53
oleic acid (monounsaturate n-9)	%	23.71
linoleic acid (omega-6)	%	15.66
alpha-linolenic acid (omega-3)	%	38.43
docosapenaenoic acid (n-3)	%	0.05
arachidonic acid (omega-6)	%	0.065
eicosapentaenoic acid (n-3)	%	0.52
eicosadienoic acid (n-6)	%	0.06
behenic acid (saturate)	%	0.21
eicosatrienoic acid	%	0.06
lignoceric acid (saturate)	%	0.24

[1]All samples were analyzed based on AOAC 996.06 method.
[2]The results are expressed as relative % of fatty acids in fat.

MATERIALS AND METHODS

Samples of ground flax seeds were subjected to chemical analysis (AACC, 1994) (Table 1). Cholesterol and fatty acids in flax seeds and yolks were determined by Gas Chromotographic method using direct saponification (AOAC, 2000) (Table 2). Lipids were saponified at high temperature with ethanolic KOH. The unsaponifiable fraction, containing cholesterol and other sterols, was extracted with toluene. Sterols were derivatized to trimethylsilyl (TMS) ethers and then quantified by gas chromatography.

Flax seeds were heat treated using large trays exposed to direct heat from a gas stove during which they were turned several times until there was a change in color to dark brown. The temperature of the seeds was supervised by checking the roasting temperature at various intervals. The result being the highest roasting temperature reached at 120°C.

During the early and growing periods, standard management practices were maintained for 300 white leghorn chicks. Starting at week 19, the birds were fed a commercial laying diet containing 17% protein. Since flax seeds are a new and unfamiliar ingredient to chicks, and in order to avoid feed rejection the dietary treatments were introduced gradually in the following pattern: During the first 2 days the birds were fed 10% treatment diets and 90% commercial layer diets. During the next 2 days, the mixture given to the birds was 20% dietary treatments and 80% commercial diets. This pattern of feeding continued in the same manner until the 22nd week of age when the birds started to be fed 100% of the experimental diets. The feeding trial continued for 12 – 28 day periods. The composition of the diets is shown in Table 3.

Each diet was supplied to five cages containing 4 birds each. There were eight 4 week dietary treatment periods and an identical control group that was not fed flax. Performance criteria such as hen-day egg production and egg weight were made, a rotated biweekly. At the end of each 2 week period, three days of egg collection were used for shell quality determination, Haugh unit (albumin height) and yolk color. Specific gravity method was used to measure the shell quality of the eggs (North, 1984). Feed was given ad-libitum daily. At the end of the experiment, 2 eggs from each replicate were collected randomly for cholesterol and fatty acid analysis.

The data was analyzed using ANOVA and SAS General Linear Model procedure (SAS, 1989). Means of the treatments were compared by Duncan Multiple Range Test (Duncan, 1955).

RESULTS AND DISCUSSION

Interaction between the level of the flax seeds in the diet and roasting treatment was significant (P < 0.05) in most of the examined parameters.

Birds fed 5 or 10 weight % roasted flax seed had a comparable production rate to the controls which had no flax, while those fed unroasted flax had a lower production rate than controls (P < 0.05) (Table 4).

Table 3. The composition of the dietary treatments.

Feed ingredients	Flax level				
	control	5%	10%	15%	20%
Yellow corn	60.1	55.5	53.0	48.5	45.0
Wheat bran	0	1.0	1.0	3.0	3.5
SBM, 48%	25.2	22.9	20.8	18.5	16.5
Fish meal, 67%	3	3.0	3.0	3.0	3.0
Choline Cl, 60%	0.4	0.27	0.25	0.1	0.09
Salt	0.4	0.4	0.4	0.4	0.4
Dl-Methionine	0.2	0.2	0.22	0.2	0.22
L-Lysine	0.1	0.08	0.1	0.1	0.2
Di-Ca P., 18%	0.6	0.6	0.6	0.6	0.6
Limestone	8.31	8.39	8.4	8.35	8.45
Alfalfa	0	1.0	1.0	1.0	1.0
Liquid fat	1.49	1.51	1.03	1.05	0.84
MINVIT [premix1]	0.2	0.2	0.2	0.2	0.2
Flax seeds	0	5.0	10.0	15.0	20.0
Total	100	100	100	100	100

[1]The multi vitamin-minerals premix provides the following, per kilogram of mix: 7000000 IU, Alfalfa vit A; 1500000 ICU, vit D3; 30000 IU, vit E; 50000 mg, vit C; 2300 mg, vit K; 1400 mg, vit B1; 5520 mg, vit B2; 2300 mg, vit B6; 12 mg, vit B12; 27600, mg Niacin; 920 mg, Folic acid; 6900 mg, PA; 92 mg, Biotin; 50000 mg, Antioxidant (BHT); 220 mg, Cobalt; 4400 mg, copper; 800 mg, Iodine; 26400 mg, Iron; 44000 mg, Manganese; 180 mg, Selenium.

Table 4. The effect of feeding different levels of roasted or un-roasted flax seed on production traits of single comb white Leghorn hens.

Source of variation	Feed consumption g/b/d	Flax intake g/b/d	Feed conversion Kg/kg	Hen-day egg production, %	Egg weight g	Egg mass g/HD
Interaction	**		*	*	NS	NS
0 T	121.3 ± 5.3	0.00 ± 0.00	2.019 ± 0.26	92.71 ± 8.51	63.91 ± 9.96	60.78 ± 6.07
5 T	121.7 ± 3.3	6.08 ± 0.17	2.041 ± 0.30	93.82 ± 9.40	63.20 ± 10.14	60.82 ± 8.76
10 T	119.6 ± 5.4	11.94 ± 0.57	2.052 ± 0.72	93.35 ± 12.61	64.95 ± 3.68	60.61 ± 9.68
15 T	118.1 ± 9.1	17.59 ± 1.75	2.114 ± 0.39	89.88 ± 13.15	63.66 ± 3.17	57.26 ± 8.98
20 T	120.7 ± 8.9	24.9 ± 1.99	2.106 ± 0.31	91.60 ± 9.90	63.55 ± 3.77	58.21 ± 7.11
0 U	122.3 ± 4.2	0.00 ± 0.00	2.036 ± 0.21	94.31 ± 7.66	64.67 ± 3.43	60.67 ± 6.07
5 U	126.6 ± 14.6	6.34 ± 0.72	2.248 ± 0.45	89.24 ± 13.26	64.61 ± 3.67	57.66 ± 8.80
10 U	123.5 ± 13.1	12.36 ± 1.38	2.182 ± 0.39	90.51 ± 10.40	63.60 ± 4.15	57.71 ± 8.07
15 U	117.4 ± 5.6	17.58 ± 1.07	2.093 ± 0.33	90.07 ± 10.56	63.62 ± 3.44	57.27 ± 7.89
20 U	115.4 ± 8.9	22.84 ± 2.50	2.076 ± 0.39	88.68 ± 13.31	64.43 ± 4.02	56.94 ± 8.46
P	0.0001	0.0001	0.0118	0.0405	0.1301	0.1574

[1]Means within columns carrying different superscripts are significantly different, $P < 0.05$. NS = Not significant, $P > 0.05$. ** Significant at 1% level of probability, LVL = 0, 5, 10, 15 and 20% of Flax seeds, TRT = T, roasted / U, Un-Roasted flax seeds.

Scheideler et al. (1995) showed that egg production was depressed with the addition of 10% flax seed to the diet, which could be attributed to anti-nutritional factors contained in full-fat seed. Whether roasting the seeds in this experiment was efficient in controlling the anti-nutritional factors in the seeds is open for speculation. Novak and Scheideler (2001) reported that addition of flaxseed to the diet of laying hens did not have any adverse effects on egg production parameters, but flaxseed supplementation can significantly alter weight of yolk solids and yolk and albumen percentages. These factors were not measured in this study.

With the exception of the 20% flax addition, feed consumption of the birds fed different levels of roasted or unroasted flax seeds decreased significantly ($p < 0.01$) with increasing amounts in the diet (Table 4). However,

Table 5. The effect of feeding different levels of roasted or unroasted flax seed on some egg characteristics and livability of single comb white leghorn hens[1].

Source of variation	Livability, %	Haugh unit	Yolk color	Specific gravity
Levels	**NS**	**NS**	**NS**	**NS**
0	100.0[a]	92.79a	3.066a	1.088a
5	100.0[a]	93.39a	2.980a	1.089a
10	99.9[a]	92.49a	3.163a	1.088a
15	100.0[a]	92.15a	3.066a	1.088a
20	99.9[a]	92.34a	3.033a	1.089a
P =	0.2887	0.4254	0.3697	0.9881
Treatments	**NS**	** ** **	**NS**	**NS**
T	99.96[a]	92.06a	3.014a	1.089a
U	99.95[a]	93.20b	3.110a	1.088a
P =	0.8696	0.0098	0.0956	0.9344
Interaction	**NS**	**NS**	**NS**	**NS**
0 T	100.0 ± 0.0	92.53 ± 4.76	3.042 ± 0.92	1.088 ± 0.006
5 T	100.0 ± 0.0	92.12 ± 3.99	2.850 ± 0.96	1.088 ± 0.007
10 T	100.0 ± 0.0	92.39 ± 3.94	3.128 ± 0.77	1.088 ± 0.006
15 T	100.0 ± 0.0	91.66 ± 4.24	3.000 ± 0.88	1.089 ± 0.007
20 T	99.8 ± 1.8	91.62 ± 4.11	3.048 ± 1.00	1.089 ± 0.006
0 U	100.0 ± 0.0	93.04 ± 4.42	3.090 ± 0.92	1.088 ± 0.006
5 U	100.0 ± 0.0	94.66 ± 16.90	3.111 ± 0.86	1.089 ± 0.006
10 U	99.8 ± 1.2	92.59 ± 4.17	3.198 ± 0.81	1.088 ± 0.007
15 U	100.0 ± 0.0	92.64 ± 5.50	3.132 ± 0.83	1.088 ± 0.006
20 U	100.0 ± 0.0	93.07 ± 4.49	3.018 ± 0.93	1.088 ± 0.006
P	0.0840	0.4800	0.5805	0.7956

[1]Means within columns carrying different superscripts are significantly different, $P < 0.05$. NS = Not significant, $P > 0.05$, ** Significant at 1 % level of probability; LVL = 0, 5, 10, 15 and 20% of Flax seeds, TRT = T, Roasted / U, Un-Roasted flax seeds.

when calculations were based on flax consumption per bird per day, there was a gradual increase in flax consumption as level of flax seeds increased in the diet. This certainly reflected on the omega-3 fatty acids content of the egg (Tables 4 and 6). Van Elswyk (1997) reported a linear increase of linolenic acid in response to increasing levels of ground and whole flax seeds. The effect of flax inclusion on egg mass, egg weight, specific gravity, yolk color, haugh unit and livability were not significant (P>0.05) (Table 5). Bean and Leeson (2003) in a long term study found that egg production, egg weight, shell weight, albumen height and shell thickness were not significantly (P<0.05) different for hens consuming 0 and 10% flax seed. The highest level of flax seed in their experiment was 10%.

Incorporation of 15% unroasted flax seed significantly increased the level of omega-3 fatty acids (DHA, DPA, EPA and alpha-linolenic acids) in the yolk fat by 194% DHA (2.9 - fold), 300% DPA (4-fold), 450% EPA (5.5-fold) and 876% alpha-linolenic acid (9.8-fold) compared with controls (Table 6). Linoleic acid (an omnipresent

omega-6 fatty acid) increased by 49% in birds fed 15% unroasted flax seeds while the linoleic:alpha-linolenic acid ratio reduced from 16:1 in the control birds to 2.4:1 in the flax fed birds (Table 6). These changes in the fatty acids profile of the egg yolk (primarily an increase in omega-3 fatty acids) were somewhat unique and very important to human health (Leaf, 1999).

Caston and Leeson (1990) reported a linoleic: alpha-linolenic ratio of 3:1 in the 10% flax fed birds compared to 37:1 in the control. Oleic acid was lower with birds fed flax diets while Arachidonic acid was better for diets containing 10 and 15% roasted or unroasted flax seeds (Table 7). Caston and Leeson (1990) reported a large increase in omega-3 fatty acids in all levels in the eggs when feeding 10, 20 and 30% flaxseed to laying hens for a 28-day period and collecting eggs for analysis in the last 3 days of the period. Cherian and Sim (1991) fed flax seed to laying hens at 8 and 16% in diets supplemented with pyridoxine (vitamin B6). They reported increased omega-3 fatty acids in the eggs and in brain tissue of embryos and chicks from the hens fed

Table 6. The unsaturated fatty acids profile of yolks of layers fed different levels of roasted or unroasted flax[1].

Source of variation	Docosahexonoic acid	Docosapento-noic acid	Eicosapentae-noic acid	Alpha-linolenic acid	Linoleic acid
Interaction	**	**	**	**	**
U * 0	0.350 ± 0.014	0.040 ± 0.000	0.020 ± 0.000	0.565 ± 0.007	8.860 ± 0.099
U * 5	0.865 ± 0.007	0.105 ± 0.007	0.050 ± 0.000	2.945 ± 0.021	13.740 ± 0.028
U * 10	0.990 ± 0.042	0.135 ± 0.007	0.085 ± 0.007	3.125 ± 0.064	11.900 ± 0.028
U * 15	1.030 ± 0.056	0.160 ± 0.000	0.110 ± 0.000	5.515 ± 0.233	13.215 ± 0.332
U * 20	0.455 ± 0.064	0.060 ± 0.000	0.035 ± 0.007	1.575 ± 0.021	10.365 ± 0.078
T * 0	0.240 ± 0.028	0.020 ± 0.000	0.010 ± 0.000	0.690 ± 0.042	8.975 ± 0.304
T * 5	0.595 ± 0.007	0.105 ± 0.007	0.050 ± 0.000	3.235 ± 0.035	13.145 ± 0.148
T * 10	0.925 ± 0.064	0.120 ± 0.000	0.060 ± 0.000	2.435 ± 0.120	12.225 ± 0.148
T * 15	0.395 ± 0.007	0.100 ± 0.000	0.070 ± 0.014	4.495 ± 0.021	13.640 ± 0.042
T * 20	0.385 ± 0.021	0.065 ± 0.007	0.040 ± 0.000	3.240 ± 0.028	11.965 ± 0.035
P	0.0001	0.0001	0.0010	0.0001	0.0001

U = Untreated flax seeds with heat, T = Treated flax seeds with heat (roasted) ** significant $P < 0.001$; 0, 5, 10, 15, 20 = percent level of flax seeds, [1]fatty acids were measured as percent of yolk fat.

Table 7. The Fatty acids profile and cholesterol of layer egg yolks fed different levels of roasted or unroasted flax seeds[1].

Source of variation	Arachidonic acid	Oleic acid	Palmitoleic acid	Palmitic acid	Cholesterol
Interaction	**	**	NS	**	NS
U * 0	0.625 ± 0.021	37.175 ± 0.02	4.650 ± 0.070	36.210 ± 0.0	0.685 ± 0.049
U * 5	0.645 ± 0.007	34.180 ± 0.14	5.065 ± 0.007	32.750 ± 0.30	0.960 ± 0.000
U * 10	0.715 ± 0.021	36.000 ± 2.23	5.030 ± 0.141	29.855 ± 0.81	0.905 ± 0.007
U * 15	0.715 ± 0.035	35.170 ± 0.38	4.320 ± 0.127	29.815 ± 0.70	1.010 ± 0.000
U * 20	0.475 ± 0.049	35.110 ± 0.18	5.285 ± 0.658	35.040 ± 1.02	0.845 ± 0.007
T * 0	0.470 ± 0.000	38.845 ± 0.83	4.110 ± 0.141	34.755 ± 1.0	0.738 ± 0.117
T * 5	0.605 ± 0.007	34.665 ± 0.09	3.895 ± 0.035	32.110 ± 0.04	0.878 ± 0.025
T * 10	0.700 ± 0.028	32.760 ± 0.66	4.165 ± 0.106	35.125 ± 0.81	0.820 ± 0.043
T * 15	0.480 ± 0.014	29.580 ± 0.44	3.570 ± 0.042	34.325 ± 0.01	0.921 ± 0.018
T * 20	0.420 ± 0.042	32.860 ± 0.03	4.215 ± 0.007	34.950 ± 0.18	0.838 ± 0.041
P =	0.0010	0.0006	0.3531	0.0001	0.1843

** Significant $P<0.001$. NS Not significant, $P > 0.05$, [1]fatty acids and cholesterol were measured as percent of yolk fat.

ground flaxseed. Aymond and Van Elswyk (1995) reported that feeding both 5 and 15% flaxseed caused increased total omega-3 fatty acids in the eggs and that ground seeds caused a greater level of these fatty acids at the 15% level of feeding than used whole seed.

It is worth noting that the alpha-linolenic acid content of the flax seed, used in this study, was much lower (38.43%) (Table 2) than what is commonly reported (52%, Kratzer and Vohra, 1996). This could be why alpha-linolenic acid did not plateau in yolk at 10% addition of flax seed, as reported by some researches. On the other hand, the subsequent observed decline in alpha-linolenate at 20% flax seed in the feed could be due to toxicity, as the consumption of flax seeds increased from 17 gm to 24 and 23 gm roasted or

unroasted flax, respectively (Table 4).

Increasing the amount of unroasted flax to 15% of feed reduced the saturate palmitic acid in the egg yolk while it increased at a 20% addition. The percentage of cholesterol in yolk fat increased significantly with increasing levels of flax, compared with controls (Table 8). However, this increase was not consistent and there was no move in this direction. On the other hand, Caston and Leeson (1990) reported that the amount of cholesterol in egg was not influenced by dietary flax.

It is conclude that adding flax seeds to the feed of layers significantly improved the fatty acid content of the egg yolk. More specifically, and importantly, it increased their omega-3 fatty acid content, an important dietary player in the prevention of cardiovascular disease

Table 8. The fatty acids and cholesterol profile of yolks of layers fed different levels of heat treated or un treated flax seeds[1].

Source of variation	Archidonic acid	Oleic acid	Palmit-oleic acid	Palmitic acid	Stearic acid	Cholesterol
Between treatments	**	**	**	**	NS	NS
U	0.635[a]	35.527[a]	4.870[a]	32.734[a]	8.967[a]	0.881[a]
T	0.535[b]	33.742[b]	3.991[b]	34.253[b]	9.506[a]	0.839[a]
P	0.0001	0.0006	0.0001	0.0003	0.2587	0.0648
Between levels	**	**	**	**	NS	**
0	0.548[c]	38.010[a]	4.380[a]	35.482[a]	9.270[a]	0.711[d]
5	0.625[b]	34.422[b]	4.480[a]	32.430[b]	8.402[a]	0.919[ab]
10	0.708[a]	34.380[b]	4.598[a]	32.490[b]	9.778[a]	0.862[bc]
15	0.598[b]	32.375[c]	3.945[b]	32.070[b]	9.438[a]	0.966[a]
20	0.448[d]	33.985[b]	4.750[a]	34.995[a]	9.295[a]	0.842[c]
P	0.0001	0.0001	0.0051	0.0001	0.4425	0.0001

** Significant P < 0.001. NS Not significant, P > 0.05, [1]fatty acids and cholesterol were measured as percent of yolk fat.

(Leaf, 1999). However, the production performance was somehow jeopardized.

ACKNOWLEDGMENT

The financial support of the Deanship of Scientific Research at King Faisal University is highly appreciated.

REFERENCES

American Association Of Cereal Chemists (AACC) (1994). Official methods of analysis St. Paul Minesota. USA).

Association of Official Analytical Chemist International, J. AOAC Int. (995) 78, 75. Revised : June 2000.

Aymond WM, Van Elswyk ME (1995). Yolk thiobarbituric acid reactive substances and n-3 fatty acids in response to whole and ground flaxseed. Poultry Sci. 74: 1388-1394.

Bean LD, Leeson S (2003). Long-term effects of feeding flaxseed on performance and egg fatty acid composition on brown and white hens. Poult. Sci. 82: 388-394.

Caston LJ, Leeson S (1990). Dietary flax and egg composition. Poult. Sci. 69: 1617 1620.

Cherian G, Sim JS (1991). Effect of feeding full fat flax and canola seeds to laying hens on fatty acid composition of eggs, embryos and newly hatched chicks. Poult. Sci. 70: 917-922

Duncan DB (1955). Multiple range and F-tests, Biometrics 11: 1-42.

Genser MV (1994). Description and composition of flax seed. Pages 9-14 in: Flax Seed, Health, Nutrition and Functionality. The Flax Council of Canada. Winnipeg, MB, Canada.

Jiang ZD, Ahn U, Sim JS (1991). Effects of feeding flax and two types of sunflower seeds on fatty acid composition of yolk lipid classes. Poult. Sci. 70: 2467-2475.Kratzer FH, Vohra P (1996). The use of flaxseed as a Poultry feedstuff. Poultry fact sheet NO. 21. Cooperative extension, University of California, Avian Sciences Department, Davis, CA 95616.

Leaf A (1999). Dietary Prevention of Coronary Heart Disease, The Lyon Diet Heart Study. American Heart Association, Inc, Circulation 99: 733-735.

Mazza G, Oomah BDB (1995). Flax seed, dietary fiber, and cyanogens. Pages 56-81 in: Flaxseed in human nutrition. S. C. Cunnane and L. U. Thompson, ed. AOCS Press, Champaign, IL.

North MO (1984). Commercial chicken production manual. 3rd ed., AVI Publishing company, Inc. Westport, Connecticut.

Novak C, Scheideler SE (2001). Long-term effects of feeding flaxseed-based diets. 1. Egg production parameters, components, and eggshell quality in two strains of laying hens. Poult. Sci. 80: 1480-1489.

Oomah BD, Mazza G, Kenaschuk EO (1992). Cyanogenic compounds in flaxseed. J. Agric. Food Chem. 40: 1346-1348.

SAS Institute, SAS/STAT® (19890. SAS User's Guide: Ver 6, 4th ed Vol 1. SAS Inst. Inc., Cary, NC.

Scheideler SE, Jaroni D, Froning G (1995). Strain and dietary oats effect on flax fed hen's egg composition and fatty acid profile. Poult. Sci. 74(suppl. 1): 165. (Abstr.).

Van Elswyk ME (1997). Nutritional and physiological effects of flax seeds in diets for laying fowl. Wolrl's Poult. Sci. J. 53: 253-264.

5

Breed characteristics in Iranian native goat populations

B. Mahmoudi[1]*, **M. Sh. Babayev**[2], **F. Hayeri Khiavi**[1], **A. Pourhosein**[1] and **M. Daliri**[3]

[1]Islamic Azad University, Meshkinshahr Branch, Meshkinshahr, Ardabil, Iran.
[2]Department of genetic, Faculty of Biology, Baku State University, Baku, Azerbijan.
[3]Department of Animal Science, Genetic Engineering and Biotechnology Institute, Tehran, Iran.

Investigation of genetic relationship among populations was traditionally based on the analysis of allele frequencies at different loci. The aim of this study was to analyze, the genetic diversity and variability of three native Iranian goat populations (Raeini, Korki Jonub Khorasan and Lori) through the use of 13 microsatellite markers. The 13 tested loci were all polymorphic in the three goat populations. Within the 13 polymorphic loci, allele frequencies, number of effective alleles (Ne), heterozygosity (He), polymorphism information content (PIC) and Nei's standard genetic distance (D) were calculated, and UPGMA phylogenetic tree was constructed based on allele frequencies. The average number of alleles was 7.57, ranging from 3 to 13 at the 13 assessed loci. The average values of Ne, He and PIC of all loci were 5.14, 0.797 and 0.757 respectively. Korki Jonub Khorasan showed the highest mean number of alleles (8.15), while the highest value for polymorphic information content was observed for Raeini population (0.78). Tests of genotype frequencies for deviation from the Hardy-Weinberg equilibrium (HWE), had been tested in the level of probability ($p<0.005$). A UPGMA diagram based on Nei's standard genetic distances, yielded relationships between populations that agreed with what is known about their origin, history and geographical distribution.

Key words: Hardy-Weinberg equilibrium (HWE), goat, diversity, microsatellite.

INTRODUCTION

The maintenance of genetic diversity in livestock species requires the adequate implementation of conservation priorities and sustainable management programs, which should be based on comprehensive information regarding the structure of the populations, including sources of genetic variability among and within breeds. Species are the most recognized and protected units of biodiversity. Yet we tend to ignore the importance of biodiversity that is fundamental to new species (Crawford and Littlejohn, 1998). Genetic diversity is shaped by past population processes and affects the sustainability of species and populations in the future (Soule, 1987). The maintenance of genetic diversity is a key to the long-term survival of most species (Hall and Bradley, 1995). Farm animal genetic diversity is required to meet current production needs in various environments, to allow sustained genetic improvement, and to facilitate rapid adaptation to

changing breeding objective (Crawford and Littlejohn, 1998; Kumar et al., 2006). Genetic variation between and within breed is described as diversity. It is essential to characterize a breed for its conservation. Microsatellites are ideal molecular markers for characterization. This study attempted to analyze the diversity of three goat populations in the Iran by using thirteen microsatellites as molecular markers, so as to help breeders to implement rational decisions for conservation and improvement of valuable germplasm.

If genetic diversity is very low, none of the individuals in a population may have the characteristics needed to cope with the new environmental conditions or challenges. Such a population could be suddenly wiped out. Low amounts of genetic diversity increase the vulnerability of populations to catastrophic events such as disease outbreaks. Low genetic diversity may also indicate high levels of inbreeding with its associated problems of expression of deleterious alleles or loss of over-dominance. Change in the distribution of the pattern of genetic diversity can destroy local adaptations and break up co-adapted gene complexes. These problems combine to lead to a

*Corresponding author. E-mail: Bizhan.mahmoudi@gmail.com

poorer 'match' of the population to its habitat increasing and eventually leading to the probability of population or species extinction. Microsatellite markers, also known as simple sequence repeats (SSRs) or short tandem repeats (STRs), are regions of DNA that exhibit short repetitive sequence motifs. Because of their high degree of polymorphism, random distribution across the genotypes, microsatellite markers have been proved to be one of the most powerful tools for evaluating genetic diversity snd estimating genetic distances among closely populations of ruminant species (Moore et al., 1991; Buchanan et al., 1994; Ellegren et al., 1997). Microsatellite are highly polymorphic and randomly markers are the simple sequence motif not more than six bases long, that is tandomly repeated for example (dC-dA)n. Microsatellite being polymorphic, they provide extremely useful markers for comparitive study of genetic variation, parentage control, linkage map analysis and could well be the marker of choice for analysis of population structure in domestic species. There are close similarities between cattle, sheep and goat chromosomes (Crawford et al., 1994; Kemp et al., 1995; Vaiman et al.,1996).

Microsatellite markers present in all three specice could be amplified with the same primer paire, so microsatellite markers developed in cattle and sheep also work in goats (Vaiman et al., 1996) and they can be used for the analysis of genetic diversity (Saitbekova et al., 1999). Indigenous livestock breeds are considered, for diverse reasons, as treasured genetic resores that tend to disappear as a result of new market demands, crossbreeding or breed replacement, and mechanized agricultural operations. There is terrible risk that most breed may perish before they have been exclusively recognized and exploited. The existance of a large gene pool is important for the potential future breeding preservation and for the development of a sustainable animal production system. Comprehensive knowlege of the existing genetic variability is the first step for the conservation and exploitation of domestic animal biodiversity (Li et al., 2002). The three distinct goat populations of Iran: Raeini, Korki Jonob Khorasan and Lori mainly distributed in Kerman, khorasan and Lorestan province respectively. The breeds were phenotypically characterized but their genetic characterization was pending. Hence, this present study is aimed to characterize Lori, Korki Jonob Khorasan and Raeini goats (Figures 1A, B and C) at molecular level using microsatellite markers with the following objectives:

1) To estimate the alleles and allelic frequencies of different microsatellite gene loci and to establish a microsatellite profile for Lori, Korki Jonob Khorasan and Raeini goat using thirteen polymorphic markers.
2) To estimate the percent heterozygosity and polymerphic information content (PIC).
3) To assess genetic variability and relationships within and among the three goat populations.

4) To analyses the population for Hardy-Weinberg equilibrium.
5) To determine time of divergence for all three populations.

MATERIALS AND METHODS

Blood samples from unrelated animals of Raeini (49) (Figure 1C), Korki Jonob Khorasan (51) (Figure 1B) and Lori (53) (Figure 1A) populations of goat were collected at random from their respective home tract. Then bleeding was transferred to laboratory (in an ice-cooled box, where they were kept under -20℃ in a deep freezer until DNA isolation) and DNA genomic was extracted by salting out method (Miller et al., 1988). We use both spectrophotometery and agarose gel (0.8%) for DNA quality definition. Hence, this study used 13 microsatellite primer pairs including MAF64, BM4621, BM121, LSCV36, TGLA122, oarJMP23, oarFCB304, oarAE133, ILSTS005, ILSTS022, ILSTS029, ILSTS033 and ILSTS34. Most of the primers used were independent and belonged to different chromosomes. These loci in prior studies had been amplified on the goat (Maudet et al., 2001; Yang et al., 1999; Hanrahan et al., 1994; Dixit et al., 2008). They showed polymorphism in the goat of world. Thirteen microsatellite markers, their sequences, type of repeat, size rang and their location is shown in Table 1. All PCR reactions were continued the following component: 200 μM dNTPs, 3.5 to 6 mM $MgCl_2$, 0.25 μM each of primer, 0.5 unit Taq DNA polymerase, 150 ng DNA. The final volume was 15 μl. Reactions were run on a thermal cycler (Biometra 96 block T-gradient, Germany). In this study annealing temperature was modified as follows: MAF64 (62.5℃), BM4621 (58℃), LSCV36 (55℃), oarFCB304 (60.5℃) and BM121 (65.5℃). The rest of PCR process is in accordance with Table 2. For oarJMP23 and TGLA122 primers used PCR program (Crawford et al., 1995), for oarAE133 used PCR program (Hanrahan et al., 1994) and For ILSTS005, ILSTS022, ILSTS029, ILSTS033 and ILSTS34 primers, the 'touchdown' PCR protocol was used. PCR products were separated on a 10% polyacrylamid gel and detected by silver staining and visualized under white light on a BIO-RAD Gel Doc XR system. The alleles and genotypic frequencies directly were identified from the gel. Hardy-Weinberg equilibrium (HWE) had been tested based on likelihood ratio for different locus-population combinations and observed number of alleles (N), effective number of alleles (Ne) and expected hetrozygosity (He) were computed by the software POPGENE (Version 3.2).

Polymorphism information content (PIC) was computed according to Botstein et al. (1980). Nei's standard genetic distance were calculated by POPGENE (Version 3.2), a phylogenetic tree was constructed by unweighted pair group method with arithmetic mean (UPGMA) method based on pair wise Nei's standard distances using the same software by a bootstrapping method.

Data analysis

Genotypes were assigned for each animal based on allele size data. On the basis of allele and genotypic frequencies, a likelihood ratio test (G^2_T) was conducted to test for deviations from Hardy-Weinberg equilibrium (Guo and Thompson, 1992). The most common measures of genetic diversity such as allelic diversity, heterozygosity and proportion of polymorphic loci were considered. The effective number of alleles (estimates the reciprocal of homozygosity) was calculated according to Hartl and Clark (1989). Nei unbiased expected heterozygosity (He = $1-\Sigma p^2_i$; where p_i is the frequency of allele (i) were estimated for all loci (Nei, 1978). These parameters were statistically analyzed using POPGENE software

Figure 1A. Lori goats.

Figure 1C. Raeini goat.

Figure 1B. Korki Jonob Khorasan goat.

package version 1.31 (Yeh et al., 1999). Polymorphism information content (PIC) (Botstein et al., 1980) values were estimated in order to assess the relevance of each locus for linkage. The Hardy-Weinberg expected hetrozygosity also defined as "gene diversity" (Weir, 1996) or polymorphism index content (PIC) (Botstein et al., 1980), was obtained from observed allele frequencies.

RESULTS

All the markers were successfully amplified in all the populations. Each 13 loci were found to be polymorphic in all populations. Korki Jonub Khorasan populations do not show the deviation of Hardy-Weinberg equilibrium (HWE). Lori and Raeini in some of loci showed the deviation of Hardy-Weinberg equilibrium (HWE). Most and least unbiased expected hetrozygosity is for K.J.K. (0.809) and Lori (0.778) respectively. The population statistics generated by the thirteen microsatellite markers in three goat populations is presented in Table 3. Yang et al. (1999) He value of oarFCB304 locus estimated 0.854 on Chinese goats but it was 0.708, 0.702 and 0.635 in Lori, K.J.K. and Raeini goats populations respectively.

The number of observed alleles for each locus ranged from 3 to 13. Highest number of allele's objective for oarJMP23 locus with the Raeini and for TGLA122 locus with the K.J.K. goats. Highest and lowest number of allele effective was 8.7 and 2.6 for oarJMP23 locus in Raeini and oarAE133 locus in K.J.K., respectively. All the average number of allele objective and effective was 7.67 and 5.14 respectively. Highest and lowest PIC value was 0.778 and 0.725 for Raeini and Lori respectively; it was between 0.746 to 0.8 in Chinese goats (Yang et al., 1999). The average expected hetrozygosity overall loci in Raeini, Lori and K.J.K. are 0.805, 0.778 and 0.809, respectively. The mean effective number of alleles, polymorphism information content and expected hetrozygosity over all populations were 5.14, 0.797 and 0.755, respectively. No significant difference in the number of alleles, Ne, He and PIC was found between these goat populations.

Among the three populations, K.J.K. populations displayed the highest values for mean He and Raeini PIC, while the Lori populations showed lower variability levels (Table 3). The standard genetic distances were calculated for these populations. The closest distance was observed between Raeini and K.J.K. (D = 0.4891) and the largest between Raeini and Lori (D = 0.6298), (Table 4; Figure 2).

DISCUSSION

The study of genetic variation plays an important role in developing rational breeding strategies for economical animal species (Maudet et al., 2002). The advantage of the use of microsatellites for estimating genetic variations among breed and among closely related populations has been investigated in farm animals such as, water buffalo (Barker et al., 1997; Moioli et al., 2001; Kumar et al., 2006), cattle (MacHugh et al., 1997), sheep (Gutierrez-Gil et al., 2006) and goat (Barker et al., 2001; Maudet et al., 2002). In the present study thirteen microsatellite loci were used to evaluate the genetic diversity within and

Table 1. Microsatellite markers, their sequences, type of repeat, size rang and location.

Locus	Primer sequence	Type of repeat	Size range	Chromosome No.
BM121	TGGCATTGTGAAAAGAAGTAAAA CTAGCACTATCTGGCAAGCA	$(TC)_{18}$	165-185	16
BM4621	CAAATTGACTTATCCTTGGCTG TGTAACATATGGGCTGCATC	$(CA)_{14}$	106-148	6
ILSTS005	GGAAGCAATGAAATCTATAGCC TGTTCTGTGAGTTTGTAAGC	$(nn)_{39}$	174-190	10
ILSTS022	AGTCTGAAGGCCTGAGAACC CTTACAGTCCTTGGGGTTGC	$(GT)_{21}$	186-202	Ann
ILSTS029	TGTTTTGATGGAACACAGCC TGGATTTAGACCAGGGTTGG	$(CA)_{19}$	148-191	3
ILSTS033	TATTAGAGTGGCTCAGTGCC ATGCAGACAGTTTTAGAGGG	$(CA)_{12}$	151-187	12
ILSTS34	AAGGGTCTAAGTCCACTGGC GACCTGGTTTAGCAGAGAGC	$(GT)_{29}$	153-185	5
LSCV36	GCACACACATACACAGAGATGCG AAAGAGGAAAGGGTTATGTCTGGA	$(CA)_{16}$	524	19
MAF64	AATAGACCATTCAGAGAAACGTTGAC CTCATCGAATCAGACAAAAGGTAGG	$(TG)_{13}$	121-125	1
oarAE133	AGCCAGTAGGCCCTCACCAGG CCAACCATTGGCAGCGGGAGTGTGG	$(TG)_{24}$	152	Ann
oarFCB304	CCCTAGGAGCTTTCAATAAAGAATCGG CGCTGCTGTCAACTGGGTCAGGG	$(CT)_{11}(CA)_{15}$	119-169	Ann
oarJMP23	GTATCTTGGGAGCCTGTGGTTTATC GTCCCAGATGGGAATTGTCTCCAC	-	-	27
TGLA122	AATCACATGGCAAATAAGTACATAC CCCTCCTCCAGGTAAATCAGC	$(CA)_{21}$	145	21

between Lori, K.J.K. and Raeini goat populations reared in Iran. The thirteen microsatellite are all polymorphic in the three goat populations. The use of microsatellites to evaluate the genetic diversity on the basis of allele frequency distribution has also been employed to differentiate between Italian, Greek and Egyptian buffalo populations (Moioli et al., 2001). The average expected heterozygosity overall loci in Lori, K.J.K. and Raeini are 0.778, 0.809 and 0.805 respectively. High value of average expects heterozygosity within the populations could be attributed to the large allele numbers detected in the tested loci (Kalinwski, 2002). The average direct count of heterozygosity overall loci in each of the three goat populations is less than the expected heterozygosity. This finding is an evidence for the presence of overall loss in heterozygosity within the three tested goat populations (allele fixation) (De Araujo et al., 2006).

Test of genotype frequencies for deviation from HWE at each locus over all populations showed, Najdi goat populations except in two loci (LSCV41 and BM121), in other loci revealed significant departure from HWE. Deviatin from HWE at microsatellite loci have also been

Table 2. PCR reaction conditions for all loci exceptional TGLA122, oarJMP23 and oarAE133 loci.

Stage	PCR process	Temperature (°C)	Time
1	Denaturation	95	2.5 min
2	Denaturation	95	30 s
3	Anealing	-	30 s
4	Extension	72	30 s
5	Final extension	72	2.5 min
6	Maintenance	4	-

Table 3. Mean numbers of alleles per locus, Ne, He, PIC and related SD (Standard division) on three goat populations.

Population	Mean number of alleles	Ne	He	PIC
Raeini	7.85(2.67)	5.38(1.65)	0.80(0.07)	0.78(0.08)
Korki Jonub Khorasan	8.15(2.51)	5.34(1.51)	0.81(0.07)	0.76(0.09)
Lori	7.00(1.78)	4.7(1.21)	0.78(0.06)	0.73(0.08)

Table 4. Nei (1978) genetic distance (D) in three goat populations.

Goat populations	Raeini	Korki Jonub Khorasan
Raeini		
Korki Jonub Khorasan	0.5031	
Lori	0.6158	0.5831

```
                +------------------------------------------------ KORKI JONOB KHORASA
        +------------1
    ---2            +------------------------------------------------ RAEENI
        !
        +--------------------------------------------------------- LORI
```

Figure 2. UPGMA of 3 native goat populations by Nei (1978) genetic

reported in various studies (Braker et al., 2001; Hassan et al., 2003; Laval et al., 2000; Luikart et al., 1999). It is known that a population is considered to be within HWE only when it is able to maintain its relative allele frequencies. Heterozygosis deficiency is one of the parameters underlying departure from HWE. Heterozygosis deficiency may results from one or more of the following reasons:

i) The presence of a null allele which is the allele that fails to multiply during PCR using a given microsatellite primer due to a mutation at the primer site (Callen et al., 1993; Pemberton et al., 1995);
ii) Small sample size, where rare genotypes are likely to be included in the samples;
iii) The Wahlund effect, that is presence of fewer heterozygotes in population than predicted on account of population subdivision;
iv) The decrease in heterozygosity due to increased consanguinity (inbreeding) (Kumar, 2006).

The information obtained in this study will aid their rational development, utilization and conservation. The result of UPGMA was consistent with the background of the origin, history and geographical location of these populations. The UPGMA tree shows that two goat populations (K.J.K. and Raeini) are distinct from the other goat population (Lori). The close kinship between Najdi and Tali might suggest some past crossing between these two geographically close populations. However, only a small number of microsatellite loci populations were analyzed. Additional markers and samples are required to increase the accuracy of the results.

ACKNOWLEDGEMENTS

This work was supported by the Islamic Azad University, Meshkinshahr branch. We wish to extend in us deep and sincere thanks for Dr. E. Asadpour, Chairperson of Islamic Azad University, Meshkinshahr branch for providing us facilities during this research project.

REFERENCES

Barker JSF, Moore SS, Evans DJS, Tan SG, Byrne K (1997). Genetic diversity of Asian water buffalo: Microsatellite variation and Comparison with protein-coding loci. Anim. Genet., 28: 103-115.

Barker JSF, Tan SG, Moore SS, Mukherjee TK, Matheson JL, Selvaraj OS (2001). Genetic variation within and relationship among populations of Asian goats (Capra hircus). J. Anim. Breed. Genet., 118: 213-233.

Botstein D, White RL, Skolnick M, Davis RW (1980). Construction of genetic linkage map in man using restriction fragment length polymorphism. Am. J. Hum. Genet., 32: 314-331.

Buchanan FC, Adams LJ, Litlejohn RP, Maddox JF, Crawford AM (1994). Determination of evolutionary relationships among sheep breeds using microsatellites. Genomics, 22: 397-403.

Callen DF, Thompson AD, Shen Y, Phillips HA, Richards RI, Mully JC, Sutherland GR (1993). Incidence and origin of "null" alleles in the (AC)n Microsatellite markers. Am. J. Hum. Genet., 52: 922-927.

Crawford AM, Dodds KG, Pierson CA, Ede AJ, Montgomery GW, Garmonsway HG, Beattie A E, Davies K, Maddox JF, Kappes SW, Stone RT, Nguyen TC, Penty JM, Lord EA, Broom JE, Buitkamp J, Schwenger W, Epplen JT, Matthew P, Matthews ME, Hulme DJ, Beh KJ, McGraw RA, Beattie CW (1995). An autosomal genetic linkage map of the sheep genome. Genetics, 140: 703-724.

Crawford AM, Littlejohn RP (1998). The use of DNA markers in deciding conservation priorities in sheep and other livestock. AGRI., 23: 21-26.

Crawford AM, Montgomery GW, Pierson CA, Brown T, Dodds KG, Sunden SLF, Henry HM, Ede AJ, Swarbrick PA, Berryman T, penty JM, Hill DF (1994). Sheep linkage mapping:Nineteen linkage groups derived from the analysis of paternal half-sib families. Genetics, 137: 573-579.

De Araujo AM, Guimaraes SEF, Machado TMM, Lopes PS, Pereira CS, Da Silva FLR, Rodrigues MT, Columbiano VDS, Da Fonseca CG (2006). Genetic diversity between herds of Alpine and Sannen dairy goats and the naturalized Brazilian Moxoto breed. Genet. Mob. Biol., 29(1): 67-74.

Dixit SP, Verma NK, Ahlawat SPS, Aggarwal RAK, Kumar S, Chander R, Singh KP (2008). Molecular genetic characterization of Kutchi breed of goat. Curr. Sci., 95(7): 946-952.

Ellegren H, Moore S, Robinson N, Byrne K, Ward W, Sheldon BC (1997). Microsatellite evolutional reciprocal study of repeat lengths at homologous loci in cattle and sheep. Mol. Biol. Evol., 14: 854-860.

Guo SW, Thompson EA (1992). Performing the exact test of Hardy-Weinberg proportions for multiple alleles. Biometrics, 48: 361-372.

Gutierrez-Gil B, Uzun M, Arranz J, Primitivo FS, Yildiz S, Cenesiz M, Bayon Y (2006). Genetic diversity in Turkish sheep. Acta Agric. Scand., Section A- Anim. Sci., 56(1): 1-7.

Hall SJG, Bradley DG (1995). Conserving livestock breed biodiversity. Trends. Ecol. Evol., 10: 267-270.

Hanrahan V, Ede AJ, Crawford AM (1994). Ovine x-chromosome microsatellite at the oarAE25 and oarAE133 loci. Anim. Genet., 25: 370-375.

Hartl DL, Clark AG (1989). Effective allele number. In: Principle of population genetics, 2nd Edn. Sinaure Associates, Stunderland, M.A.(Eds.), p. 125.

Hassan AA, Abou MHA, Oraby HA, De Hondt El, Nahas SM (2003). Genetic Diversity of three Sheep Breed in Egypt Based on Microsatellite Analysis. J. Eng. Biotechnol., (NRC) 1(1): 141-150.

Kalinwski ST (2002). How many alleles per locus should be used to estimated genetic distances. Heredity, 88: 62-65.

Kemp SJ, Hishida O, Wambugu J, Rink A, Longeri ML, Ma RZ, Da J, Lewin HA, Barendse W, Teale AJ (1995). A panel of polymorphic bovine, ovine and caprine microsatellite markers. Anim. Genet., 26: 299-306.

Kumar S, Gupta T, Kumar N, Dikshit K, Navani N, Jain P, Nagarajan M (2006). Genetic variation and relationships among eight Indian riverine buffalo breeds. Mol. Ecol., 15: 593-600.

Laval G, Iannuccelli N, Legault C, Milan D, Groenen MAM, Giuffra E, Andersson L, Nissen PH, Jorgensen CB, Beeckmann P, Geldermann H, Foully JL, Chevalet C, Olivier L (2000). Genetic diversity of eleven European pig breeds. Genet. Select. Evol., 32: 187-203.

Li MH, Zhao SH, Bian C, Wang HS, Wei H, Liu B, Yu M, Fan B, Chen SL, Zhu MJ, Li SJ, Xiong TA, Li K (2002). Genetic relationships among twelve Chinese indigenous goat populations based on microsatellite analysis. Gen. Select. Evol., 34: 729-744.

Luikart G, Biju-Duval MP, Ertugral O, Zagdsuren Y, Maudet C, Taberlet P (1999). Power of 22 microsatellite markers in fluorescent multiplexes for parentage testing in goats (Capra hircus). Anim. Genet., 30: 431-438.

MacHugh DE, Shriver MD, Loftus RT, Cunningham P, Bradley DG (1997). Microsatellite DNA variation and the evolution, domestication and phylogeography of Taurine and Zebu cattle (Bos taurus and Bos indicus). Genetics, 146: 1071-1086.

Maudet C, Luikart G, Tberlet P (2001). Development of Microsatellite Multiplexes for wild goats using Primers Designed from Domestic Bovidea. Genet. Select. Evol., 33: 193-203. ISSN: 0999-193X, cote INIST: 1441893540001032389.0110.

Maudet C, Miller C, Bassano B, Breitenmoser-Wursten C, Gauthier D, Obexer-Ruff G, Michallet B, Taberlet P, Luikart G (2002). Microsatellite DNA and recent statistical methods in wildlife conservation management: Applications in Alpine Ibex (Capra ibex). Mol. Ecol., 11: 421-436.

Miller SA, Dykes DD, Polesky HF (1988). A simple salting out procedure for extracting DNA from human nucleated cells. Nucl. Acids Res., 16: 1215-1215.

Moioli B, Georgoudis A, Napolitano F, Catillo G, Giubilei E, Ligda CH, Hassnane M (2001). Genetic diversity between Italian, Greek and Egyptian buffalo populations. Livest.Prod. Sci., 70: 203-211.

Moore SS, Sarageant LL, King TJ, Mattick JS, Georges M, Hetzel DJS (1991). The conservation of dinucleotid microsatellites among mammalian genomes allows the use of heterologous PCR primer pairs in closely related species. Genomics, 10: 645-660.

Nei M (1978). Estimation of average heterozygosity and genetic distance from a small number of individuals. Genetics, 89: 583-590.

Pemberton JM, Slate J, Bancroft DR, Barrett JA (1995). Non amplifying alleles at microsatellite loci: A caution for parentage and population studies. Mol. Ecol., 4: 249-252.

Saitbekova N, Gaillard C, Obexer-Ruff G, Dolf G (1999). Genetic diversity in Swiss goat breeds based on microsatellite analysis. Anim. Genet., 30: 36-41.

Soule M (1987). Viable populations for conservation. Cambridge Univ. press, Cambridge.

Vaiman D, Schibler L, Bourgeois F, Oustry A, Amigues Y, Cribiu EP (1996). A genetic linkage map of the male goat genome. Genetics, 144: 279-305.

Yang L, Zhao SH, Li K, Peng ZZ, Montgomery GW (1999). Determination of Genetic Relationships among five Indigenous Chinese Goat Breeds with Six Microsatellite Markers. Anim. Genet., 30: 452-455.

Yeh FC, Yang R, Boyle T (1999). POPGENE Version 3.1, Microsoft Window–based Freeware for Population Genetic Analysis, University of Alberta. Edmonton, AB, Canada.

6

The effects of sildenafil ciltrate on the medial geniculate body of adult Wistar rats (*Rattus norvegicus*)- A histological study

A. O. Eweka[1]* and A. B. Eweka[2]

[1]Department of Anatomy, School of Basic Medical Sciences, College of Medical Sciences, University of Benin, Benin City, Edo State, Nigeria.
[2]School of Nursing, University of Benin Teaching Hospital, Benin City, Edo State, Nigeria.

The histological effect of oral administration of sildenafil citrate (Viagra), commonly used as an aphrodisiac and for the treatment of erectile dysfunction on the medial geniculate body of adult Wistar rat was carefully studied. The rats of both sexes (n=24), average weight of 202 g were randomly assigned into three treatment (n=18) and control (n=6) groups. The rats in the treatment groups 'A', 'B' and 'C' received respectively, 0.25, 0.70 and 1.43 mg/kg body weight of sildenafil citrate base dissolved in distilled water daily for 30 days, through orogastric feeding tube, while that of the control group D, received equal volume of distilled water daily during the period of the experiment. The rats were fed with growers' mash obtained from Edo Feeds and Flour Mill Ltd, Ewu, Edo State, Nigeria and were given water liberally. The rats were sacrificed on day thirty-one of the experiment. The medial geniculate body was carefully dissected out and quickly fixed in 10% formal saline for histological studies. The histological findings after H&E method indicated that the treated section of the medial geniculate body showed some decreased cellular population, degenerative changes, cellular hypertrophy, with some vacuolations appearing in the stroma, with the group that received higher doses of sildenafil citrate (1.43 mg/kg) more severe. These findings indicate that sildenafil citrate consumption may have some deleterious effects on the medial geniculate body of adult Wistar rats at higher doses and by extension may affect the functions of the medial geniculate body and this may probably have some adverse effects on auditory sensibilities by its deleterious effects on the cells of the medial geniculate body of adult Wistar rats. It is therefore recommended that further studies aimed at corroborating these observations be carried out.

Key words: Sildenafil citrate, medial geniculate body, degenerative changes, Wistar rats.

INTRODUCTION

Sildenafil citrate is widely used as an effective and safe oral treatment for erectile dysfunction of various etiologies (Goldstein et al., 1998; Cheitlin et al., 1999; Benchekroun et al., 2003). It is a potent and selective inhibitor of phosphodiesterase type 5 enzymes that acts to break down cyclic guanosine monophosphate (cGMP) (Boolell et al., 1996). The medication amplifies the effect of sexual stimulation by retarding the degradation of this enzyme. Sildenafil has been found effective in several subpopulations of men with erectile dysfunction, including sufferers from diabetes (Basu and Ryder, 2004), hypertension (Feldman et al., 1999), spinal cord injuries (Hultling et al., 2000; Deforge et al., 2006), multiple sclerosis (Fowler et al., 2005), depression (Seidman et al., 2001; Rosen et al., 2004; Tignol et al., 2004; Fava et al., 2006), PTSD (Orr et al., 2006), and schizophrenia (Aviv et al., 2004; Gopalakrishnan et al., 2006), men after resection of the prostate or radical prostatectomy (Nandipati et al., 2006), after renal transplant (Sharma et al., 2006), men on dialysis (Dachille et al., 2006), and men aged 65 years and older (Wagner et al., 2001;

*Corresponding author. E-mail: andreweweka@yahoo.com.

Carson, 2004).

Psychogenic erectile dysfunction (ED) patients are excellent candidates for sildenafil citrate therapy due to the intact neurovascular pathway. Nevertheless, the drug has been reported to be effective only in about 78% of patients with psychogenic ED (McMahon et al., 2000). It is likely that performance anxiety and sympathetic over-tone are the cause of this unresponsiveness to sildenafil citrate during awakening, though data supporting this assumption are lacking (Rosen, 2001). The drug has been found to be effective and well tolerated in men with mild to moderate erectile dysfunction of no clinically identifiable organic cause (Eardley, 2001).

With the presence of PDE5 in choroidal and retinal vessels sildenafil citrate increase choroidal blood flow and cause vasodilation of the retinal vasculature. The most common symptoms are a blue tinge to vision and an increased sensitivity to light (Kerr and Danesh-Meyer, 2009). Adverse effects include headache, visual and retinal disturbances, dizziness and pupil-sparing third nerve palsy (Monastero et al., 2001). There have been reports of non-arteritic anterior ischaemic optic neuro-pathy and serous macular detachment in users of PDE5 inhibitors, although a causal relationship has not been conclusively shown. Despite the role of cGMP in the production and drainage of aqueous humor these medications do not appear to alter intraocular pressure and are safe in patients with glaucoma. All PDE5 inhibi-tors weakly inhibit PDE6 located in rod and cone photoreceptors resulting in mild and transient visual symptoms that correlate with plasma concentrations. Psychophysical tests reveal no effect on visual acuity, visual fields or contrast sensitivity; however, some studies show a mild and reversible impairment of blue-green colour discrimination. PDE5 inhibitors transiently alter retinal function on electroretinogram testing but do not appear to be retinotoxic. Despite the role of cyclic nucleotides in tear production there is no detrimental effect on tear film quality. Based on the available evi-dence PDE5 inhibitors have a good ocular safety profile (Kerr and Danesh-Meyer, 2009).

It has been reported that sildenafil citrate significantly improves nocturnal penile erections in sildenafil non-responding patients with psychogenic erectile dysfunction (Abdel-Naser et al., 2004). Several pharmacological and physiological properties of sildenafil have been described (Cheitlin et al., 1999; Aviv et al., 2004; Galie et al., 2005; Hoeper et al., 2006).

In Nigeria, most individuals often use sildenafil citrate indiscriminately for sexual arousal. There is a growing apprehension that it could be harmful or injurious to the body. Though sildenafil is currently being used to treat erectile dysfunction in patients with multiple sclerosis, Parkinson disease, multisystem atrophy, and spinal cord injury by improving their neurologically related erectile dysfunction, conversely, it has been implicated in a number of neurological problems, such as intracerebral hemorrhage, migraine, seizure, transient global amnesia,

nonarteritic anterior ischemic optic neuropathy, macular degeneration, branch retinal artery occlusion, and ocular muscle palsies. Thus, preclinical and very limited clinical data suggest that sildenafil may have therapeutic potential in selected neurological disorders. However, numerous reports are available regarding neurological adverse events ascribed to the drug. Although sildenafil shows some promise as a therapeutic agent in selected neurological disorders, well-designed clinical trials are needed before the agent can be recommended for use in any neurological disorder (Farooq et al., 2008).

The inferior colliculus and medial geniculate body con-stitute the intracranial auditory relay centres. The medial geniculate body is the target of ascending projections from the inferior colliculus and descending input from the auditory cortex; this is the obligatory synaptic target in the thalamus for hearing. It contains interleaved and overlapping tonotopic and aural bands (Fall, 1999). The cerebral cortex strongly affects the medial geniculate body through descending projections which are thought to consist primarily of small areas with slow conduction velocities (Winner, 1996). Cerebral nuclei such as the medial and lateral geniculate bodies, inferior and superior colliculi have higher glucose utilization than other structures. There is also a correlation between functional activity and metabolic rate such as in the visual and auditory system (Siesjo, 1978). The effects of sildenafil citrate on the intracranial auditory relay centre may not have been documented, but there have been reports that it may be implicated in varied symptoms of dizziness, vomiting, headaches, diarrhea, tinnitus, increase hearing loss, macular rash, neutronpenia, migraine, seizure, transient global amnesia, nonarteritic anterior ischemic optic neuropathy, macular degeneration, branch retinal artery occlusion, and ocular muscle palsies.

It is probable that the adverse effects of sildenafil citrate on hearing such as tinnitus may be due to direct effect of sildenafil citrate on this auditory relay centre. This present study was to elucidate the histological effects of sildenafil citrate on the medial geniculate body of adult Wistar rats.

MATERIALS AND METHODS

Animals

Twenty-four (24) adult Wistar rats of both sexes with average weight of 202 g were randomly assigned into four groups A, B, C and D of (n = 6) in each group. Groups A, B, and C of (n = 18) serves as treatments groups while group D (n = -6) was the control. The rats were obtained and maintained in the Animal holdings of the Department of Anatomy, School of Basic Medical Sciences, University of Benin, Benin city, Nigeria. They were fed with grower's marsh obtained from Edo feed and flour mill limited, Ewu, Edo state, and were given water liberally. The rats were allowed to gain maximum acclimatization before the actual commencement of the experiment. Sildenafil citrate tablet were obtained from the University of Benin Teaching Hospital Pharmacy, Benin City, Edo state, Nigeria.

Sildenafil citrate administration

The rats in the treatment groups (A, B, and C) received respectively, 0.25, 0.70 and 1.43 mg/kg body weight of sildenafil citrate base dissolved in distilled water daily for 30 days, through orogastric feeding tube, while that of the control group D, received equal volume of distilled water daily during the period of the experiment. The rats were sacrificed by cervical dislocation on day thirty-one of the experiment. The skulls were opened using bone forceps to expose the brain of the rat, and the medial geniculate body was quickly dissected out and fixed in10% formal saline for routine histological techniques.

Histological study

The tissue was dehydrated in an ascending grade of alcohol (ethanol), cleared in xylene and embedded in paraffin wax. Serial sections of 7 microns thick were obtained using a rotatory microtome. Some of the deparaffinized sections were stained routinely with hematoxylin and eosin (H&E) method (Drury, 1967). The digital photomicrographs of the desired sections were made in the Department of Anatomy research laboratory, University of Benin, Nigeria for further observations.

RESULT

Photomicrographs of the sections of the medial geniculate (MGB) from the control group (D) showed normal histological features, with the neurons appearing distinct and the glial cells normal without vacuolation in the stroma (Figure 1).

The sections of the medial geniculate body from the treatment (A, B, and C) groups showed some decrease in cellular population, degenerative changes, cellular hypertrophy and vacuolations appearing in the stroma (Figures 2, 3 and 4).

DISCUSSION

The results (H & E) revealed that administration of sildenafil citrate showed some varied degree of cellular degenerative changes, cellular hypertrophy, decrease cell population and intercellular vacuolations appearing in the stroma of the treatment groups compared to the control section of the medial geniculate body of the adult Wistar rat. Neuronal degeneration has been reported to result in cell death, which is of two types, namely apoptotic and necrotic cell death. These two types differ morphologically and biochemically (Wyllie, 1980). Pathological or accidental cell death is regarded as necrotic and could result from extrinsic insults to the cell such as osmotic, thermal, toxic and traumatic effects (Farber, 1981). It was reported that cell death in response to neurotoxins might trigger an apoptotic death pathway within brain cells (Waters, 1994).

The process of cellular necrosis involves disruption of the membranes structural and functional integrity. Cellular necrosis is not induced by stimuli intrinsic to the cells as in programmed cell death (PCD), but by an abrupt

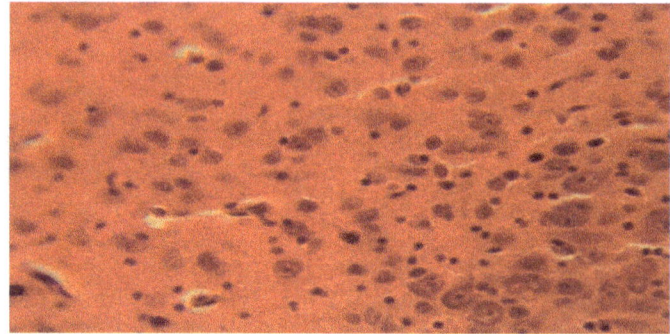

Figure 1. Group D: Control section of the medial geniculate body (Mag. x400).

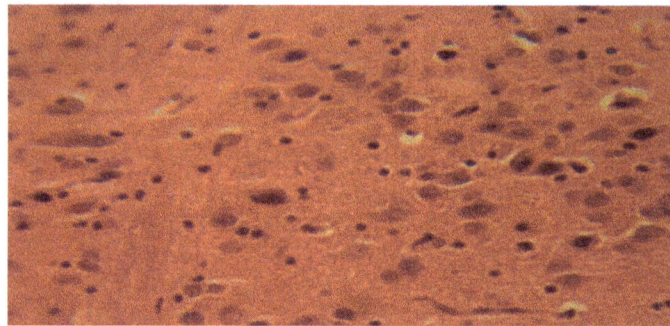

Figure 2. Photomicrograph of treatment section of the medial geniculate body of rats that received 0.25 mg/kg of sildenafil citrate base dissolved in distilled water daily for 30 days (Mag. X400).

environmental perturbation and departure from the normal physiological conditions (Martins, 1978). There is the need to further investigate the actual mechanism by which sildenafil citrate induced neuronal degeneration in the medial geniculate body of adult Wistar rat in this study.

Extensive cell death in the central nervous system is present in all neurodegenerative diseases (Waters, 1994). The type of nerve cell loss and the particular part of the brain affected dictate the symptoms associated with an individual disease (Waters, 1994). In this study sildenafil citrate may have acted as toxin to the cells of the medial geniculate body, affecting their cellular integrity and causing defect in membrane permeability and cell volume homeostasis.

In cellular necrosis, the rate of progression depends on the severity of the environmental insults. The principle holds true for toxicological insult to the brain and other organs (Martins, 1998). The prime candidates for inducing the massive cell destruction observed in neurodegeneration are neurotoxins (Waters, 1994). The latter when present at a critical level can be toxic to the brain cells they normally excite (Waters, 1994). It is inferred from this results that prolonged and high dose of sildenafil citrate resulted in increased toxic effects on the medial geniculate body.

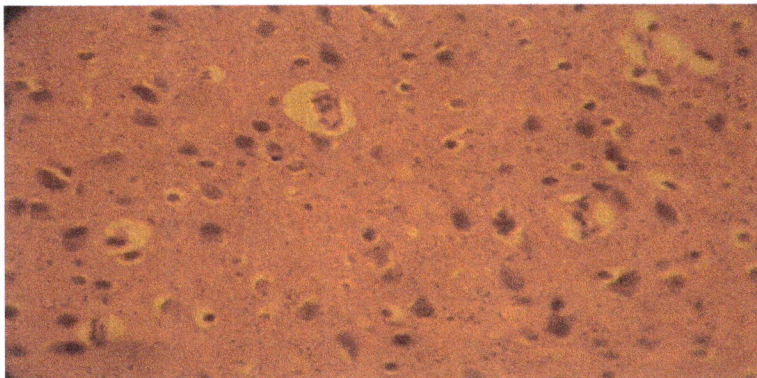

Figure 3. Photomicrograph of treatment section of the medial geniculate body of rats that received 0.70 mg/kg of sildenafil citrate base dissolved in distilled water daily for 30 days (Mag. X400).

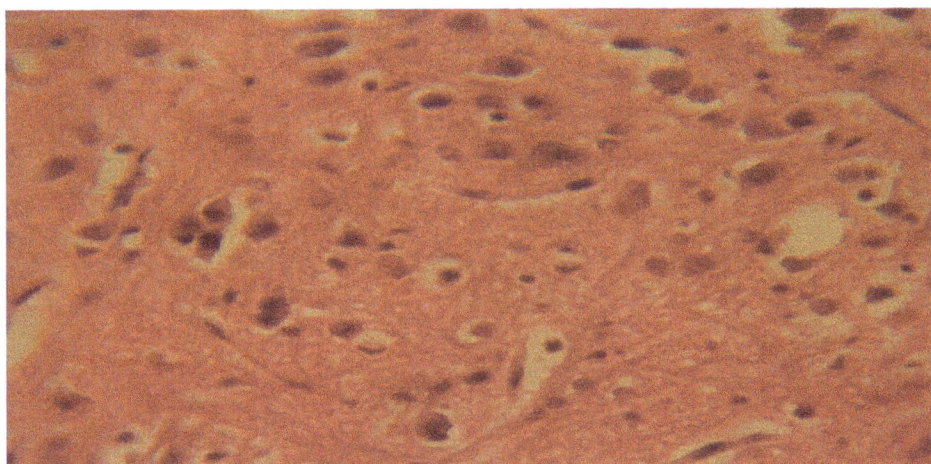

Figure 4. Photomicrograph of treatment section of the medial geniculate body of rats that received 1.43 mg/kg of sildenafil citrate base dissolved in distilled water daily for 30 days (Mag. X400).

The vacuolations observed in the stroma of the medial geniculate body in this experiment may be due to sildenafil citrate interference. The cellular hypertrophy observed in this experiment may be due to the adverse effects of sildenafil citrate on the medial geniculate body. This study may underlie the possible neurological symptoms such as dizziness and tinnitus. Sildenafil citrate has been implicated as a possible cause of blindness—diagnosed as nonarteritic anterior ischemic optic neuropathy (Cunningham and Smith, 2001; Pomeranz et al., 2002; Pomeranz and Bhavsar, 2005).

the medial geniculate body of adult Wistar rats. These results may probably affect the functions of the medial geniculate body in auditory sensibility in adult Wistar rats. It is recommended that further studies be carried out to examine these findings.

CONCLUSION

Our study revealed that high doses and long term administration of sildenafil citrate caused some varied degree of cellular degenerative changes, cellular hypertrophy, clustering of cells and intercellular vacuolations in

REFERENCES

Abdel-Naser MB, Imam A, Wollina U (2004). Sildenafil citrate significantly improves nocturnal penile erections in sildenafil non-responding patients with psychogenic erectile dysfunction. Int. J. Impot. Res., 16, 552–556.

Aviv A, Shelef A, Weizman A (2004). An open-label trial of sildenafil addition in risperidone-treated male schizophrenia patients with erectile dysfunction. J. Clin. Psychiatry, 65:97–103.

Basu A, Ryder RE (2004). New treatment options for erectile dysfunction in patients with diabetes mellitus. Drugs, 64: 2667–2688.

Benchekroun A, Faik M, Benjelloun S (2003). A baseline-controlled, open-label, flexible dose-escalation study to assess the safety and efficacy of sildenafil citrate (Viagra) in patients with erectile dysfunction. Int. J. Impot. Res., 15(Suppl. 1):S19–S24

Boolell M, Allen MJ, Ballard SA (1996). Oral sildenafil: an orally active type 5 cyclic GMP specific phosphodiesterase inhibitor for the treatment of penile erectile dysfunction. Int J Impot Res; 8: 47–52.

Carson CC (2004). Erectile dysfunction: evaluation and new treatment options. Psychosomatic Med., 66: 664–671.

Cheitlin MD, Hutter AM, Brindis RG, Kaul S, Russell RO, Zusman RM (1999). Use of sildenafil (Viagra) in patients with cardiovascular disease. Circulation, 99: 168–177.

Cunningham AV, Smith KH (2001). Anterior ischemic optic neuropathy associated with Viagra. *Journal of Neuro-Ophthalmology, 21,* 22–25.

Dachille G, Pagliarulo V, Ludovico GM (2006). Sexual dysfunction in patients under dialytic treatment. Minerva Urologica e Nefrologica, 58: 195–200.

Deforge D, Blackmer J, Garritty C (2006). Male erectile dysfunction following spinal cord injury: a systematic review. Spinal Cord, 44: 465–473.

Drury RAB, Wallington EA, Cameron R (1967). Carleton's Histological Techniques: 4th ed., Oxford University Press NY. U.S.A. pp. 279-280.

Eardley I (2001). Efficacy and safety of sildenafil citrate in the treatment of men with mild to moderate erectile dysfunction. British J. Psychiatry. 178: 325-330

Fall. Mammalian Neuroanatomy MCB 163: 1999

Farber JL Chein KR, Mittnacht S (1981). The pathogenesis of Irreversible cell injury in ischemia. Am. J. Pathology; 102: 271-281.

Farooq MU, Naravetla B, Moore PW, Majid A, Gupta R, Kassab MY (2008). Role of Sildenafil in Neurological Disorders Clin. Neuropharmacol., 31(6): 353-362.

Fava M, Nurnberg HG, Seidman SN (2006). Efficacy and safety of sildenafil in men with serotonergic antidepressant-associated erectile dysfunction: results from a randomized, double-blind, placebo-controlled trial. J. Clin. Psych. 67: 240-246.

Feldman R, Meuleman EJ, Steers W (1999). Sildenafil citrate (VIAGRA) in the treatment of erectile dysfunction: analysis of two flexible dose-escalation studies. Sildenafil Study Group. Int. J. Clin. Pract., 53(102): 10-12.

Fowler CJ, Miller JR, Sharief MK (2005). A double blind, randomised study of sildenafil citrate for erectile dysfunction in men with multiple sclerosis. J. Neuro. Neurosurg. Psych., 76: 700-705.

Galie N, Ghofrani HA, Torbicki A, Barst RJ, Rubin LJ, Badesch D, Fleming T, Parpia T, Burgess G, Branzi A, Grimminger F, Kurzyna M, Simonneau G (2005). Sildenafil citrate therapy for pulmonary arterial hypertension. N. Engl. J. Med.. Nov. 17(353): 2148-57.

Goldstein I, Lue TF, Padma-Nathan H, Rosen RC, Steers WD, Wicker PA (1998). Oral sildenafil in the treatment of erectile dysfunction. N Engl. J. Med., 338: 1397-1404.

Gopalakrishnan R, Jacob KS, Kuruvilla A (2006). Sildenafil in the treatment of antipsychotic-induced erectile dysfunction: a randomized, double-blind, placebo-controlled, flexible-dose, two-way crossover trial. Am. J. Psych., 163: 494-499.

Hoeper MM, Welte T, Izbicki G, Rosengarten D, Picard E, Kuschner WG, Galiè N, Rubin LJ, Simonneau G (2006). Sildenafil Citrate Therapy for Pulmonary Arterial Hypertension. N. Engl. J. Med., 354: 1091-1093.

Hultling C, Giuliano F, Quirk F (2000). Quality of life in patients with spinal cord injury receiving Viagra (sildenafil citrate) for the treatment of erectile dysfunction. Spinal Cord, 38: 363-370.

Kerr NM, Danesh-Meyer HV (2009). Phosphodiesterase inhibitors and the eye. Clin. Exp. Ophthalmol., 37(5): 514-523

Martins LJ, Al-Abdulla NA, Kirsh JR, Sieber FE, Portera-Cailliau C (1998). Neurodegeneration in excitotoxicity, global cerebral ischaemia and target deprivation: A perspective on the contributions of apoptosis and necrosis. Brain Res. Bull., 46(4): 281-309.

McMahon CG, Samali R, Johnson H (2000). Efficacy, safety and patient acceptance of sildenafil citrate as treatment for erectile dysfunction. J. Urol., 164: 1192-1196.

Monatero R, Pipia C, Camarda LKC, Camarda R (2001). Intracerebral haemorrhage associated with sildenafil citrate, J. Neurol., 248(2): 141-142.

Nandipati KC, Raina R, Agarwal A, Zippe CD (2006). Erectile dysfunction following radical retropubic prostatectomy: epidemiology, pathophysiology and pharmacological management. Drugs Aging, 23: 101-117.

Orr G, Weiser M, Polliack M (2006). Effectiveness of sildenafil in treating erectile dysfunction in PTSD patients: a double-blind, placebo-controlled crossover study. J. Clin. Psychopharmacol., 26: 426-430.

Pomeranz HD, Smith KH, Hart WM, Egan RA (2002). Sildenafil-associated nonarteric anterior ischemic optic neuropathy. Ophthalmology, 109: 584-587.

Pomeranz HD, Bhavsar AR (2005). Nonarteric ischemic optic neuropathy developing soon after use of sildenafil (Viagra): A report of seven new cases. J. Neuro-Ophthalmol., 25: 9-13.

Rosen RC (2001). Psychogenic erectile dysfunction. Classification and management. Urol. Clin. North Am., 28: 269-278.

Rosen RC, Seidman SN, Menza MA (2004). Quality of life, mood, and sexual function: a path analytic model of treatment effects in men with erectile dysfunction and depressive symptoms. Int. J. Impot. Res., 16: 334-340.

Seidman SN, Roose SP, Menza MA (2001). Treatment of erectile dysfunction in men with depressive symptoms: results of a placebo-controlled trial with sildenafil citrate. Am. J. Psych., 158: 1623-1630.

Sharma RK, Prasad N, Gupta A, Kapoor R (2006). Treatment of erectile dysfunction with sildenafil citrate in renal allograft recipients: a randomized, double-blind, placebo-controlled, crossover trial. Am. J. Kidney Dis., 48: 128-133.

Siesjo BK (1978). Utilization of substrates by brain tissues, (In) Brain energy metabolism. John Wiley & Sons, USA. pp. 101-130.

Tignol J, Furlan PM, Gomez-Beneyto M (2004). Efficacy of sildenafil citrate (Viagra) for the treatment of erectile dysfunction in men in remission from depression. Intl. Clin. Psychopharmacol., 19: 191-199.

Wagner G, Montorsi F, Auerbach S, Collins M (2001). Sildenafil citrate (VIAGRA) improves erectile function in elderly patients with erectile dysfunction: a subgroup analysis. J. Gerontol. Biol. Sci. Med. Sci., 56: M113-M119.

Waters CM (1994). Glutamate induced apoptosis of striatal cells in rodent model for Parkinsonism. Neuroscience, 63: 1-5

Winner JA, Saint Marie RL, Larue DT, Oliver DL (1996). The cerebral cortex strongly affects the medial geniculate body through descending projections. Proc. Nat. Acad. Sci, USA, 93: 8005-8010.

Wyllie AH (1980). Glucocorticoid-induced thymocyte apoptosis is associated with endogenous endonuclease activation. Nature: London, 284: 555-556.

7

Toxicity evaluation of dextran-spermine polycation as a tool for genetherapy *in vitro*

Fatemeh Abedini[1], Maznah Ismail[1,4]*, Hossein Hosseinkhani[2], Tengku azmi[1,3], Abdolrahman Omarb[1,3], Chong PeiPei[4], Norsharina Ismail[1], Ira-Yudovin Farber[5] and Abraham J. Domb[5]

[1]Institute of Bioscience, Faculty of Medicine, University of Putra Malaysia, 43400 UPM Serdang, Selangor Darul Ehsan, Malaysia.
[2]School of Biomedical Engineering, National Yang Ming University, Taipei 112, Taiwan.
[3]Faculty of Veterinary, University of Putra Malaysia, 43400 UPM Serdang, Selangor Darul Ehsan, Malaysia.
[4]Faculty of Medicine, University of Putra Malaysia, 43400 UPM Serdang, Selangor Darul Ehsan, Malaysia.
[5]Department of Medicinal Chemistry and Natural Products, School of Pharmacy, the Hebrew University-Hadassah Medical School, Jerusalem, Israel.

Cationic polymers are a leading class of nonviral self-assembled nucleic acid delivery systems. Cationic polymers have been shown to condence the DNA so that the entrapped DNA is protected from contact with DNase. The objective of the present study is to evaluate the effect of cationic dextran on the proliferation rate, morphological changes and biosynthetic activities *in vitro*. Cationic dextran was prepared by means of reductive-amination between oxidized dextran and the natural oligoamine, spermine. Four kinds of biological evaluations including cell proliferation assay, ultrastructural changes of cells using transmission electron microscopy (TEM), acridine orange/Propidium Iodide and cell cycle were studied. Our results clearly indicated that the toxicity of cationic dextran is dose depended and it is not toxic at low concentration and tolerable by the cells, and it can be used as a tool for gene delivery.

Key words: Dextran-spermine, genetherapy, nonviral vectors, toxicity.

INTRODUCTION

Genetic therapy includes gene therapy by DNA, inhibition or silencing of gene expression either by antisense oligonucleotides or by siRNA (Ye, 1998; Shuey, 2002). The gene transfection with naked plasmid DNA always shows low efficiency *in vivo* although it is simple and safe (Kawabata, 1995; Yu et al., 2001). On the other hand, when a plasmid DNA is complexed with a vector, the molecular size and surface charge of the complex affect the body fate of the plasmid DNA (Kircheis et al., 2001).

The success of gene therapy is greatly dependant on the development of a vector or vehicle that can selectively and efficiently deliver a gene or siRNA to target cells with minimal toxicity. The safety concerns regarding

the use of virus in humans make nonviral delivery systems an attractive alternative (Pouton and Seymour, 1998). Although at present, the *in vivo* expression levels of the synthetic molecular gene vectors are lower compared to viral vectors and gene expression is transient, these vehicles are likely to present several advantages including safety, low-immunogenicity, capacity to deliver large genes, and large-scale production at low cost (Schmidt-Wolf and Schmidt-Wolf, 2003). The two leading classes of synthetic gene delivery systems that have been mostly investigated involve the use of either cationic lipids or cationic polymers (Schatzlein, 2001). Polycations used for gene complexation are polyamines that become cationic at physiologic conditions. The most studied polycations used for gene complexation and delivery are the branched/linear polyethylene imine (PEI) (Abdallah et al., 1996; Ferrari et al., 1997) poly(L-Lysine), (Vanderkerken et al., 2000) poly(dimethyl aminoethyl metha-

---
*Corresponding author. E-mail: maznahis@putra.upm.edu.my.

crylate, pDMAEMA), (van de Wetering et al., 1998) poly(trimethyl aminoethyl methcryalte, pTMAEMA), (Wolfert et al., 1996) poly(vinylpyridine), (Yaroslavov et al., 1996) Chitosan, (MacLaughlin et al., 1998) and diethylami-noethyl dextran (DEAE-dextran). (Takai and Ohmori, 1990) Among different polycations were prepared starting from various polysaccharides of different molecular weights (Jolly et al., 1999) only the dextran-spermine (D-SPM) polycations of defined molecular weights were found to be active in transfection of genes (Azzam et al., 2003a; Azzam et al., 2004b; Eliyahu et al., 2003).

The reason for the transfection of D-SPM conjugate was attributed to spermine residues, which play a crucial role in cell transfection (Li and Huang, 2000). Dextran-spermine is a water-soluble and biodegradable cationic polymer. Dextran-spermine cationic polysaccharide was prepared by means of reductive-amination between oxidized dextran and the natural oligoamine, spermine. The formed schiff_base imine-based conjugate was reduced with borohydride to obtain the stable amine-based conjugate (Hosseinkhani et al., 2006a, Hosseinkhani et al., 2006e). As non viral vector, we have previously reported that the dextran-spermine based polycations found to be highly effective in transfecting many different cells in vitro (Azzam et al., 2002, Hosseinkhani et al, 2005c; Hosseinkhani et al., 2006b, Hosseinkhani et al., 2005d, Hosseinkhani et al., 2005f). Transfection efficacy was highly dependent upon the charge ratio.

Histopathological assessment of dextran-spermine in vivo revealed mild toxicity in the muscle and no abnormal findings in liver or lung. No systemic toxicity decrease in WBC counts, thrombocytopenia, and no increase in levels of serum transaminases were found (Eliyahua et al., 2006). In the present study, the rate of proliferation, morphological and biosynthetic activities on HT29 cells (human colonic adenocarcinoma cell) and MCF7 cells (human breast cancer cell) after treatment with different doses of dextran expermine were assessed for its safety and tolerance by the cell lines. We confirmed the biological evaluation of cationic dextran by a different method such as MTS, transmission electron microscopy, AO/PI staining and cell cycle flow cytometery.

MATERIALS AND METHODS

HT29 (Human colon adenocarcinoma), HeLa (Human epithelial carcinoma), and MCF7 (human breast cancer) cell lines were purchased from American Type Culture Collection (ATCC), USA. Dextran-spermine (FI-4/12A) was received as a generous gift from Prof Abaraham J. Domb (The Hebrew University-Hadassah Medical School, Jerusalem, Israel). RPMI 1640 (Cat # 125K83551) was purchased from GIBCO, UK., 4% Glutaraldehyde (Cat # R1010) UK, Sodium Cacodylate Buffer (Cat # R1010), Osmium Tetroxide (Cat # R1010), Agar 100 resin (Cat # R1044), Dodecenyl succinic anhydric (Cat # R1051), Benzyldimethylamine (Cat # R1061), and Methyl Nadic Anhydride (Cat # R1081) were purchased from Agar Scientific Co., UK. Foetal bovine serum (PAA Cat # A11-105, Tripsin-EDTA (1X) (PAA Cat no: L11-004), MTS kit (Promega Cat #

G3580), RNase, Acridine orange (Cat: 235474-5G), Propidium Iodide (Cat: 1056-1) were purchased from Sigma Aldrich, USA.

Synthesis of dextran-spermine conjugate

Dextran-spermine based conjugate was prepared as described elsewhere (Azzam et al., 2002). Briefly, a solution of oxidized dextran (1 g) in 100 ml DDW (6.9 mmole aldehyde groups) was slowly added during 5 h to a basic solution containing 1.25 equimolar amount of oligoamine (to aldehyde) dissolved in 50 ml borate buffer (0.1 M, pH 11). The mixture was stirred at room temperature for 24 h and excess of sodium borohydride (1 g) was added and stirring was continued for 48 h at room temperature. The reduction was repeated with additional portion of NaBH4 (1 g) and stirring for 24 h under the same conditions. The resulting light-yellow solution was dialyzed against DDW (6 x 6L) applying 3,500 cutoff cellulose tubing (Membrane Filtration Products Inc., San Antonio, TX, USA) followed by lyophilization to obtain a yellowish reduced amine-based conjugate in 50% overall yield (to polysaccharide). 1H-NMR (D2O): 1.645 (m, 4H, dextran-CH2NH (CH2)3NHCH2CH2CH2CH2NH (CH2)3NH2), 1.804 (m, 4H, dextran-CH2NHCH2CH2CH2NH (CH2)4NHCH2CH2 2.815 (m, 14H, dextran-CH2NHCH2CH2NHCH2 (CH2)2CH2NHCH2CH2CH2NH2), 3.52-4.19 (m, sugar hydrogens) and 5.02 ppm (m, 1H, anomeric hydrogen). %N=10.90 ± 0.5% (equal to ~50% substitution). TNBS = 1.33 ± 0.15 μmole/mg (primary amino content). Mw = 10000 ± 1500 Da (n=10).

Cell culture

HT29 (passage 4), HeLa (passage 50) and MCF7 (passage 35) cells were grown in RPMI 1640 medium supplemented with 10% foetal bovine serum, 2 mM L-glutamine and 10% penicillin-streptomycin. Cells were incubated at 37ºC and 5% CO_2 separately, and subcultured after 3 days with trypsin-EDTA (1X).

Serial dilution on MCF7 cells and cell proliferation assay on HT29 cells

Dextran spermine (1mg/ml) was dissolved in PBS (pH 7.4). MCF7 cells were treated with different dilution of dextran spermine from 1:2, 1:4, 1:8, 1:16 and finally to 1:36 for 72 h. Dextran spermine (D-SPM) conjugates in phosphate buffer, pH 7.4, are stable for at least 1 week (6).

In the proliferation assay, HT-29 cells were seeded at a density of 4×10^3 cells/well onto 96-well cell culture plates and allowed to adhere for 24 h. After 24 h, cells were treated with dextran-spermine with concentration of 30 μg/ml for 24, 48 and 72 h. An aliquot of 20 μl of combined solution of a tetrazolium compound MTS [3-(4,5-dimethylthiazol-2-yl)-5-(3-carboxymethoxyphenyl)-2-(4-sulfophenyl)-2htetrazolium, inner salt} was added to each well. After incubation for 4 h at 37ºC and 5% CO_2, the absorbance at 490 nm was measured on an enzyme-linked immunosorbent assay (ELISA) plate reader.

Cell cycle analysis

HT29, HeLa, and MCF7 cells were plated in a small flask at a concentration of 1×10^5 cells in RPMI. The cells were treated with dextran-spermine with concentration of 5 μg/ml for 72 h. For cell cycle analysis, the cells were trypsinized and washed with PBS. Cells were fixed with 70% ethanol (500 μl) for at least 2 h at -20ºC. The fixed cells were then washed with PBS and incubated with 500 μl RNase/PI on ice and keep in dark place for 30 min. The cells were subjected to fluorescence-activated cell sorting analysis Dakkocytomation.

Morphological and organelles characterization of cells

HT29 cells were treated with dextran-spermine at different doses. After 72 h, cells were fixed in fixative 4% Glutaraldehyde for 12-24 h at 4ºC for further processing. When processing resumed, the cells were washed with 0.1 M sodium cacodylate buffer for 10 min repeatedly for three times. The cells were post fixed in 1% osmium tetroxide for 2 h at 4ºC. The cells were washed again with 0.1 M sodium cacodylate buffer to repeat the above process. Then they were dehydrated in a series of acetone (35% for 10 min, 50% for 10 min, 75% for 10 min, 95% for 10 min, and 100% for 15 min. The specimens were infiltrated with acetone and resin mixture. Specimens were put in mixture of acetone and resin proportionally for 1 h transferred into mixture of acetone and resin proportionally 1 to 3 for 2 h. 100% resin was made by mixing of Agar 100 resin 10 ml, dodecenyl succinic anhydric 5.5 ml, benzyldimethylamine 0.5 ml, methyl nadic anhydride 6 ml. Specimens were put into 100% resin overnight and resin 100% was changed and specimen were kept for 2 h. The specimens were placed into beam capsules and topped up, and then polymerized in oven at 60ºC for 24-48 h. After sectioning, the specimens were stained with toluidine blue, and were examined under transmission electron microscope (TEM) LEO 912AB.

Acridine orange and propidium iodide (AO/PI) staining

HT29 cells were plated in 6 well plates at the density 1×10^5. HT29 cells were treated with 5 µg/ml dextran spermine at 48 h. After 48 h, cells were tripsinized with tripsin-EDTA and then were washed twice with PBS and centrifuged. Cell suspension (95 µl) was mixed with 5 µl of dye mixture containing 100 µg/L AO with 100 µg/l PI in PBS. Cells were visualized immediately under fluorescent microscope (LEICA DM LB, filter cube 13).

Statistic

All the experiment was conducted in triplicate. ANOVA single factor test was used to indicate statistical significance.

RESULTS

Serial dilution and cell viability assay on MCF7 and HT29 cells

The results of cytotoxcity evaluation of dextran-spermine on MCF7 breast cancer cells at different interval time are shown in Figures 1-E. The concentration of dextran-spermine in this experiment was selected at the highest amount (1000 □g/ml) to evaluate the level of the toxicity of dextran-spermine. The morphology of the cells was also evaluated after treatment with dextran-spermine. As shown in Figure 1-B and C, MCF7 cells were lysed at a concentration of 1000 and 500 □g/ml. At the concentration of 250 and 125 □g/ml the morphology of the cells were changed (Figure 1-D and E) and the proliferation of the cells was still lower than the control group (Figure 1-A; untreated cells). At the concentration of 60.5 µg/ml

the morphology of the cells and proliferation were normal (Figures 1-E). Viability assay was done in HT29 cells by MTS assay at concentration of 30 µg/ml of dextran spermine for 24, 48 and 72 h. The percentages of viability in HT29 after 24, 48 and 72 h were 0.72± 0.01, 0.74± 0.01 and 0.69± 0.02 µg/ml of dextran spermine (Figure 1-F).

Transmission electron microscopy to observe morphology of cells and organelles

Transmission electron microscopy of HT29 cell line (Figure 2A, and B) showed that the cells were round and have microvilli on the surface, nucleus with nucleolus inside, peripheral heterochromatin and have abundant euchromatin in nucleus which is representative of gene expression by the cell in control group. In the cytoplasm there are many oval mitochondria with transverse crista that show the activity of the cell also as well as a lot of endoplasmic reticulum. Treated HT29 cells with 60 µg/ml of dextran-spermine after 72 h resulted macrovilli on cell surface were swollen and inclusion body in cells were observed (Figure 2-C), but treatment of HT29 cells with 15 µg/ml of dextran-spermine led to the same morphology of control group (Figure 2-D-E). Round cells with tiny microvilli on the cell surface, nucleolus, heterochromatin and euchromatin were observed in nucleus. Mitochondria with transverse crista, and golgi apparatus were observed in cytoplasm. The result demonstrated that dextran-spermine has no morphological change of the cell membrane and organelles in HT29 cells at concentration of 15 µg/ml.

Acridine orange and propidium iodide (AO/PI) staining

Cell death occurs by two methods, necrosis and apoptosis. In this experiment Acridine orange and propidium iodide (AO/PI) were used to visualize living and dead HT29 cells simultaneously after treatment with 5 µg/ml of dextran-spermine. Propidium Iodide (PI) intercalates into double-stranded nucleic acids. It is excluded by viable cells but can penetrate cell membranes of dying or dead cells and acridine orange (AO) to stain viable cells. In this experiment based on ANOVA single factor test calculated F=7.708 (P = 0.09) as P>0.05, we conclude that dextrane spermine did not affect on necrosis or apoptosis of HT29 cells at concentration of 5 µg/ml (Figure 3-A and B).

Cell cycle

Cell division is a vital process through well defined phases of the cell cycle. These cell cycle phases include an initial

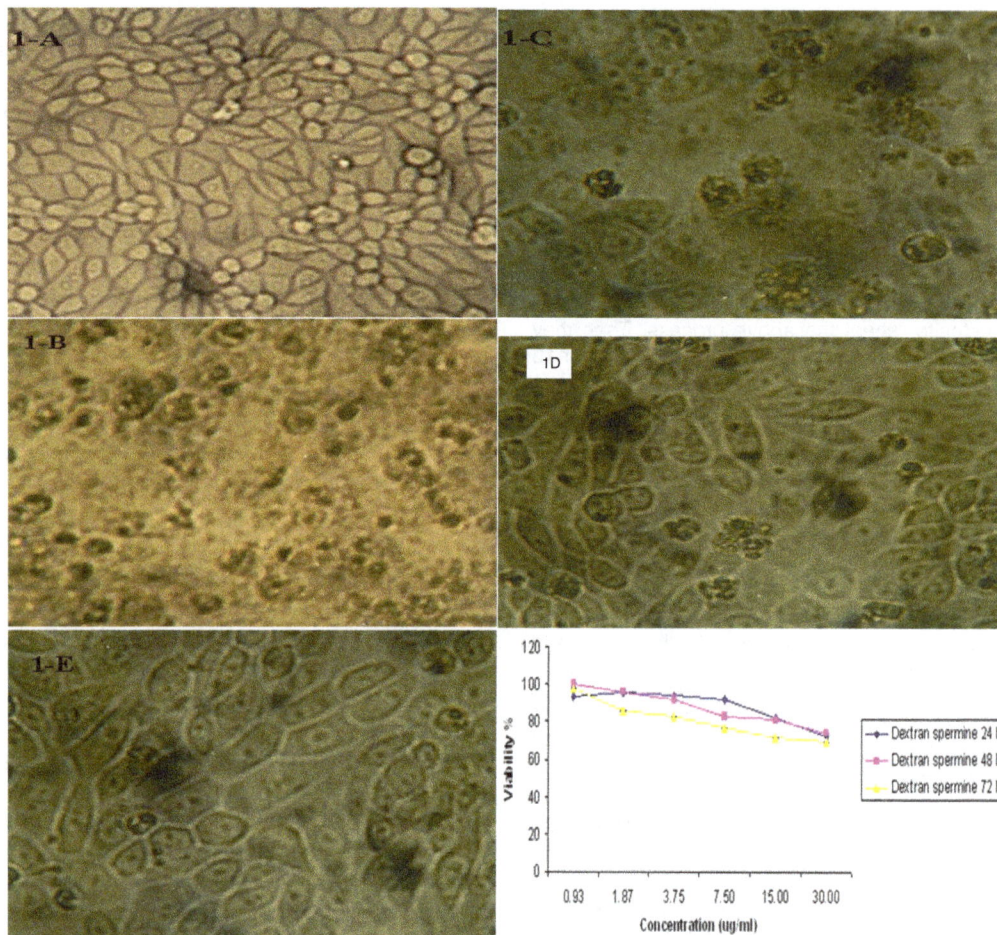

Figure 1. A: Serial dilution, MCF7 cells were treated with different dilution of dextran spermine for 72 h. A; MCF7 cells, control, **B:** MCF7 cells treated with dextran spermine stock solution (1mg/ml), C;MCF7 cells treated with dextran spermine at concentration of (500 µg/ml, 1:2 dilution of stock solution, **D:** MCF7 cells treatment with dextran spermine at concentration of (250 µg/ml, 1:4 dilution of stock solution), **E:** MCF7 cells treated with dextran spermine at concentration of (120 µg/ml, 1:8 dilution of stock solution). **F:** Cell viability was done in HT29 cells by MTS assay at concentration of 30 µg/ml dextran spermine after 24, 48 and 72 h.

gap (G1), synthesis of DNA (S), a second gap (G2) and the final mitotic nuclear and cellular division (M) to result in two identical daughter cells. Cell cycle analysis was performed for HT29, MCF7 and HeLa cells at 5 □g/ml of dextran spermine. Based on ANOVA single factor test for G0/G1 of HT29, MCF7 and HeLa cells was calculated F=32.28 (P=0.78), as P>0.05, we conclude that there was no influence on cell cycle after treatment with dextran spermine at concentration of 5 □g/ml.

DISCUSSION

Toxicity is still an obstacle to the application of non-viral vectors to gene therapy (Jamesonet et al., 1998). Cationic lipids and cationic polymers are the most probable alternative to viral delivery systems and are increasingly being used in vitro and in vivo (Taira et al.,

2005). Administration of cationic substances may be toxic to tissue due to their positive charge. Comparison between the cytotoxicity of dextran-spermine with well-established polyplexes found that cytotoxicity at the concentration range used for in vitro transfection (equivalent to 1-2 µg cationic carrier/well of 96-well plates) was minimal for all cationic carriers studied (Branched-PEI, Linear-PEI, D-spm, and DOTAP/cholesterol). Above 10 µg cytotoxicity was obtained, with D-spm being the least cytotoxic. DOTAP/ cholesterol showed higher cytotoxicity and was the most toxic cationic carrier tested (Eliyahu et al., 2005).

Polyplexes protect DNA by sterically blocking the access of nucleolytic enzymes. Naked plasmid DNA is degraded by DNase within minutes, whereas plasmid DNA in polyplexes is stable for hours (Abdelhady et al., 2003). Gene-delivery vectors bind and condense DNA into small, compact structures through electrostatic

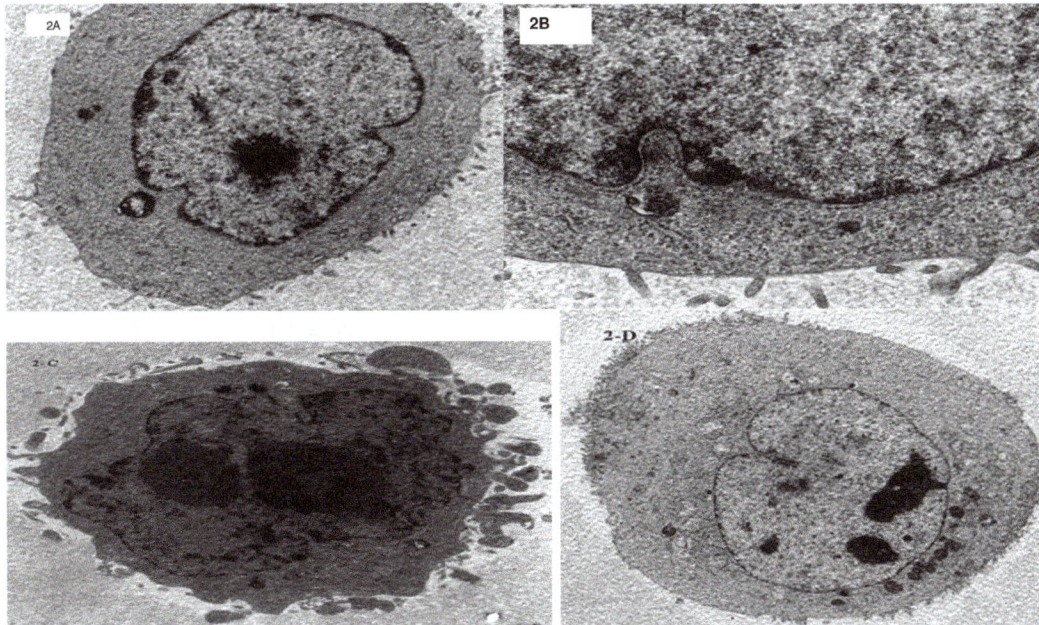

Figure 2. Transmission Electron Microscopy (TEM) observation of HT29 cells, control; A and B control groups after 72 h. C: HT29 cells treated with 60 µg/ml of dextran spermine after 72 h. D and E: HT29 cells treated with dextran-spermine at concentration of 15 µg/ml after 72 h.

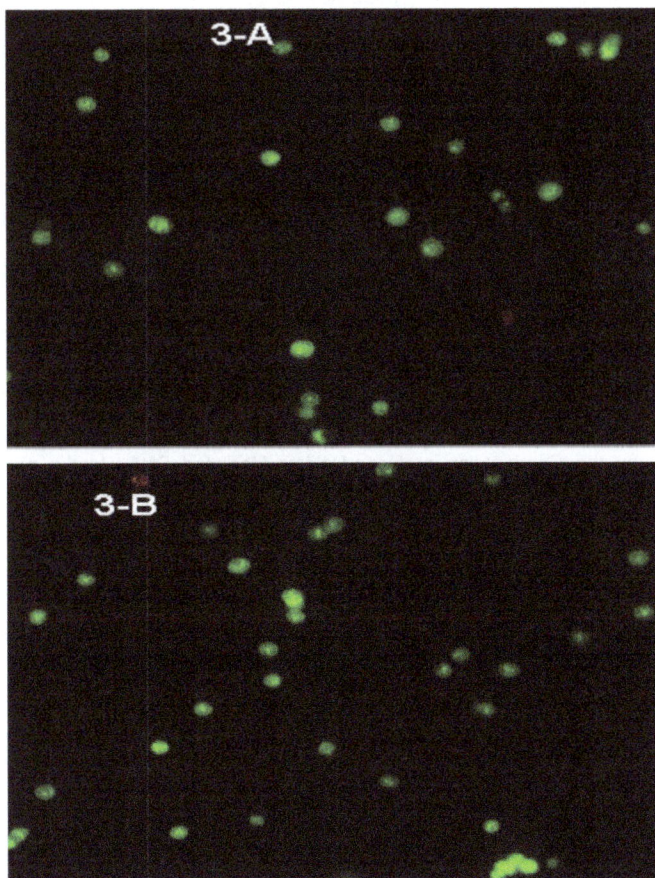

Figure 3. HT29 cells stained with AO/PI. A; HT29 cells, control, B; HT29 cells treated with 5 µg/ml of dextran spermine after 72 h.

interactions between the negative phosphates along the DNA backbone and positive charges displayed on the vector material. The structure of the polycation can affect DNA binding and condensation. Previous studies have reported that a minimum of six to eight charges in a polyplex are required for efficient DNA condensation (Schaffer et al., 2000; Hosseinkhani et al., 2006). Since positively charged complexes interact with the cell surface by an electrostatic interaction with anionic substances on the cell surface, such as sialic acid, proteoglycan and also with anionic microtubules or motor proteins inside cytoplasm. Motor proteins are the driving force behind most active transport of proteins and vesicles in the cytoplasm. Kinesins and dyneins play essential roles in intracellular transport such as axonal transport and in the formation of the spindle apparatus and the separation of the chromosomes during mitosis and meiosis (Ruponen et al., 1999). To test the hypothesis that dextran spermine could affect on cell membrane and organelles inside cells, we assessed the effect of toxicity dextran-spermine on MCF7, HeLa, and HT29 cells. The presence of D-spm at the high concentration (1000 and 500 ug/ml) was lysed cells and stop proliferation and at 250 and 125 ug/ml were changed the morphology and low proliferation. The percentage of viability in HT29 after 24, 48 and 72 h were 0.72 ± 0.01, 0.74 ± 0.01 and 0.69 ± 0.02 at 30 □g/ml of dextran spermine. We evaluated the effect of dextran spermine on HT29 cells by transmission electron microscopy at 60 and 15 µg/ml in each small flask. At 60 µg/ml of D-spm after 72 h was observed few inclusion body in some cells and was changed the cell membrane and nuclear membrane and the microvillis were swollen. The nuclear to cytoplasm ratio (N/C) was high (Figure 2-C). Cell cycle analysis in the G0/G1 phase showed F=32.28 (P=0.78), that there was no different of accumulation of cells in the G0/G1 phase after treatment with 5 ug/ml dextran spermine between HT29, HeLa and MCF7 control and treatment cells after 72 h. Based on the result of cell cycle and AO/PI staining we concluded that positively charge dextran spermine at 5 ug/ml has no affect on cell membrane, nuclear membrane, motor proteins, spindle apparatus and mitosis. These results suggest that dextran spermine induce a dose-dependent toxicity in HT29, CT26.WT, HeLa and MCF7 cells *in vitro*.

Conclusion

In the present study we assessed the toxicity of dextran spermine at different doses on HT29, MCF7 and HeLa cells *in vitro*. The toxicity of dextran spermine is dose dependent, whereby at dose lower than 5 µg/ml it is safe and has no toxicity effects as observed by MTS, cell cycle, TEM observation as well as AO/PI staining.

REFERENCES

Abdallah B, Hassan A, Benoist C, Goula D, Behr JP, Demeneix BA (1996). A powerful nonviral vector for *in vivo* gene transfer into the adult mammalian brain. Polyethylenimine. Human Gene Therapy. 7(16): 1947-1954.

Abdelhady HG, Stephanie A, Martyn CD, Clive JR, Saul JBT, Philip MW (2003). Direct real-time molecular scale visualisation of the degradation of condensed DNA complexes exposed to DNase I. Nucleic Acids Res. 31: 4001-4005.

Azzam T, Eliyahu H, Makovitzki A, Linial M, Domb AJ (2004). Hydrophobized dextran-spermine conjugate as potential vector for in vitro gene transfection, J. Controlled Release. 96: 309-323.

Azzam T, Eliyahu H, Makovitzki A, Domb AJ (2003). Dextranspermine conjugate: an efficient vector for gene delivery, Macromol. Symp. 2003; 195: 247-261.

Eliyahu H, Joseph A, Azzam T, Barenholz Y, Domb AJ (2006). Dextran-spermine-based polyplexes-Evaluation of transgene expression and of local and systemic toxicity in mice. Biomaterials. 27: 1636-1645.

Eliyahu H, Makovitzki A, Azzam T, Zlotkin A, Joseph A, Gazit D (2005). Novel dextran-spermine conjugates as transfecting agents: comparing water-soluble and micellar polymers. Gene Ther. 12: 494-503.

Ferrari S, Moro E, Pettenazzo A, Behr JP, Zacchello F, Scarpa M (1997). ExGen 500 is an efficient vector for gene delivery to lung epithelial cells in vitro and in vivo. Gene Therapy. 4(10): 1100-1106.

Jameson LJ (1998). Principles of Molecular Medicine. : Humana Press Inc. 1998; 65-72.

Jolly D (1995). in The Internet Book of Gene Therapy: Cancer Therapeutics, eds. Sobol RE & Scanlon KJ. (Appleton & Lange, East Norwalk, CT). 1995; 3-16.

Hosseinkhani H, Yamamoto M, Inatsugu Y, Hiraoka Y, Inoue S, Shimokawa H, Tabata Y (2006). Enhanced ectopic bone formation using a combination of plasmid DNA impregnation into 3-D scaffold and bioreactor perfusion culture. Biomaterials. 27: 1387-1398.

Hosseinkhani H, Azzam T, Kobayashi H, Hiraoka Y, Shimokawa H, Domb AJ, Tabata Y (2006). Combination of 3-D tissue engineered scaffold and non-viral gene enhance *in vitro* DNA expression of mesenchymal stem cells. Biomaterials. 27: 4269-4278.

Hosseinkhani H, Inatsugu Y, Hiraoka Y, Inoue S, Shimokawa H, Tabata Y (2005). Impregnation of plasmid DNA into three-dimensional scaffolds and medium perfusion enhance *in vitro* DNA expression of mesenchymal stem cells. Tissue Eng. 11: 1459-1475.

Hosseinkhani H, Inatsugu Y, Inoue S, Hiraoka Y, Tabata Y (2005). Perfusion culture enhances the osteogenic differentiation of rat mesenchymal stem cells in collagen sponge rein forced with poly (glycolic acid) fiber. Tissue Eng. 11: 1476-1488.

Hosseinkhani H, Yamamoto M, Inatsugu Y, Hiraoka Y, Inoue S, Shimokawa H, Tabata Y (2006). Enhanced ectopic bone formation using a combination of plasmid DNA impregnation into 3-D scaffold and bioreactor perfusion culture. Biomaterials. 27: 1387-1398.

Hosseinkhani H, Azzam T, Kobayashi H, Hiraoka Y, Shimokawa H, Domb AJ, Tabata Y (2006). Combination of 3-D tissue engineered scaffold and non-viral gene enhance *in vitro* DNA expression of mesenchymal stem cells. Biomaterials. 27: 4269-4278.

Hosseinkhani H, Inatsugu Y, Hiraoka Y, Inoue S, Shimokawa H, Tabata Y (2005). Impregnation of plasmid DNA into three-dimensional scaffolds and medium perfusion enhance *in vitro* DNA expression of mesenchymal stem cells. Tissue Eng. 11: 1459-1475.

Kawabata K, Takakura Y, Hashida M (1995). The Fate of Plasmid DNA after Intravenous-Injection in Mice - Involvement of Scavenger Receptors in Its Hepatic-Uptake. Pharmaceut. Res. 12(6): 825-830.

Kircheis R, Blessing T, Brunner S, Wightman L, Wagner E (2001). Tumor targeting with surface-shielded ligand-polycation DNA complexes. J. Controlled Release. 72(1-3): 165-170.

Liu F, Huang L (2002). Development of non-viral vectors for systemic gene delivery. Journal of Controlled Release. 78(1-3): 259-266.

MacLaughlin FC, Mumper RJ, Wang J, Tagliaferri JM, Gill I, Hinchcliffe M, Rolland AP (1998). Chitosan and depolymerized chitosan oligomers as condensing carriers for in vivo plasmid delivery. J. Controlled Release. 56(1-3): 259-272.

Pouton CW. Seymour LW (1998). Key issues in non-viral gene delivery. Adv. Drug Delivery Reviews. 34(1): 3-19.

Ruponen M, Yla-Herttuala S, Urtti A (1999). Interactions of polymeric and liposomal gene delivery systems with extracellular glycosaminoglycan:physicochemical and transfection studies. Biochem Biophys. Acta. 1415: 331-341.

Schatzlein AG (2001). Non-viral vectors in cancer gene therapy: principles and progress. Anti-Cancer Drugs. 12(4): 275-304.

Schaffer DV, Fidelman NA, Dan N, Lauffenburger DA (2000). Vector unpacking as a potential barrier for receptormediated polyplex gene delivery. Biotechnol. Bioeng. 67: 598-606.

Schmidt-Wolf GD, Schmidt-Wolf IGH (2003). Non-viral and hybrid vectors in human gene therapy: an update. Trends Mole. Med. 9(2): 67-72.

Shuey DJ, McCallus DE, Giordano T (2002). RNAi: gene-silencing in therapeutic intervention. Drug Discov Today. 104(7): 104-1046.

Taira T, Kataoka K, Niidome T (2005). Non-viral gene therapy. springer-verlag Tokyo. 35-45.

Takai T, Ohmori H (1990). DNA Transfection of Mouse Lymphoid-Cells by the Combination of Deae-Dextran-Mediated DNA Uptake and Osmotic Shock Procedure. Biochem. Et Biophys. Acta. 1048(1): 105-109.

Vanderkerken S, Vanheede T, Toncheva V, Schacht E, Wolfert MA, Seymour L, Urtti A (2000). Synthesis and evaluation of poly(ethylene glycol)-polylysine block copolymers as carriers for gene delivery. J. Bioactive Compatible Polymers. 15(2): 115-138.

van de Wetering P, Cherng JY, Talsma H, Crommelin DJ, Hennink WEI (1998). 2-(dimethylamino)ethyl methacrylate based (co)polymers as gene transfer agents. J. Controlled Release. 53(1-3): 145-153.

Wolfert MA, Schacht EH, Toncheva V, Ulbrich K, Nazarova O, Seymour LW (1996). Characterization of vectors for gene therapy formed by self- assembly of DNA with synthetic block co-polymers. Human Gene Ther. 7(17): 2123-2133.

Yaroslavov AA, Sukhishvili SA, Obolsky OL, Yaroslavova EG, Kabanov AV, Kabanov VA (1996). DNA affinity to biological membranes is enhanced due to complexation with hydrophobized polycation. Febs Letters. 384(2): 177-180.

Ye S, Cole-Strauss AC, Frank B, Kmiec EB (1998). Targeted gene correction: a new strategy for molecular medicine. Mol. Med. Today. 4: 431-437.

Yu L, Suh H, Koh JJ, Kim SW (2001). Systemic administration of TerplexDNA system: Pharmacokinetics and gene expression. Pharmaceut. Res. 18(9): 1277-1283.

8

Isolation and characterization of breast cancer stem cells from malignant tumours in Vietnamese women

Pham Van Phuc[1]*, Tran Thi Thanh Khuong[1], Le Van Dong[2], Truong Dinh Kiet[3], Tran Tung Giang[4] and Phan Kim Ngoc[1]

[1]Laboratory of Stem cell Research and Application, University of Science, VNU-HCM, Vietnam.
[2]Military Medical University, Ha Noi, Vietnam.
[3]University of Medicine - Pharmacy, HCM city, Vietnam.
[4]The University of New South Wales, Sydney, Australia.

Cancer stem cells are the origin of tumors and have been isolated successfully from different kinds of tumors. Breast cancer stem cells have been recently identified in breast carcinoma with markers CD44+/CD24-/dim. This population can cause tumor and display stem cell-like properties. However, direct evidences that breast cancer stem cells can be propagated *in vitro* is still lacking. This research was carried out to isolate and propagate *in vitro* breast cancer stem cells from tumor biopsy. Breast tumor biopsy was used to isolate breast cancer cells by primary tissue culture. As such, breast cancer stem cells were isolated from breast cancer cells by catcher-tube based cell sorter on flow cytometter machine. These cells were propagated by an *in vitro* culture in a free serum specific medium. The results showed that the CD44+CD24-/dim cell population that were maintained, were capable of self-renewal and extensive proliferation as clonal non-adherent spherical clusters. Interestingly, cultured cells were CD44+CD24-/dim expressed by the putative stem cell marker Oct-4, resistant with verapamil at 50 μg/ml, and gave rise to new tumors when as few as 1000 cells were injected into the mammary fat pad of immune-deficient mice. This population was a suitable *in vitro* model to study breast cancer stem cells and develop therapeutic strategies to treat breast cancer.

Key words: Breast cancer, breast tumor, cancer stem cell, CD44+CD24-/dim.

INTRODUCTION

Cancer is a disease that causes cells to grow unlimitedly, thus, forming tumors in the body. Tumors are groups of cells that contain many kinds of cells with different biological characteristics (Reya et al., 2001; Campbell and Polyak, 2007). Many publications showed evidences for existence of cancer stem cells (CSCs) in malignant tumors. CSCs have been identified in many solid tumors, including brain, prostate, pancreatic, liver, colon, head and neck, lung, and skin tumors (Anton Aparicio et al., 2007; Ceder et al., 2008; Eramo et al., 2008; Ferrandina et al., 2007; Glinsky, 2007; Li et al., 2007; Prince et al., 2007; Seo et al., 2007). Indeed, this idea was first postulated by Rudolph Virchow and Julius Cohnheim in

the nineteenth century (Bignold et al., 2006; Huntly et al., 2005). Virchow's embryo rest hypothesis noted the similarities between fetal tissue and cancer cells (Sell, 2004). Later, Cohnheim and Durante extended this hypothesis that there exist embryonic remnants immature organs, and Beard hypothesized that cancer arises either from germ cells or from placental tissue.

The concept about tumor containing heterogeneous populations of cells was demonstrated firstly by Lapidot and colleagues (Lapidot et al., 1994) in leukemia. They showed that $CD34^+CD38^-$ cells isolated from acute myeloid leukemia patients developed a tumor when injected into nonobese diabetic/severe combined immuno-deficiency (NOD/SCID) mice, while injection of even larger numbers of the more differentiated cells, $CD34^+CD38^+$, did not initiate tumor formation. Moreover, the tumors formed by injection of the $CD34^+CD38^-$ cells were similar in morphology compare to original tumors. And this concept also was demonstrated in solid tumors,

*Corresponding author. E-mail: pvphuc@hcmuns.edu.vn, phamvanphuc2308@gmail.com.

especially breast cancer by Al-Hajj (Al-Haii et al., 2003). Following the results of Al-Hajj et al. (2003), the cells expressed protein CD44 and weakly or unexpressed protein CD24 could established the new tumors when allografted in mice. A lot of other researches demonstrated that 200 cells with this phenotype could cause tumors in NOD/SCID mouse models. While 20,000 cells not been expressed these markers could not cause tumor after transplantation in mice. These tumors contained a lot of different kinds of cells and CSCs. These CSCs derived from this tumor could continuously cause tumor when injected in immunodeffiency mice. This capacity of CSCs demonstrated that they could undergo self-renewal for a long time.

In Vietnam, there are about 20.3 breast cancer patients per 100,000 people. This disease is the most common cancer in Vietnamese women. Up to date, there is not any publication about existing breast CSCs in their tumors. Therefore, this research is aimed at demonstrating the existence of breast CSCs in Vietnam. Moreover, this research further confirmed the hypothesis of breast CSCs in breast cancer.

Three methods are commonly used for the isolation of CSCs. These methods include:

(1) The isolation by sorting of a side population (SP) based on Hoechst dye efflux.
(2) Sorting on the basis of cell surface marker expression.
(3) Sphere culture (mammosphere culture).

CSCs were obtained from three difference methods, in degrees of enrichment of CSCs as well as advantages and limitations. Despite the isolation methodology, the cells were named CSCs when they passed some assays, including tumorigenicity, self-renewal, and the ability to histologically recapitulate the tumor of origin.

In this research, breast cancer stem cells were firstly isolated by sorting on the basic of cell surface CD44 and CD24 marker expression and then they were cultured in mammosphere type. The propagation of breast cancer stem cells was carried out by cell suspension techniques in free-serum medium. After that, these CSC candidates were checked about tumorigenicity, self-renewal, and the ability to histologically recapitulate the tumor of origin.

MATERIALS AND METHODS

Primary tissue culture

Isolation and *in vitro* expansion of stem cells was carried out from breast tumor specimens. Tumor specimens were obtained from consenting patients, whereas tumor biopsy, obtained from a hospital, was transferred to the laboratory. Biopsy was washed 3 to 4 times by PBS (phosphate buffer saline) supplemented with antibiotic-antimycotic to remove blood and then the fat binding in the biopsy as well as in the necrotic part. After that, biopsy was cut into some small fragments of about 1 to 2 mm^2 in square. These fragments were seeded in a tissue dish that is 35 mm in diameter

(Nunc, Germany), in medium DK-SFM (Gibco, Invitrogen), supplemented with FGF (fibroblast growth factor) and EGF (epidermal growth factor) bought from Sigma (Sigma-Aldrich, St Louis, MO). Dishes were incubated in 37°C, 5% CO_2 condition and the medium was changed at a 3-day interval.

Isolation of candidate breast cancer stem cells

Candidate breast cancer stem cells were isolated by sorting CD44+CD24-/dim cell population by catcher-tube based cell sorter in combination with the flow cytometter (Facscalibur, BD Bioscience). The cells that were obtained from the primary tissue culture were stained with 20 µl anti-body CD44 and 20 µl anti-body CD24 (all bought from BD Bioscience) in 5 ml tube in 10^7 cells/ml concentration. Tubes had been incubated in the dark, in room temperature for 45 min and more tubes were added to the FACSflow solution (1 ml) in 10^7 cells/ml concentration. In CellQuest Pro software, CD44+CD24-/dim cell population was identified by quadrant analysis. CD44+CD24-/dim cells were positive with CD44 and negative or weakly positive with CD24. In Figure 1, this population was R1 and it was sorted into a 50 ml tube coated with BSA (bovine serum albumin) 1 mg/ml initially. Candidate cells were harvested by centrifuging at 3,000 rpm for 5 min.

Culture of breast cancer stem cells

After sorting, candidate breast cancer stem cells were plated at 1,000 cells/mL in serum-free DMEM-F12, supplemented with 10 ng/mL basic fibroblast growth factor (bFGF), 20 ng/mL epidermal growth factor (EGF), 5 ng/mL insulin and 0.4% bovine serum albumin (all from Sigma). Cells grown in these conditions as non-adherent spherical clusters of cells (usually named "spheres" or "mammospheres") were enzymatically dissociated every 3 days by incubation in a 0.25% trypsin-EDTA solution (Sigma, St Louis, MO) for 2 min at 37°C. Conversely, differentiation was induced by culturing mammosphere derived cells for 8 days on collagen-coated dishes in DMEM-F12, supplemented with 5% fetal bovine serum (Sigma, St Louis, MO) without growth factors.

Sphere formation assay

Primary spheres were dissociated as previously described and 100 cells per well were plated in 96-well culture dishes in 200 µL growth medium. As such, 25 µL of the medium per well was added every 2 days. However, the number of spheres for each well was evaluated after 7 days of culture.

In vivo injection of mammosphere cells

Spheres were collected, enzymatically dissociated, washed in PBS, and kept at 4°C until they were injected into the layer that is under the skin (subcutaneous) of the 5-week-old SCID mice. Mice received an estradiol supplementation (0.4 mg/kg s.c., Progynon Depot, Schering-Plough, Kenilworth, NY) every 10 days for 40 days after cell injection and were inspected for tumor appearance, by observation and palpation, for 15 weeks following cell injection. After this time interval, all mice were sacrificed by cervical dislocation and the presence of each tumor nodule was confirmed by necropsy.

Oct-4 gene expression assay

Oct-4 gene expression was evaluated by RT realtime PCR. In each

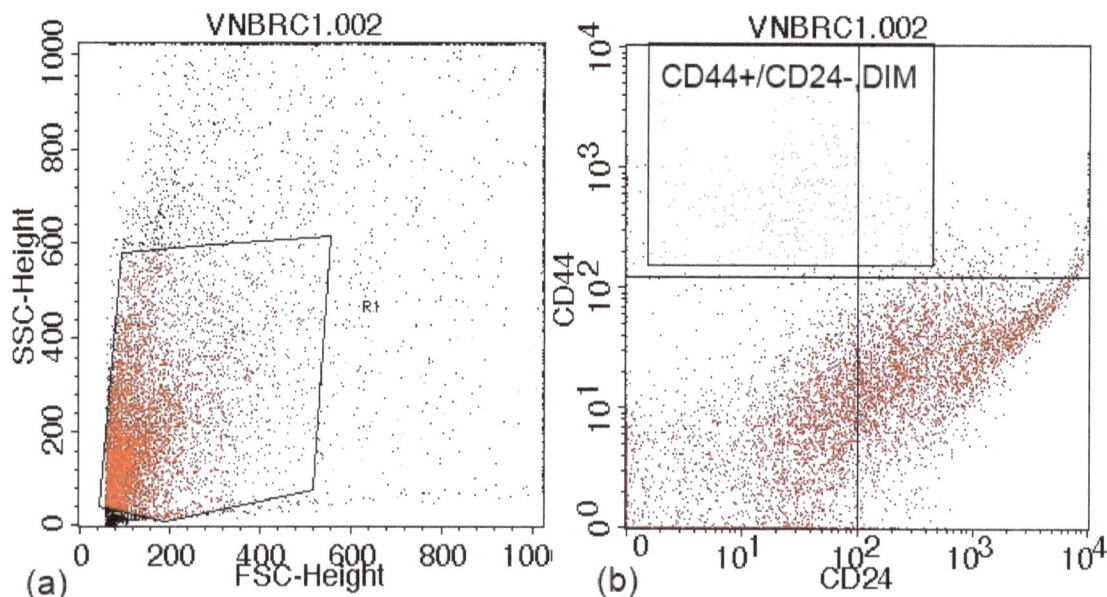

Figure 1. Cell population with suitable size was gated in the R1 region (a). Breast cancer stem cells were identified by expression of CD44 and an un-expression or weak expression of CD24 in quadrant analysis (b).

tube, the following volumes which amounted to a total volume of 25 µl were added: 12.9 µl of PCR master mix, 9.6 µL of nuclease-free water, 0.5 µL of forward primer (gene of interest), 0.5 µL of reverse primer (gene of interest) and 1.5 µl of RNA template. After complete addition of all the components, tubes were centrifuged at 3,000 rpm, in 4 °C for 3 min. Tubes were then loaded into the Mastercycler epgradient S real-time PCR machine (realplex4, Eppedorf, Germany) according to the template created using the realplex® programme. As such, the cells were analyzed repeatedly for 3 times.

Anti-cancer drug assay

Breast cancer stem cell (CD44+CD24-/dim) and breast cancer cells were at a density of 0.4 and 10^4 cells per well in 24-well plate (Nunc), respectively in DMEMF12/10% FBS. After 24 h culture for confluence, the cells were treated with 50 µg/ml verapamil (Sigma). The cells were observed under inverted phase contrast microscope (Carl-Zeiss, Gemany) after 24, 48 and 72 h treatment. The apoptosis of two populations were investigated by flow cytometry using Annexin-V and PI (BD Bioscience, USA).

RESULTS AND DICUSSION

Existence of breast cancer stem cells in primary tissue culture

Establishment of primary cultures of mammary gland precursors

The study was carried out to culture primarily 21 tumors from 21 different patients. About 15 samples of the 21 tumors had grown out with a lot of single cells surrounding the tissues. The cells from the 15 samples

were propagated until they reach the 80% cell confluence. In almost all the samples, single cells appeared around the fifth day, with the earliest on the third day (Figure 2). Then cells proliferated rapidly and got clonal combination on the fifteenth day. There were two kinds of cell shape in the primary culture. These are epithelial cells with bean shape and big nucleus; and stromal cells like fibroblast having small nucleus and long shape.

Existence of CD44+CD24-/dim cell sub-population in primary cell lines

When analyzing 15 primary cell samples with two markers, CD44 and CD24, all of them had a small population of cells that was positive with CD44 and negative or weakly positive with CD24. This population got 3.59 ± 1.65% in average in the total number of cells derived from the primary culture with the lowest ratio at 1.25% and the highest at 7.12%. The result also showed that there were about 50% of the cells that were positve with marker CD24. However, there were more than 90% of the cells that were negative with marker CD44. By flow cytometry analysis, almost all the cells that were positive with CD44 were nagetive or weakly positive with CD24. As such, CD44+CD24-/dim cell population have characteristics of cancer stem cells after culture.

Expression of breast cancer stem cell markers

Following the results of Clarke et al. (2006), xenotransplan-

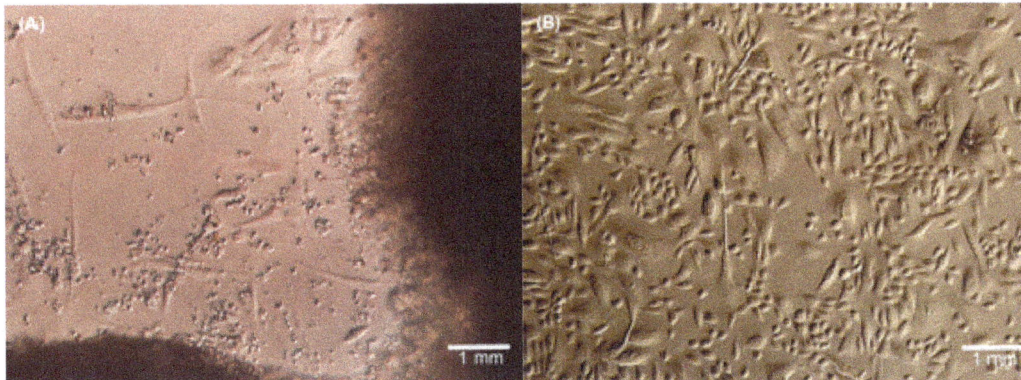

Figure 2. Results of the primary tissue culture. (A) Migration of cells out of the tissue after 7 days, (B) Two kinds of cells appearing in the primary culture (epithelial cells and stromal cells).

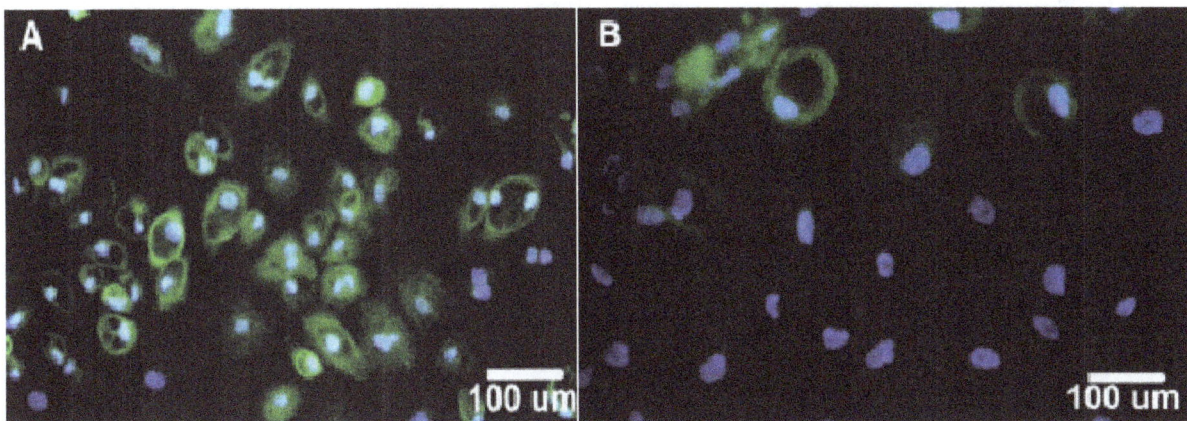

Figure 3. Expression of CD44 (A) and CD24 (B) was identified by immunohistochemistry. Candidate cancer stem cells were stained with anti CD44-FITC and anti CD24-FITC, and were counterstained with Hoescht 33342 (stained nucleus).

tation was used to isolate the population of cells that have tumorigenicity potential in NOD/SCID mice (Al-Hajj et al., 2003). This cell population expressed CD44 protein, but unexpressed or weakly expressed protein CD24 in the cell surface. This result was confirmed by flow cytometry and immunohistochemistry (Figure 3 and 4. They demonstrated that 200 cells with these pheno-type could cause tumors when injected into NOD/SCID mice, while 20,000 cells without these phenotype could not cause tumors *in vivo*. However, the population that was isolated by this research expressed CD44+CD24-/dim phenotype.

In vitro self renewal

When cultured in the free serum medium, CD44+CD24-/dim cells could not adhere into the flask surface, but floated in the medium with mammosphere style (Figure 5a). Mammosphere creating capacity demonstrated that sorted cells had been self renewed. This chacracteristic was similliar to stem cells which were isolated from the

mamary gland. It was demonstrated that cells derived from the mamary gland which could form mammosphere *in vitro* could form *de novo* human mamary gland in mice after xenotransplantation (Dontu et al., 2003), in that sorted cells had been cultured in the free serum medium. After 15 days, a lot of colony were formed in the so called mammosphere and were found floating in the medium. As such, the size of the mammosphere grew bigger and bigger, owing to time. However, the mammospheres were rather condensed when observed through an inverted microscope.

In vivo tumor formation

Tumorigenicity is an important characteristic of cancer stem cells. Many researches showed that new tumors could be formed by injecting a little cancer stem cells into immunodeficient mice. In this research, 1 thousand breast cancer stem cells were used to create tumor in mice (Figure 5b). The number of cells needed for tumorigenicity was relatively big when compared to some

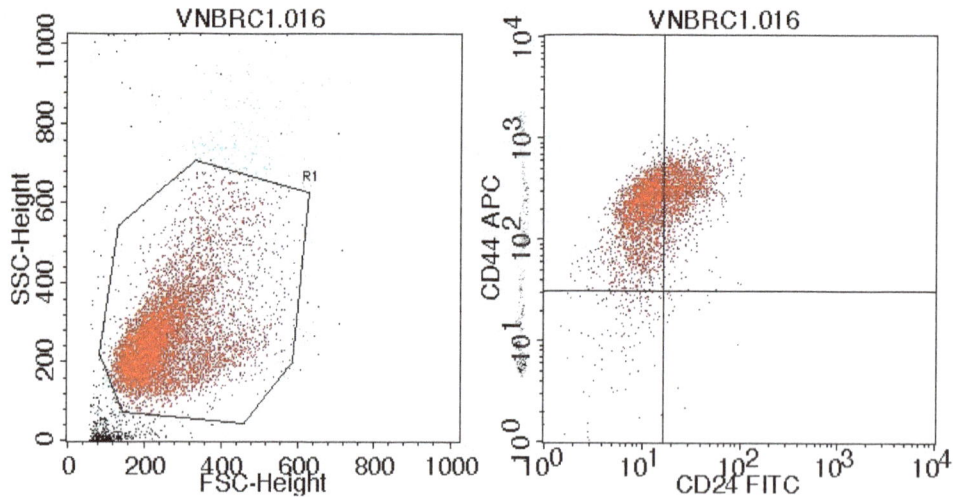

Figure 4. The breast cancer stem cells were established with CD44 positive expression and CD24 negative/dim expression.

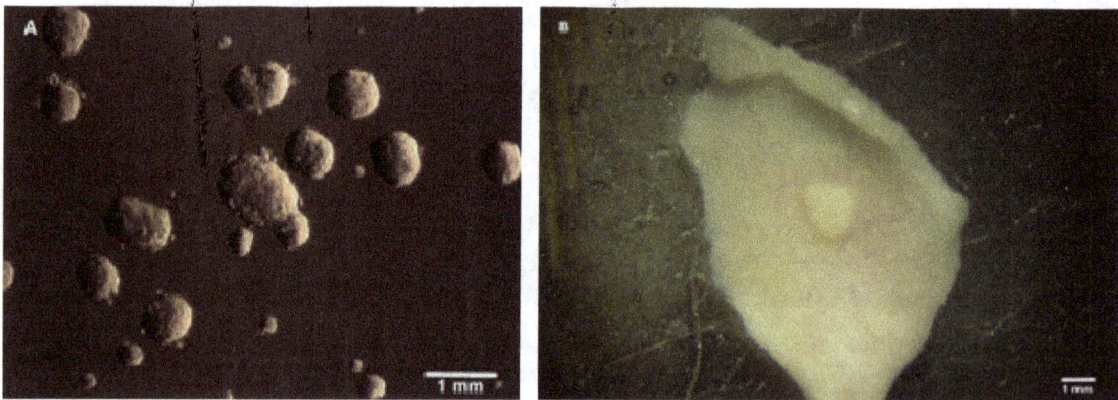

Figure 5. Mammosphere formed in the serum free medium after 10 culture days (A). The tumor formed subcutenously after injection of 106 breast cancer stem cells (B).

others. However, in this research, immunodeficient mice were used instead of NOD/SCID mice, and it was observed that the mice's immune system in this research could not be inhibited completely like the NOD/SCID mice.

Breast cancer stem cells expressed by Oct-4

Oct-4 gene is a specific gene marker for embryonic stem cells and the expression of its pluripotent potential is related to stem cells. However, its expression was also found in cancer cells. Previous researches have confirmed that cells or stem cells, expressing Oct-4 gene, could cause tumors after they have been injected in immunodeficient mice. Oct-4 expression in the candidate breast cancer stem cells demonstrated that the confirmed cells isolated from this procedure were breast cancer stem cells (Figure 6).

CD44+CD24-/dim cell population are resistant with anti-cancer drugs

Verapamil was known as an anti-cancer drug for inhibition of drug efflux pump proteins such as P-glycoprotein. They were used in clinical treatment as well as in the research for growth inhibition of tumor or induction to apoptosis. In this research, verapamil was used to test the resistance of the anti-cancer drug of two subpopulations (CD44+CD24-/dim) and breast cancer cell population. The cells were seeded at 6.10^4 cells/cm^2 density at the first day in serum DMEM media. When the cells got to confluence, 50 µg/ml verapamil was supplied with media for culture in a 2 day experiment. The cell numbers of breast cancer cells significantly decreased after 48 h of verapamil treatment. Verapamil possessed strong effect to these cells which led to cell apoptosis, whereas breast cancer stem cells had growth normally under verapamil effect. The data indicated that breast

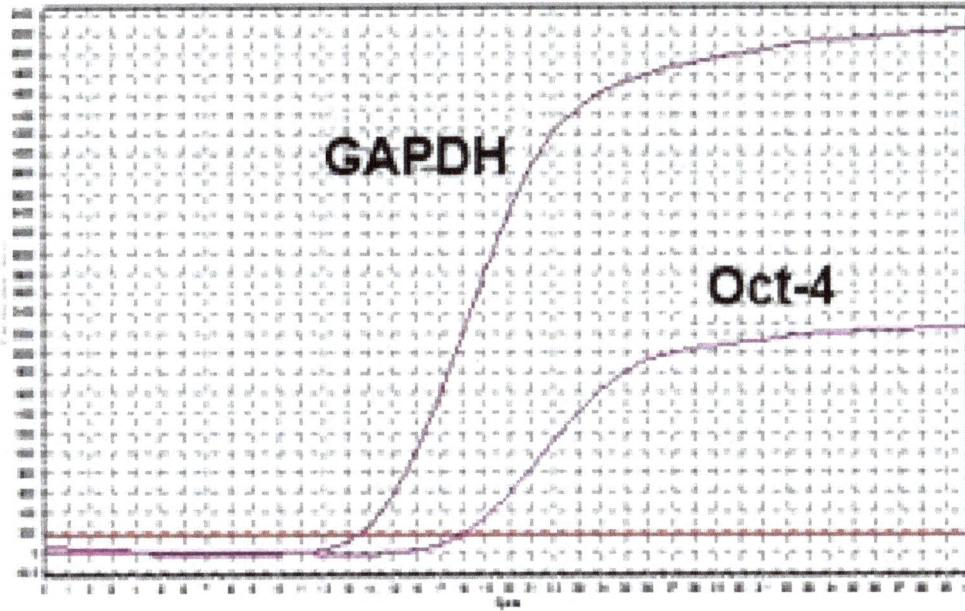

Figure 6. Result of RT realtime PCR for genes GAPDH (house keeping gene) (red line) and Oct-4 (pink line). The CD44+CD24-/dim were positive with GAPDH and Oct-4. The Ct values of GAPDH and Oct-4 genes are 13.5 and 18.0, respectively.

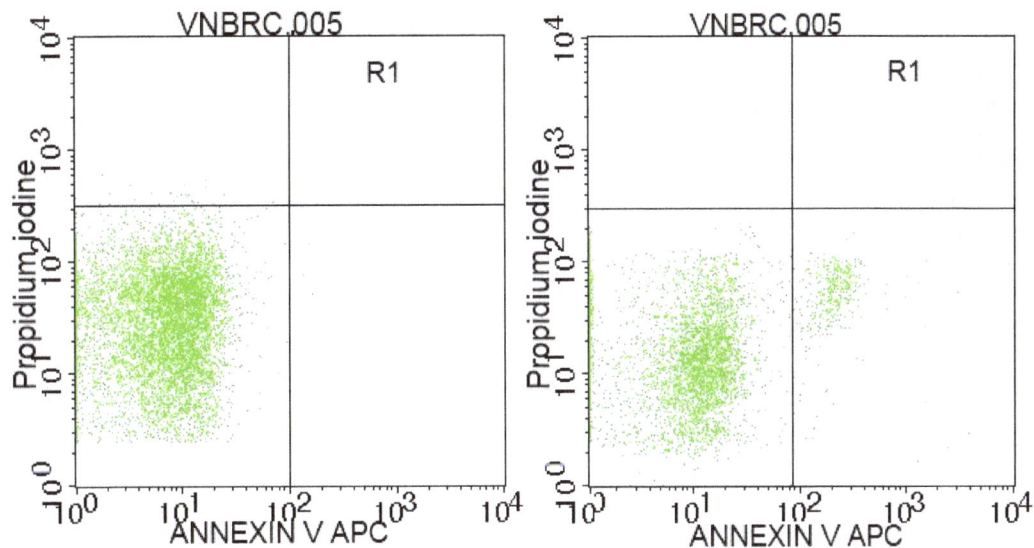

Figure 7. Results of apoptosis analysis by flow cytometry techniques using kit annexin-V and PI. The breast cancer stem cells were resistant with verapamil after 2 days treatment (left). Breast cancer cells contained a apopototic cell population after 2 days treatment in verapamil medium (right).

cancer stem cell population has extreme resistance of anti cancer drug when compared to breast cancer cells (Figure 7). As such, this property remains a serious situation in cancer treatment clinically. This result was also similar with the apoptosis analysis results.

Our results are a 100% similar to that of Al-Hajj and colleagues, who showed that breast cancer tumors at stage II and III contained a small population of CD44+CD24-/dim. Contrary to these results, the results of Honeth et al. (2008) showed that only 31% of their tumors contained breast cancer stem cells. This discordance could depend on our study involving mainly metastatic tissues, while they used both metastatic and primary tumors. However, another study demonstrated 59% of CD44+/CD24- cells in human breast tumors (Mylona et al., 2008).

Conclusion

In this study, breast cancer stem cells can isolated from maglinant tumors in Vietnamese women. We retrospectively confirmed that CD44+/CD24dim/- breast cancer cells, which have been prospectively identified as tumorigenic cells, display stem/progenitor cell properties. They are positive with CD44 protein and negative or weakly positive with CD24. Also, they have self-renewal capacity via mammosphere assay and a tumor causing capacity *in vivo*.

To our knowledge, for the first time, we showed that breast tumorigenic cells with stem/progenitor cell properties can be propagated *in vitro* as non-adherent mammospheres from breast cancer tissue in Vietnamese women, in keeping with similar findings obtained by normal mammary stem/progenitor cells. This experimental system may represent a suitable *in vitro* model to study breast cancer-initiating cells and to challenge them with molecularly targeted agents specifically interfering with the self-renewal and survival of breast cancer-initiating cells.

REFERRENCES

Anton Aparicio LM, Cassinello Espinosa J, García Campelo R, Gómez Veiga F, Diaz Prado S, Aparicio Gallego G (2007). Prostate carcinoma and stem cells. Clin. Transl. Oncol. 9(2): 66-76.

Bignold LP, Coghlan BL, Jersmann HP (2006). Hansemann, Boveri, chromosomes and the gametogenesis-related theories of tumours. Cell Biol. Int. 30(7): 640-644.

Campbell LL, Polyak K (2007). Breast tumor heterogeneity: cancer stem cells or clonal evolution? Cell Cycle 6(19): 2332-2338.

Ceder JA, Jansson L, Ehrnström RA, Rönnstrand L, Abrahamsson PA (2008). The characterization of epithelial and stromal subsets of candidate stem/progenitor cells in the human adult prostate. Eur Urol. 53(3):524-31.

Clarke MF, Dick JE, Dirks PB, Eaves CJ, Jamieson CH, Jones DL, Visvader J, Weissman IL, Wahl GM (2006). Cancer stem cells – perspectives on current status and future directions: AACR Workshop on Cancer Stem Cells. Cancer Res. 66: 9339-9344.

Dontu G, Abdallah WM, Foley JM, Jackson KW, Clarke MF, Kawamura MJ, Wicha MS (2003). In vitro propagation and transcriptional profiling of human mammary stem/progenitor cells. Genes Dev. 17(10): 1253-70.

Eramo A, Lotti F, Sette G, Pilozzi E, Biffoni M, Di Virgilio A, Conticello C, Ruco L, Peschle C, De Maria R (2008). Identification and expansion of the tumorigenic lung cancer stem cell population. Cell Death Differ 15: 504-514.

Ferrandina G, Legge F, Mey V, Nannizzi S, Ricciardi S, Petrillo M, Corrado G, Scambia G (2007). A case of drug resistant clear cell ovarian cancer showing responsiveness to gemcitabine at first administration and at re-challenge. Cancer Chemother. Pharmacol. 60(3): 459-61.

Glinsky GV (2007). Stem cell origin of death-from-cancer phenotypes of human prostate and breast cancers. Stem Cell Rev. 3(1): 79-93.

Huntly BJ, Gilliland DG (2005). Leukaemia stem cells and the evolution of cancerstem cell research. Nat. Rev. Cancer 5: 311-321.

Lapidot T, Sirard C, Vormoor J, Murdoch B, Hoang T, Caceres-Cortes J, Minden M, Paterson B, Caligiuri MA, Dick JE (1994). A cell initiating human acute myeloid leukaemia after transplantation into SCID mice. Nature. 367: 645-648.

Li C, Heidt DG, Dalerba P, Burant CF, Zhang L, Adsay V, Wicha M, Clarke MF, Simeone DM (2007). Identification of pancreatic cancer stem cells. Cancer Res. 67: 1030-1037.

Misra S, Hascall VC, Berger FG, Markwald RR, Ghatak S (2008). Hyaluronan, CD44, and cyclooxygenase-2 in colon cancer. Connect Tissue Res. 49: 219-224.

Prince ME, Sivanandan R, Kaczorowski A, Wolf GT, Kaplan MJ, Dalerba P, Weissman IL, Clarke MF, Ailles LE (2007). Identification of a subpopulation of cells with cancer stem cell properties in head and neck squamous cell carcinoma. Proc. Natl. Acad. Sci. USA 104: 973-978.

Reya T, Morrison SJ, Clarke MF, Weissman IL (2001). Stem cells, cancer, and cancer stem cells. Nature 414: 105-111.

Sell S (2004). Stem cell origin of cancer and differentiation therapy. Crit. Rev. Oncol. Hematol. 51: 1-28.

Seo DC, Sung JM, Cho HJ, Yi H, Seo KH, Choi IS, Kim DK, Kim JS, El-Aty AM A, Shin HC (2007). Gene expression profiling of cancer stem cell in human lung adenocarcinoma A549 cells. Mol. Cancer 6: 75.

9

Effect of dietary supplementation of probiotic and prebiotic on growth indices and serum biochemical parameters of broiler chickens

A. Ashayerizadeh[1]*, N. Dabiri[1,2], K. H. Mirzadeh[1] and M. R. Ghorbani[1]

[1]Department of Animal Science, Ramin Agricultural and Natural Resources University, Ahvaz, Iran.
[2]Faculty of Agricultural, Animal Science Department, Islamic Azad University, Karaj Branch, Karaj, Iran.

This experiment was conducted to compare the effects of antibiotic, probiotic, prebiotic and mixture of probiotic and prebiotic as dietary growth promoter on serum biochemical parameters, energy and protein efficiency of broiler chickens. Three hundred day old Ross 308 broilers were equally distributed into 30 floor pens and reared for 42 days. A basal diet was formulated according to the recommendations of NRC for starter (1 to 21 days) and grower (22 to 42 days) periods and considered as control diet. Four tested diets were formulated by supplementing the basal control diet with antibiotic, Flavomycin; probiotic, Primalac; prebiotic, Biolex-MB and mixture of probiotic plus prebiotic (synbiotic). Six replicates were used for each treatment. Specific growth rate and growth efficiency were highest in birds under prebiotic and synbiotic treatments in starter and total rearing period, respectively. Broilers that were fed diet containing synbiotic had the highest energy efficiency ratio which was significant as compared to control group. Protein efficiency ratio followed the same trend and improved when synbiotic was used in the diet (p < 0.05). At 21 and 42 day of age, dietary supplementation with probiotic and synbiotic decreased (p < 0.05) serum cholesterol concentration, when compared with birds fed Flavomycin diet. The results suggested that the mixture of probiotic and prebiotic could be effective as antibiotic to improve the health and performance of broiler chickens.

Key words: Broiler, performance, cholesterol, probiotic, prebiotic

INTRODUCTION

Nowadays, the efficiency of poultry to convert the feed into meat plays a key role In economics of broiler industry. Therefore, it is highly essential to improve feed efficiency of poultry to produce meat economically and also food safety is more seriously considered than before. On the other hand, economy of food production is also a factor that can not be ignored. A huge amount of antibiotics have been used to control diseases and improve performances in livestock. However, due to growing concerns about antibiotic resistance and the potential for a ban for antibiotic growth promoters in many countries in the world, there is an increasing interest in finding alternatives to antibiotics in poultry production.

One choice could be directed microbials (DFM), also called probiotics, which are live microbial feed supplements that beneficially affect the host animal by improving its intestinal health (Fuller, 1989). Primalac is a kind of commercial probiotic that contains at least 1×10^8 CFU g^{-1} Lactobacillus casei, Lactobacillus acidophilus, Bifidobacterium thermophilum, and Enterococcus faecium (Chichlowski et al., 2007a,b). Prebiotics are non-digestible food ingredient that beneficially affects the host by selectively stimulating the growth and/or activity of one or a limited number of bacteria in the colon (Gibson and Roberfroid, 1995). Biolex-MB is a commercial prebiotic of the mannan-oligosaccharides family, which is obtained by extraction from the outer cell wall of the yeast Saccharomyces cerevisiae.

Various findings on the effect of different probiotics and prebiotics on the health and growth responses of broiler chickens was reported (Kabir et al., 2004; Piray et al.,

*Corresponding author. E-mail: amin.ashayerizadeh@yahoo.com.

2007). Most recently, considerable attention has been paid to test the potency of growth promotants on altering lipid metabolism. Probiotic supplementation has been shown to reduce the cholesterol concentration in egg yolk (Abdulrahim et al., 1996; Haddadin et al., 1996) and serum in chicken (Mohan et al., 1996; Jin et al., 1998). Recent report suggested that feeding of chicory beta fructans, a prebiotic, reduced the serum cholesterol and abdominal fat of broiler chicken (Yusrizal and Chen, 2003). One of the most important compounds of feeds for farm animals is the protein source. It is reported that protein efficiency could be affected by intestinal microflora. Pathogenic bacteria increase the breakdown of proteins to nitrogen and reduce the efficiency of dietary protein (Mikulec et al., 1999). The aim of this study was comparing the effects of the antibiotic, probiotic, prebiotic and the probiotic plus prebiotic mixture on blood biochemical parameters and efficiency of dietary energy and protein of broiler chickens.

MATERIALS AND METHODS

Birds and housing

In this study, 300 broiler chickens of the commercial Ross 308 strain were used in a randomized completely design with 5 treatment (6 replicates in each treatment and 10 birds/replicates) and reared on the floor pens for 42 days. A basal diet was formulated and considered as control according to recommendation of NRC, 1994 for starter (1 to 21 days), and grower (22 to 42 days) diets. Four tested diets were formulated by supplementing the basal control diet with antibiotic (Flavomycin, 650 g ton[-1]), probiotic (Primalac, 900 g ton[-1]), prebiotic (Biolex-MB, 2000 g ton[-1]), and mixture of probiotic (900 g ton[-1]) plus prebiotic (2000 g ton[-1]), respectively (Table 1). From day 1 to 42 of the study, water and experimental diets were given to the birds, ad-libitum. Broilers and feed intake were weighed weekly. Energy and protein efficiency ratios and specific growth rate were calculated as follows:

Specific growth rate (SGR) = 100 × (ln FBW - ln IBW) / t

Where, FBW is final body weight (g), IBW is initial body weight (g) and t is time in days.

GE = WG / LW

Where, GE is growth efficiency for time period, WG is weight gain for specific time period and LW is initial weight as a covariate.

Protein intake (PI) = total feed intake × (CP% of diet / 100)

Protein efficiency ratio (PER) = weight gain/total protein intake

Energy intake (EI) = (kcal ME of per kg diet × feed intake) /1000

Energy efficiency ratio (EER) = weight gain × 100 / total ME intake

Biochemical serum data collection and analysis

At 21 and 42 days of age, 4 ml of blood was collected from wing vein from 6 birds in each treatment. In order to prevent clotting, blood was collected in heparinized test tubes and centrifuged (at

2,000 rpm for 10 min), and the serum was separated, then stored at -20°C until assayed to measuring blood parameters (cholesterol, triglycerides and high density lipoprotein (HDL) cholesterol) using commercial kits (Pars Azmoon) according to the manufacturer's protocols. Very low density lipoprotein (VLDL) cholesterol was calculated from triglycerides by dividing the factor 2.2. The low density lipoprotein (LDL) cholesterol was calculated by using the formula:

LDL cholesterol = Total cholesterol − HDL cholesterol − VLDL cholesterol.

Statistical analysis

All data were analyzed using the One-Way Anova procedure of SAS® (SAS, 1998) for analysis of variance. Significant differences among treatments were identified at 5% level by Duncan's multiple range tests (Duncan, 1955).

RESULTS AND DISCUSSION

Growth performance

The results of feed additives on broilers performance are presented in the Table 2. Supplementation of prebiotic significantly increased the SGR and GE as compared to the control and probiotic treatments during the starter period (p < 0.05). Moreover, these indices were highest in broilers fed synbiotic than those of control and probiotic groups from 1-42 d of experiment (p < 0.05). At the starter (1 to 21 d) and total (1 to 42 d) periods of the experiment, the energy intake and protein intake were not affected by dietary treatments (p > 0.05). Energy efficiency ratio (EER) and protein efficiency ratio (PER) followed the same trend. All the growth promoting additive treatments had better PER and EER than control birds (p < 0.05). The highest value of PER and EER was shown by broilers under synbiotic treatment.

In the present study, the beneficial effects of a probiotic, prebiotic and synbiotic products on broiler performance are in agreement with previous studies (Zulkifli et al., 2000; Thitaram et al., 2005; Nayebpor et al., 2007; Falaki et al., 2010). In contrast, Gunal et al. (2006), Zhang et al. (2005) and Willis et al. (2007) reported that using these additives shed in the broiler ration had no significant effects on growth performance of broiler chickens. It is reported that the use of mixture of probiotic plus prebiotic because of their synergistic effect could reduce the count of pathogenic bacteria and increase the population of useful microflora in gut (Fairchild et al., 2001; Rada et al., 1995). So, it could be concluded by removing pathogenic bacteria that can adhere to the gastrointestinal track wall, Immune system may be less stimulated and a favorable medium is provided for the use of energy and nutrients by birds (Savage and Zakrzewska, 1996; Fairchild et al., 2001). Also, the use of prebiotics by increasing in length of the intestinal mucosa, increases the absorption areas and improves the birds energy and protein efficiency ratio

Table 1. Ingredient composition (as percent of dry matter) and calculated analysis of the basal diets.

Grower (22 – 42 days)	Starter (1 – 21 days)	Ingredient
Corn	61	58.7
Soybian meal	29	30
Wheat bran	5	5
Fish meal	0	2
Soybian oil	2	1
Oister shell meal	1	1.2
DCP	1.07	1
Vitamin and mineral perimix	0.5	0.5
DL- Methionine	0.13	0.1
L-lysine	0.15	0.25
Salt	0.25	0.1
Coccidiostat	0	0.05
Total	100	100
Nutrient content		
ME(Kcal/Kg)	2850	2950
Crude protein (%)	20.48	18.44
Crude fiber (%)	3.89	3.81

Vitamin and mineral provided per kilogram of diet: vitamin A, 360000 IU; vitamin D3, 800000 IU; vitamin E, 7200 IU; vitamin K3, 800 mg; vitamin B1, 720 mg; vitamin B9, 400 mg; vitamin H2, 40 mg; vitamin B2, 2640 mg, vitamin B3, 4000 mg; vitamin B5, 12000 mg; vitamin B6, 1200 mg; vitamin B12, 6 mg; choline chloraid, 200000 mg, manganeze, 40000 mg, iron, 20000 mg; zinc, 40000 mg, copper, 4000 mg; iodine, 400 mg; selenium, 80 mg

(Santin et al., 2001). Furthermore, the effect of probiotics and prebiotics on reduction of pathogenic bacteria could reduce the breakdown of proteins to nitrogen. In this way, the utilization of proteins (amino acids) is improved, particularly from food that does not contain them in optimum quantities (Mikulec et al., 1999). Finally, each of the above mentioned reasons may lead to better growth responses of broiler chickens.

Serum lipid concentrations

The effect of feed additives on blood constituents is presented in Table 3. At 21 day of age, no significant differences were observed in triglycerides, HDL, LDL and VLDL levels between treatments ($p > 0.05$). In 21 day old brids, dietary supplementation with probiotic decrease cholesterol concentration ($p < 0.05$), when compared with birds fed the control, prebiotic and antibiotic diets. Also, at 42 day of age, HDL and LDL levels were not affected by dietary treatments, while the synbiotic and probiotic supplemented groups had a lower cholesterol and triglycerides concentrations ($p < 0.05$) compared with those of control and antibiotic supplemented groups, respectively. However, the cholesterol and triglycerides concentration did not show significant difference between the non-antibiotic additives groups. Moreover, in the birds under probiotic treatment the serum VLDL was lower

than those under the control and antibiotic treatments ($p < 0.05$).

In agreement with our findings, it is reported that the probiotic supplementation significantly reduces the serum cholesterol level of the chickens (Panda et al., 2001; Kalavathy et al., 2003; Jin et al., 1998). Kannan et al. (2005) have reported that the use of 0.5 g kg⁻¹ mannanoligosaccharide obtained from yeast in the ration of broiler chickens, significantly reduced the serum cholesterol level on day 35 as compared with the control ($p < 0.05$). Tizard et al. (1989) reported that mannans and other similar carbohydrates (such as fructans) prevent cholesterol absorption in gastrointestinal tract. In contrast, Yalcinkaya et al. (2008) reported that the use of MOS in broilers diet could not significantly reduce the serum cholesterol and triglycerides levels as compared with the control group. Synthesis of bile acids from cholesterol in the liver is the most important way of cholesterol excretion (Wilson et al., 1998). The use of probiotics and prebiotics can disintegrating bile salts and de-conjugate production of enzymes by the activity of lactic acid bacteria, as well as reduction of the pH in the intestinal tract can be effective in reducing the cholesterol concentration. Solvability of non-conjugate bile acids is lowered at a low pH and consequently, they are absorbed less from the intestine and are excreted more in the feaces (Klaver and Van der Meer, 1993). Consequently, the liver, for re-establishment of the hepatic cycle of bile

Table 2. The main effects of treatments on growth performance broiler chickens.

SGR (% d^{-1})	Control	Antibiotic	Probiotic	Prebiotic	Synbiotic
1 - 21	12.30^c	12.44^{abc}	12.39^{bc}	12.67^a	12.64^{ab}
1 - 42	18.52^b	18.76^a	18.64^{ab}	18.69^{ab}	18.78^a
GE (g/g)					
1 - 21	12.27^c	12.66^{abc}	12.51^{bc}	13.33^a	13.24^{ab}
1 - 42	47.92^b	50.40^a	49.23^{ab}	49.69^{ab}	50.64^a
Energy intake (Kcal)					
1 - 21	2986.09	2916.37	2932.89	2973.58	2973.76
1 - 42	13149.5	13163.5	13154.2	13067.4	13165.6
EER (g/g)					
1 - 21	17.12^b	18.34^a	17.95^{ab}	18.80^a	18.86^a
1 - 42	15.18^b	16.17^a	15.76^{ab}	15.95^a	16.30^a
Protein intake (g)					
1 - 21	214.57	209.56	210.75	213.68	213.69
1 - 42	821.96	822.83	822.25	816.82	822.96
PER (g/g)					
1 - 21	2.38^b	2.55^a	2.49^{ab}	2.61^a	2.62^a
1 - 42	2.42^b	2.58^a	2.52^{ab}	2.55^a	2.60^a

^{a,b,c} Means in each row with different superscripts are significantly different ($p < 0.05$). EER = energy efficiency ratio, PER = protein efficiency ratio, SGR = specific growth rate, GE = growth efficiency.

Table 3. The main effects of treatments on serum lipid concentratios (as mmol/l) of broiler chickens.

21 day of age	Control	Antibiotic	Probiotic	Prebiotic	Synbiotic
Cholesterol	3.89^a	4.00^a	3.32^c	3.76^{ab}	3.47^{bc}
Triglycerid	1.04	1.06	0.91	1.12	1.03
HDL	1.87	1.99	2.02	2.11	2.28
LDL	1.54	1.53	0.88	1.13	0.71
VLDL	0.47	0.48	0.42	0.51	0.47
42 day of age					
Cholesterol	4.15^{ab}	4.32^a	3.71^{bc}	3.77^{abc}	3.58^c
Triglycerid	0.97^a	0.91^a	0.62^b	0.76^{ab}	0.72^{ab}
HDL	2.02	2.16	1.67	1.81	1.93
LDL	1.68	1.74	1.75	1.60	1.31
VLDL	0.44^a	0.41^a	0.28^b	0.34^{ab}	0.33^{ab}

^{a,b,c} Means in each row with different superscripts are significantly different ($p < 0.05$).

acids, coverts more cholesterol concentration into the tissues and therefore their concentrations in the blood is reduced (Ros, 2000). In the growing birds, VLDL is the most important triglycerides carrier. A reduction in the serum triglycerides level may be due to an increase in the population of lactic acid bacteria in the gastrointestinal tract. Santose et al. (1995) have reported that supplementation of *Bacillus subtilis* to the ration of broiler chickens, in addition to reducing the carcass fat, reduces the triglycerides concentration in the serum, the liver and the carcass and suggest that this bacterium can be effective in reducing the activity of acetyl coenzyme A carboxylase (the enzyme limiting the synthesis rate of fatty acids).

Conclusion

The use of probiotic plus prebiotic due to the improvement in growth indices; EER, PER and reduction in serum cholesterol was more effective than separately probiotic or prebiotic supplementation to improve the gastrointestinal health and performance of boilers. Thus, this plays an important role in increasing the economic efficiency and conserving the health of consumers. Furthermore, according to the results, synbiotic could be introduced as a safe and natural alternative to antibiotic growth promoters in broiler diets.

REFERENCES

Abdulrahim SM, Haddadinm MSY, Hashlamoun EAR, Robinson RK (1996). The influence of Lactobacillus acidophilus and bacitracin on layer performance of chickens and cholesterol content of plasma and egg yolk. Br. Poult. Sci., 37: 341-346.

Chichlowski M, Croom J, McBride BW, Daniel L, Davis G, Koci MD (2007a). Direct-fed Microbial primalac and salinomycin modulate whole-body and intestinal oxygen consumption and intestinal mucosal cytokine production in the broiler chick. Poult. Sci., 86: 1100-1106.

Chichlowski M, Croom J, Edens FW, McBride BW, Qui R, Chiang CC, Daniel LR, Havenstein GB, Koci MD (2007b). Microarchitecture and spatial relationship between bacreria and ileal, cacal, and colonic epithelium in chicks fed a direct-fed microbial, primalac, and salinomycin. Poult. Sci., 86: 1121-1132.

Duncan DB (1955). Multiple range test and multiple F-tests. Biometrics., 11: 1-42.

Falaki M, Shams Shargh M, Dastar B, Zrehdaran S (2010). Effects of different levels of probiotic and prebiotic on performance and carcass characteristics of broiler chickens. J. Anim. Vet. Adv., 9(18): 2390-2395.

Fairchild AS, Grimes JI, Jones FT, Wineland MJ, Edens FW, Sefton AE (2001). Effects of Hen Age, Bio-Mos, and Flavomycin on Poult Susceptibility to Oral Escherichia coli Challenge. Poult. Sci., 80: 562-571.

Fuller R (1989). Probiotic in man and animal. J. Appl. Bacteriol., 66: 365-378.

Gibson GR, Roberfroid MB (1995). Dietary modulation of human colonic microbiota: Introducing the concept of prebiotic. J. Nutr., 125: 1401-1412.

Gunal M, Yayli G, Kaya O, Karahan N, Sulak O (2006). The effects of antibiotic growth promoter, probiotic or organic acid supplementation on performance, intestinal microflora and tissue of broiler. Int. J. Poult. Sci., 5: 149-155.

Haddadin MSY, Abdulrahim SM, Hashlamoun EAR, Robinson RK (1996). The effect of Lactobacillus acidophilus on the production and chemical composition of hen's eggs. Poult. Sci., 75: 491-494.

Jin LZ, Ho YW, Abdullah N, Jalaludin S (1998). Growth performance, intestinal microbial populations, and serum cholesterol of broilers fed diets containing Lactobacillus cultures. Poult. Sci., 77: 1259-1265.

Kabir SML, Rahman MM, Rahman MB, Rahman MM, Ahmed SU (2004). The dynamics of probiotics on growth performance and immune response in broilers. Int. J. Poult. Sci., 3: 361-364.

Kalavathy R, Abdullah N, Jalaludin S, Ho YW (2003). Effects of lactobacillus cultures on growth performance, abdominal fat deposition, serum lipids and weight of organs of broiler chickens. Br. J. Poult. Sci., 44: 139-144.

Kannan M, Karunakaran R, Balakrishnan V, Prabhakar TG (2005). Influence of prebiotics supplementation on lipid profile of broilers. Int. J. Poult. Sci., 4(12): 994-997.

Klaver FAM, Van Der Meer R (1993). The assumed assimilation of cholesterol by lactobacilli and Bifidobacterium bifidum is due to their bile salt-deconjugating activity. Appl. Environ. Microbiol., 59: 1120-1124.

Mohan B, Kadirvel R, Natarajan A, Bhaskaran M (1996). Effect of probiotic supplementation on growth, nitrogen utilization and serum cholesterol in broilers. Br. Poult. Sci., 37: 395-401.

Mikulec Z, Serman V, Mas N, Lukac Z (1999). Effect of probiotic on production results of fattened chickens fed different quantities of protein. Vet. Arhiv., 69: 199-209.

Nayebpor M, Farhomand P, Hashemi A (2007). Effect of different levels of direct fed microbial (Primalac) on the growth performance and humoral immune response in broiler chickens. J. Anim. Vet. Adv., 6: 1308-1313.

NRC (National Research Council), (1994). Nutrient Requirements of Poultry. 9th. Rev.(ed).National Academy Press, Washington, D. C.

Panda AK, Reddy MR, Praharaj NK (2001). Dietary supplementation of probiotic on growth, serum cholesterol and gut microflora of broilers. Indian. J. Anim. Sci., 71: 488-490.

Piray AH, Kermanshahi H, Tahmasbi AM, Bahrampour J (2007). Effects of Cecal Cultures and Aspergillus Meal Prebiotic (Fermacto)on Growth Performance and Organ Weights of Broiler Chickens. Int. J. Poult. Sci., 6(5): 340-344.

Rada V, Marounek M, Rychly I, Santruckova D, Vorisek K (1995). Effect of Lactobacillus salivarius administration on microflora in the crop and caeca of broiler chickens. J. Anim. Feed Sci., 4: 161-170.

Ros E (2000). Intestinal absorption of triglyceride and cholesterol. Dietary and pharmacological inhibition to reduce cardiovascular risk. Atherosclerosis., 51: 357-379.

Santin E, Maiorka A, Macari M, Grecco M, Sanchez JC, Okada TM, Myasaka AM (2001). Performance and intestinal mucosa development of broiler chickens fed diet containing Sccharomyces cerevisiae cell wall. J. Appl. Poult. Res., 10: 236-244.

Santose U, Tanaka K, Othani S (1995). Effect of dried Bacillus subtilis culture on growth, body composition and hepatic lipogenic enzyme activity in female broiler chicks. Br. J. Nutr., 74: 523-529.

Savage TF, Zakrzewska EI (1996). The performance of male turkeys fed a starter diet containing a mannanoligosaccharide(Bio-Mos) from day old to eight weeks of age. In: Biotechnology in the Feed Industry. Proceedings of Alltech's Twelfth Annual Symposium. Lyons TP and Jacques KA, ed. Nottingham University Press, Nottingham, UK., pp. 47-54.

(SAS) Statistic Analysis System (1998). Painless windows, a handbook for SAS users. 2nd ed. Guelph (ON): Jodie Gilmore.

Thitaram SN, Chung CH, Day DF, Hinton A, Bailey JS, Siragusa GR (2005). Isomaltooligosaccharide increases cecal bifidobacterium population in young broiler chickens. Poult Sci., 84: 998-1003.

Tizard IR, Carpenter RH, McAnalley BH, Kemp MC (1989). The biological activities of mannans and related complexcarbohydrates. Mole. Biother., 1: 290-296.

Willis WL, Isikhuemhen OS, Ibrahim A (2007). Performance assessment of broiler chickens given mushroom extract alone or in combination with probiotic. Poult. Sci., 86: 1856-1860.

Wilson TA, Nicolosi RJ, Rogers EJ, Sacchiero R, Goldberg DJ (1998). Studies of cholesterol and bile acid metabolism, and early atherogenesis in hamsters fed GT16-239, a novel bile acid sequestrant (BAS), Atherosclerosis., 40: 315-324.

Yalcinkaya H, Gungori T, Bafialani M, Erdem E (2008). Mannan oligosaccharides (MOS) from Saccharomyces cerevisiae in broilers: effects on performance and blood biochemistry. Turk. J. Vet. Anim. Sci., 32(1): 43-48.

Yusrizal N, Chen TC (2003). Effect of adding chicory fructons in feed on broiler growth performance, serum cholesterol and intestinal length. Int. J. Poult. Sci., 2: 214-219.

Zhang AW, Lee BD, Lee SK, Lee KW, An GH, Song KB, Lee CH (2005). Effects of Yeast (Saccharomyces cerevisiae) Cell Components on Growth Performance, Meat Quality, and Ileal Mucosa Development of Broiler Chicks. Poult. Sci., 84: 1015-1021.

Zulkifli I, Abdullah N, Azrin NM, Ho YW (2000). Growth performance and immune response of two commercial broiler strains fed diets containing Lactobacillus cultures and oxytetracycline under heat stress conditions. Br. Poult. Sci., 41: 593-597.

10

Effects of dietary inclusion of several biological feed additives on growth response of broiler chickens

A. Ashayerizadeh[1*], N. Dabiri[1,2], Kh. Mirzadeh[1] and M. R. Ghorbani[1]

[1]Department of Animal Science, Ramin Agricultural and Natural Resources University, Ahvaz, Iran.
[2]Faculty of Agricultural, Animal Science Department, Islamic Azad University, Karaj Branch, Karaj, Iran.

The present study aimed at evaluating the effect of the probiotic Primalac, prebiotic Biolex-MB, a combination of the two supplements (probiotic plus prebiotic) and one growth promoter antibiotic (Flavomycin) on performance results of broiler chickens. Three hundred day old Ross 308 broilers were equally distributed into 30 floor pens and reared for 42 day. A basal diet was also supplemented with Flavomycin (650 g/ton^{-1}), Primalac (900 g/ton^{-1}), Biolex-MB (2000 g/ton^{-1}) and mixture of Primalac plus Biolex-MB (synbiotic) for starter (1 to 21 days) and grower (22 to 42 days) periods, resulting 5 dietary treatments were prepared including control group. Each dietary treatment was fed ad-libitum to six replicate group of 10 birds at the beginning of rearing period. Birds and feed were weighed weekly. The daily body weight gain and feed efficiency were significantly ($P<0.05$) more improved by the synbiotic treatments compared with the control broilers. The birds under antibiotic treatment had higher breast and thigh percent as compared to control birds ($p<0.05$). Also, supplementation of broiler's diet with prebiotic significantly ($p<0.05$) increased gizzard percent. The results of present study showed that probiotic and prebiotic can be used as non-antibiotic growth promoter feed additives to improve broiler chickens growth indices.

Key words: Antibiotic, prebiotic, probiotic, performance.

INTRODUCTION

Antimicrobials have been used as feed supplement for more than 50 years in poultry feed to enhance the growth performance and to prevent diseases in poultry. However, in recent years great concern has arisen about the use of antibiotics as supplement at sub-therapeutic level in poultry feed due to emergence of multiple drug resistant bacteria (Wray and Davies, 2000). As a consequence, it has become necessary to develop alternatives using either beneficial microorgansims or nondigestible ingredients that enhance microbial growth. A probiotic is a culture of a single bacteria strain, or mixture of different strains, that can be fed to an animal to improve some aspect of its health (Griggs and Jacob, 2005). Probiotics are also referred to as direct fed microbials (DFM). On the other hand, a prebiotic was defined as nondigestible food ingredient that beneficially affects the host,

selectively stimulating the growth or activity, or both, of one or a limited number of bacteria in the colon (Gibson and Roberfroid, 1995).

The efficacy of probiotics may be potentiated by several methods: the selection of more efficient strains, gene manipulation, the combination of several strains, and the combination of probiotics and synergistically acting components. This approach seems to be the best way of potentiating the efficacy of probiotics and is widely used in practice. A way of potentiating the efficacy of probiotic preparations may be the combination of both probiotics and prebiotics as synbiotics, which may be defined as a mixture of probiotics and prebiotics that beneficially affects the host by improving the survival and implantation of live microbial dietary supplements in the gastrointestinal (GI) tract. Those effects are due to activating the metabolism of one or a limited number of health-promoting bacteria or by selectively stimulating their growth, which improved the welfare of the host, or both (Gibson and Roberfroid, 1995). Lactobacilli and enterococci are among the wide variety of microbial species that have been used extensively as probiotics

*Corresponding author. E-mail: amin.ashayerizadeh@yahoo.com.

(Patterson and Burkholder, 2003). Primalac is a kind of commercial probiotic that contains at least 1×10^8 CFU g^{-1} *Lactobacillus casei*, *Lactobacillus acidophilus*, *Bifidobacterium thermophilum*, and *Enterococcus faesium* (Chichlowski et al., 2007a, b). After feeding of probiotics, improvements in growth performance and feed efficiency have been reported in broiler chickens (Cavazzoni et al., 1998; Zulkifli et al., 2000; Kabir et al., 2004; Samli et al., 2007). Two of the most commonly studied prebiotic oligosaccharides are fructooligosaccharides (FOS) and mannanoligosaccharides (MOS). FOS can be found naturally in some cereal crops and onions (Bailey et al., 1991). MOS is obtained from the cell wall of yeast (*Saccharomyces cerevisiae*). Biolex–MB is a commercial prebiotic of the mannanoligosaccharides family, which is obtained by extraction from the outer cell wall of the yeast *Saccharomyces cerviciae*. Several typical probiotics contain either of these oligosaccharides, thereby comprising a synbiotic. The combination of a probiotic and prebiotic in one product has been shown to confer benefits beyond those of either on its own (Gallaher and Khil, 1999).

The search for new additives effective on animal growth and free from harmful side effects on consumers health is still continuing. The aim of this study was comparing the effects of the antibiotic flavomycin and primalac and Biolex-MB as alternatives for the growth stimulating antibiotics on growth performance and carcass characteristics of broiler chickens.

MATERIALS AND METHODS

Birds and housing

Three hundred 1 day old broiler chicks (Ross 308) were obtained from a commercial hatchery. The birds were randomly divided into 5 groups (60 birds/group) and housed in pens of identical size (120 × 100 cm) in a deep litter system with a wood shaving floor. Each group had 6 replicates (10 birds/pen). The birds had free access to water and feed. Environmental temperature in the first week of life was 34 °C and decreased to 20 °C until the end of the experiment.

Diets

A basal diet was formulated and considered as control according to recommendation of NRC, 1994 for starter (1 to 21 days), and grower (22 to 42 days) diets. Four tested diets were formulated by supplemented the basal control diet with antibiotic Flavomycin (650 g/ton^{-1}), probiotic (Primalac, 900 g/ton^{-1}), prebiotic (Biolex-MB, 2000 g/ton^{-1}), and mixture of Primalac (900 g/ton^{-1}) plus Biolex-MB (2000 g/ton^{-1}) as synbiotic, respectively. The composition of the diets are shown in Table 1.

Growth performance traits

All birds were weighed individually after their arrival from the hatchery to the experimental farm (initial weight) and on d 21 and

42. Daily weight gain for each dietary treatment was calculated by difference between two initial and final weight of birds in each growth period (kg)/ number of days in the growth period. Feed consumption was recorded in the course of the whole starter and grower periods for each treatment, and the feed efficiency ratio were calculated subsequently.

Carcass measurements

At the end of the experiment, 12 birds per treatment were randomly taken to study carcass characteristics. Chicks were fasted for approximately 12 h, and then individually weighed, slaughtered, feathered and eviscerated. Weights of breast, thigh, heart, liver, spleen and gizzard were recorded. The percentage (% of live body weight) of carcass parts and organs was calculated.

Statistical analysis

Based a randomized complete design, all data were analyzed using the One-Way Anova procedure of SAS® (SAS, 1998) for analysis of variance. Significant differences among treatments were identified at 5% level by Duncan's multiple range tests (Duncan, 1955).

RESULTS AND DISCUSSION

Growth performance

The results of performance are presented in the Table 2. At the starter period of the experiment (1 to 21 days), birds supplemented with prebiotic and synbiotic had a greater (P<0.05) daily body weight gain (BWG) compared with control group. The daily BWG (from day 1 to 42) were increased (P<0.05) for birds supplemented with antibiotic and synbiotic compared with control treatment. The inclusion of all treatments, except probiotic, in the feed also promoted feed efficiency (1 to 21 and 1 to 42 days) as compared to the control group (p<0.05). Daily feed intake during the experimental period did not differ between treatments (p>0.05).

In general the positive effects of experimental additives on performance are in agreement with the results reported previously. Fairchild et al. (2001), who showed that flavomycin as antibiotic has favorable effects on the weight gain of broiler chickens. Esteve-garcia et al. (1997), observed that addition of flavomycin to a wheat-based ration could improve significantly the chickens feed efficiency ratio in all breeding periods (1 to 21 days and 22 to 42 days). In the present study, the beneficial effects of a probiotic, prebiotic and synbiotic products on broiler performance parameters including daily BWG and feed efficiency are in agreement with previous studies (Zulkifli et al., 2000; Pelicano et al., 2003; Cavit, 2004; Thitaram et al., 2005; Kermanshahi and Rostami, 2006; Nayebpor et al., 2007; Falaki et al., 2010). In contrast, Gunal et al., (2006), Zhang et al., (2005), Jamroz et al., (2004) and Willis et al., (2007), reported that using these additives shed in the broiler ration had no significant effects on growth performance of broiler chickens. Variance among

Table 1. Ingredient composition (as percent of dry matter) and calculated analysis of the basal diets.

Ingredients	Starter	Grower
	(1-21 days)	(22-42 days)
Corn	58.7	61
Soybian meal	30	29
Wheat bran	5	5
Fish meal	2	0
Soybian oil	1	2
Oister shell meal	1.2	1
DCP	1.07	1
Vitamin and mineral perimix	0.5	0.5
DL- Methionine	0.13	0.1
L-lysine	0.15	0.25
Salt	0.25	0.1
Coccidiostat	0	0.05
Total	100	100
Nutrient content		
ME (Kcal/Kg)	2850	2950
Crude protein (%)	20.48	18.44
Crude fiber (%)	3.89	3.81

Vitamin and mineral provided per kilogram of diet: vitamin A, 360000 IU; vitamin D3, 800000 IU; vitamin E, 7200 IU; vitamin K3, 800 mg; vitamin B1, 720 mg; vitamin B9, 400 mg; vitamin H2, 40 mg; vitamin B2, 2640 mg, vitamin B3, 4000 mg; vitamin B5, 12000 mg; vitamin B6, 1200 mg; vitamin B12, 6 mg; Choline chloraid, 200000 mg, Manganeze, 40000 mg, Iron, 20000 mg; Zinc, 40000 mg, coper, 4000mg; Iodine, 400 mg; Selenium, 80 mg.

Table 2. The main effects of treatments on growth performance of broiler chickens.

	Control	Antibiotic	Probiotic	Prebiotic	Synbiotic
Daily BWG (g)					
1-21	24.35 [b]	25.45 [ab]	25.07 [ab]	26.55 [a]	26.70 [a]
1-42	47.53 [b]	50.65 [a]	49.28 [ab]	49.47 [ab]	51.08 [a]
Daily FI (g)					
1-21	49.89	48.72	49.00	49.68	49.68
1-42	106.13	106.24	106.16	105.46	106.26
Feed efficiency (g/g)					
1-21	0.48 [b]	0.52 [a]	0.51 [ab]	0.53 [a]	0.53 [a]
1-42	0.44 [b]	0.47 [a]	0.46 [ab]	0.47 [a]	0.48 [a]

[a,b] means in each row with different superscripts are significantly different ($p<0.05$).
BWG= Body Weigh Gain, FI= Feed Intake.

reports of researchers could be related to differences in the type of probiotics and prebiotics, management and environmental conditions that exist in various experiments. It's suggested that under benefit management and/or environmental conditions, the effect of such feed additives may be worthless. On the other hand, the responses of breeding bird in a warm climate condition similar to the present environmental in south west of Iran to these growth promoters may be better than an ideal condition. The results of some studies shown that growth stimulating antibiotics, increase the growth of broiler chickens by an increase in the uptake of nutrients (especially fatty acids and glucose), fixation of nitrogen and reduction in excretion of fat in the feces and microbial urea (Anderson et al., 1999). Also, antibiotics reduce the number of bacteria, toxins and their

Table 3. The main effects of treatments on carcass characteristic (as percent of live body weight) of 42d broiler chickens.

	Control	Antibiotic	Probiotic	Prebiotic	Synbiotic
Breast	18.91 [b]	20.20 [a]	20.41 [a]	18.82 [b]	19.92 [ab]
Thigh	19.28 [c]	20.87 [a]	19.99 [abc]	19.59 [bc]	20.55 [ab]
Heart	0.45	0.46	0.45	0.41	0.41
Spleen	0.15	0.18	0.16	0.20	0.15
Liver	2.24	2.04	2.07	2.29	2.19
Gizzard	1.71 [ab]	1.63 [ab]	1.57 [b]	1.88 [a]	1.85 [ab]
Abdominal Fat pad	1.98 [a]	1.91 [a]	1.74 [ab]	1.62 [b]	1.86 [ab]

[a,b,c] means in each row with different superscripts

secondary products in the GI tract (Gunal et al., 2006). Our findings showed that consumption of synbiotic, like that of antibiotics, had a positive effect on the body weight gain when compared to control treatment. The reason may be ascribed to the synergism of probiotic and prebiotic following the concurrent action of prebiotics fermentation by lactic acid bacteria in the GI tract and production of some acids by this group of bacteria, the pH of the GI tract is further reduced (Fuller, 1989). Reduction in pH is effective in controlling the population of pathogenic bacteria. During the infections due to pathogenic bacteria, lymphocytes crowd up to kill them, and after inflammation the thickness of the muscular layer increases (Gunal et al., 2006). It seems that in our study, due to the synergism between probiotic and prebiotic, followed by absorption of nutrients by the GI system, the birds under synbiotic treatment had the best feed efficiency in both start (1 to 21 days) and whole (1 to 42 days) of experimental periods.

Carcass composition

The effects of dietary treatments on carcass characteristics and some internal organs of 42 days old broilers are shown in Table 3. The efficiency of breast was higher (p<0.05) in the antibiotic and probiotic treatments than in the control group, and it in the prebiotic and synbiotic groups did not differ significantly as compared to the control treatment. The antibiotic supplemented group had a greater (P<0.05) thigh percent compared with the control and prebiotic supplemented groups. However, the thigh percent did not show significant differences between the probiotic and prebiotic groups with the control treatment. Also, there was no significant difference in mean percentage yield of heart, spleen and liver between the fed additives treatments and control group. The birds under prebiotic treatment, had a higher gizzard percent as compared to the probiotic treatment, whereas the percent of abdominal fat pad was lower than in the antibiotic and control treatments (p<0.05).

The benefits of the antibiotic use on carcass characteristics have been reported by other researchers (Woodward et al., 1988; Elwinger et al., 1998; Fidler et al., 2003). Kabir et al. (2004) have reported that adding 2 g probiotic per each liter of water consumed by broiler chickens, would increase the efficiency in their thigh and breast as compared with the control treatment. In Ammerman et al. (1989) study, adding 0.375% oligofructose to the bird's ration, on day 47, decreased the percent of abdominal fat. However, our findings on carcasse composition were in contrast to those of Pelicia et al. (2004); Pelicano et al. (2003); Willis et al. (2007) and Kannan et al. (2005). As pointed out before, these differences between reported results could be related to the mode of action of those feed additives is quite different, particularly their antimicrobial activity, the similar physiologic pattern was probably exerted by modifying intestinal pH, altering the composition and balance of intestinal flora, enhancing nutrient digestibility and improving growth rate and carcass characteristics. Also, by observing a reduction in the abdominal fat level of birds fed by prebiotic, it is suggested that this product may interfere in the accessibility to fat for formation of fat tissue in the birds.

Conclusions

Supplementation of broiler diets with the feed growth promoters improved the growth responses compared with the unsupplemented control treatment. Also, based on obtaining better results for birds fed diets containing growth promoters (synbiotic), particularly in growth response, when compared to control treatment, it is concluded that by using non-antibiotic additives particularly mixing of both probiotic and prebiotic could obtained the advantages of antibiotic (performance) without their disadvantages.

REFERENCES

Ammerman E, Quarles C, Twining PV (1989). Evaluation of

fructo-ligosaccharides on performance and carcass yield of male broilers. Poult. Sci., 68: 167.

Anderson DB, Mc Cracken JJ, Aminov RI, Simpson JM, Mackie RI, Verstegen MWA, Gaskins HR (1999). Gut microbiology and growth-promoting antibiotics in swine. Pig News Inform., 20: 115N-122N.

Bailey JS, Blankenship LC, Cox NA (1991). Effect of fructooligosaccharide on Salmonella colonization of the chicken intestine. Poult. Sci., 70: 2433-2438.

Cavazzoni V, Adami A, Cstrivilli C (1998). Performance of broiler chickens supplemented with Bacillus coagulans as probiotic. Br. Poult. Sci., 39: 526-529.

Cavit, A (2004). Effect of dietary probiotic supplementation on growth performance in the rock partridge (Alectoris graeca). Turk. J. Vet. Anim. Sci., 28: 887-891.

Chichlowski M, Croom J, McBride BW, Daniel L, Davis G, Koci MD (2007a). Direct-fed Microbial primalac and salinomycin modulate whole-body and intestinal oxygen consumption and intestinal mucosal cytokine production in the broiler chick. Poult. Sci., 86: 1100-1106.

Chichlowski M, Croom J, Edens FW, McBride BW, Qui R, Chiang CC, Daniel LR, Havenstein GB, Koci MD (2007b). Microarchitecture and spatial relationship between bacreria and ileal, cacal, and colonic epithelium in chicks fed a direct-fed microbial, primalac, and salinomycin. Poult. Sci., 86: 1121-1132.

Duncan DB (1955). Multiple range test and multiple F-tests. Biometrics., 11: 1-42.

Elwinger K, Berndtson E, Engstrom B, Fossum O, Waldenstedt L (1998). Effect of antibiotic growth promoters and anticoccidials on growth of clostridium prefringenes in the caeca and on performance of broiler chickens. Acta. Vet. Scand., 39: 433-441.

Esteve-garcia E, Brufau J, Perez-vendrell A, Miquel A, Duven K (1997). Bioefficacy of enzyme preparations containing b-glucanase and xylanase activities in broiler diets based on barley or wheat, in combination with flavomycin. Poult. Sci., 76: 1728-1737.

Fairchild AS, Grimes JI, Jones FT, Wineland MJ, Edens FW, Sefton AE (2001). Effects of Hen Age, Bio-Mos, and Flavomycin _on Poult Susceptibility to Oral Escherichia coli Challenge. Poult. Sci., 80: 562-571.

Falaki M, Shams Shargh M, Dastar B, Zrehdaran S (2010). Effects of different levels of probiotic and prebiotic on performance and carcass characteristics of broiler chickens. J. Anim. Vet. Adv., 9(18): 2390-2395.

Fidler DJ, George B, Quarles CL, Kidd MT (2003). Broiler performance and carcass traits as affected by dietary liquid saccharopoly spora solubles concentrate. J. Appl. Poult. Res., 12: 153-159.

Fuller R (1989). Probiotic in man and animal. J. Appl. Bacteriol., 66: 365-378.

Gallaher DD, Khil J (1999). The effect of synbiotics on colon carcinogenesis in rats. J. Nutr., 129(7): 1483S-1487S.

Gibson GR, Roberfroid MB (1995). Dietary modulation of human colonic microbiota: Introducing the concept of prebiotic. J. Nutr., 125: 1401-1412.

Griggs JP, JacobJP (2005). Alternatives to Antibiotics for Organic Poultry Production. J. Appl. Poult. Res., 14: 750-756.

Gunal M, Yayli G, Kaya O, Karahan N, Sulak O (2006). The effects of antibiotic growth promoter, probiotic or organic acid supplementation on performance, intestinal microflora and tissue of broiler. Int. J. Poult. Sci., 5: 149-155.

Jamroz D, Wiliczkiewicz A, Orda J, Wertelecki T, Skorupin' ska J (2004). Response of broiler chickens to the diets supplemented with feeding antibiotic or mannan-oligosaccharides. Electr. J. Polish Agric. Univ., Ser., Anim. Husb., 7.

Kabir SML, Rahman MM, Rahman MB, Rahman MM, Ahmed SU (2004). The dynamics of probiotics on growth performance and immune response in broilers. Int. J. Poult. Sci., 3: 361-364.

Kannan M, Karunakaran R, Balakrishnan V, Prabhakar TG (2005). Influence of prebiotics supplementation on lipid profile of broilers. Int. J. Poult. Sci., 4(12): 994-997.

Kermanshahi H, Rostami H (2006). Influence of supplemental dried whey on broiler performance and cecal flora. Int. J. Poult. Sci., 5: 538-543.

Nayebpor M, Farhomand P, Hashemi A (2007). Effect of different levels of direct fed microbial (Primalac) on the growth performance and humoral immune response in broiler chickens. J. Anim. Vet. Adv., 6: 1308-1313.

NRC (National Research Council), (1994). Nutrient Requirements of Poultry. 9th. Rev.(ed).National Academy Press, Washington, D. C.

Patterson JA, Burkholder KM (2003). Application of prebiotics and probiotics in poultry production. Poult. Sci., 82: 627-631.

Pelicano ERL, Souza PA, Souza HBA, Oba A, Norkus EA, Kodawara LM, Lima TMA (2003). Effect of different probiotics on broiler carcass and meat quality. Brazil J. Poult. Sci., 5: 207-214.

Pelicia K, Mendes AA, Saldanha ESPB, Pizzolante CC, Takahashi SE, Moreira J, Garcia RG, Quinterio RR, Paz LCLA, Komiyama CM (2004). Use of prebiotics and probiotics of bacterial and yeast origin for free-range broiler chickens. Brazil J. Poult. Sci., 6: 163-169.

Samli HE, Senkoylu N, Koc F, Kanter M, Agma A (2007). Effects of Enterococcus faecium and dried whey on broiler performance, gut histomorphology and microbiota. Arch. Anim. Nutr., 61: 42-49.

Statistic Analysis System (1998). Painless windows, a handbook for SAS users. 2nd ed. Guelph (ON): Jodie Gilmore.

Thitaram SN, Chung CH, Day DF, Hinton A, Bailey JS, Siragusa GR (2005). Isomaltooligosaccharide increases cecal bifidobacterium population in young broiler chickens. Poult. Sci., 84: 998-1003.

Willis WL, Isikhuemhen OS, Ibrahim A (2007). Performance assessment of broiler chickens given mushroom extract alone or in combination with probiotic. Poult. Sci., 86: 1856-1860.

Woodward SA, Hams RH, Miles RD, Janky DM, Ruiz N (1988). Research note: Influence of virginiamycin on yield of broilers fed four levels of energy. Poult. Sci., 67: 1222-1224.

Wray C, Davies RH (2000). Competitive exclusion-An alternative to antibiotics. Vet. J., 59: 107-108.

Zhang AW, Lee BD, Lee SK, Lee KW, An GH, Song KB, Lee CH (2005). Effects of Yeast (Saccharomyces cerevisiae) Cell Components on Growth Performance, Meat Quality, and Ileal Mucosa Development of Broiler Chicks. Poult. Sci., 84: 1015-1021.

Zulkifli I, Abdullah N, Azrin NM, Ho YW (2000). Growth performance and immune response of two commercial broiler strains fed diets containing Lactobacillus cultures and oxytetracycline under heat stress conditions. Br Poult Sci., 41: 593-597.

11

Beneficial effects of canola oil on breast fatty acids profile and some of serum biochemical parameters of Iranian native turkeys

R. Salamatdoust Nobar[1]*, A. Gorbani[1], K. Nazeradl[1], A. Ayazi[2], A. Hamidiyan[2], A. Fani[2], H. Aghdam Shahryar[1], J. Giyasi ghaleh kandi[1] and V. Ebrahim Zadeh Attari[3]

[1]Department of Animal Science, Islamic Azad University, Shabestar Branch, Shabestar, Iran.
[2]Agriculture and Natural Resources Animal Science Department, East Azerbaijan Research Center, Tabriz, Iran.
[3]Department of Biochemistry and Nutrition, Tabriz University of Medical Science, Tabriz, Iran.

During many years, the main objective of the poultry meat industry was to improve body weight and feed efficiency of the birds. However, in the modern poultry industry, there are other parameters that need to be taken into consideration such as low cholesterol and improved fatty acid profile. For this purpose, an experiment was conducted to evaluate canola oil effects on the Iranian native Turkey's serum parameters and breast meat fatty acid. Ninety male turkey chicks were randomly distributed into three experimental (0, 2.5 and 5%) with three replicate for each group. Diets were isonitrogenous and isoenergetic were given to turkey chicks throughout four periods of breeding (4 - 8, 8 - 12, 12 - 16 and 16 - 20th). The blood sample was taken at the end of the breeding period and serum parameters calculated by Friedewald method. Two pieces from each pen randomly selected and slaughtered with cutting the neck vessels and experimental samples from each breast meat sample prepared and pattern of fatty acids of breast samples was determined by gas chromatography. Serum values were not found to be significantly different ($P < 0.05$) in triglycerides and VLDL and in CHOL, LDL and HDL ($P < 0.05$) was significantly different compared to the control group. n-3 Fatty acids used as α-linoleic acid using canola oil had positive effect on the values and amount of this fatty acid in the control of 3.5562% reached to 6.7994 and 8.2447%, respectively in the experimental treatments ($P < 0.05$). Finally, our results illustrated that canola oil had significant impact on lipid metabolism in native turkey and could improve their serum lipid profile.

Key words: Turkey, canola oil, cholesterol, triglyceride, HDL, LDL.

INTRODUCTION

Turkeys are raised all over the world to produce meat and their meat production is currently developed in Iran. Oils are commonly been used as energy sources in the diets of poultry especially in grower and finisher. Studies have shown that type of dietary lipids of poultry can drastically alter the fatty acids profile of meat (Balnave, 1970; Scaife et al., 1994; Hrdinka et al., 1996; Ló'pez-Ferrer et al., 1999a, b; Salamatdoust et al., 2007). Thus, alter the biochemical parameter such as cholesterol, triglyceride, and LDL and HDL content very important to human health. Canola oil has been recognized as the rich

plant source of linolenic acid (C18:3). Linolenic acid can be converted to longer chain omega-3 fatty acids, such as Eicosapentaenoic (EPA, C20:5), Docosahexaenoic (DPA, C22:5) and docosahexaenoic (DHA, C22:6) acids in poultry through an elongation and desaturation pathway, thus enriching the broiler meat with omega-3 fatty acids (Sim, 1995; Crespo and Esteve-Garcı´a, 2001, 2002a,b; Hrdinka et al., 1996). Omega-3 fatty acids have many health benefits including the ability to cardiovascular disease (Cherian and Sim, 1991; Grobas et al., 2001), antithrombic (Herod and Kinsella, 1986) and rheumatoid arthritis. Health recommendations have encouraged a reduction in the consumption of total lipids, saturated fatty acid and cholesterol but to increase the proportion of mono unsaturated and polyunsaturated fatty

*Corresponding author. E-mail: r.salamatdoust@gmail.com.

acids (PUFA) in human diets (Walsh et al., 1975; Temple, 1996; Grundy, 1980) found that dietary mono- unsaturated fatty acids (e.g. oleic acid) were very effective in lowering blood cholesterol concentration and may be important in preventing coronary heart disease (Howard et al., 2006). Grundy (1980) found that dietary mono-unsaturated fatty acids (e.g. oleic) were very effective in lowering blood cholesterol concentration and may be important in preventing coronary heart disease. Poultry species, age and breeding condition is known to affect cholesterol deposition (Hargis, 1988; Halle, 1996, 2001). The objective of this research was to determine the effects of feeding canola oil on serum biochemical parameters and breast meat fatty acids profile of Iranian native turkeys.

MATERIAL AND METHODS

Animal and diet

The investigation was performed on 90 male native Iranian turkeys in their fattening period (from 4th to 20th week of age). The turkey chicks with completely randomized design of 3 treatments, with 3 repetitions and 10 chicks in each box were fed experimental diets containing 0% CO(T1) , 2.5% CO(T2) and 5%CO (T3) in the fattening period. The experimental diets formulated isonitrogenouse and isoenergetic, accordance with the 1994 recommendations of the National Research Council (NRC). The birds were given access to water and diets ad-libitum. The composition and calculated nutrient composition of the treatment diet is shown in Table 1. Four birds in 20th week of age from each replicate after two hour fasting were taken blood and after separate serum, translated to the lab for analyses a cholesterol and triglyceride content. At the end of the growing period the number of two pieces from each pen randomly selected and slaughtered with cutting the neck vessels and experimental samples from each breast meat samples prepared and sent to the laboratory at temperature - 20°C below zero were stored.

Biochemical serum analysis

Total serum cholesterol, triglycerides and High density lipoprotein cholesterol was assayed using a commercial kit supplied by (Pars azmoon Co., Ltd.) and detected by (Alison 300) autoanalyser system. Very low density lipoprotein cholesterol is estimated as [Triglycerides/5] (Friedewald et al., 1972). Low density lipoprotein cholesterol is estimated using the Friedewald equation [Low density lipoprotein cholesterol = Total cholesterol – [High density lipoprotein cholesterol – Trigylcerides/5] (Friedewald et al., 1972).

Gas chromatography of fatty acids methyl esters

Sample preparation

Fatty acids: Total lipid was extracted from breast and thigh according to the method of Folch et al. (1957). Approximately 0.5 g of meat weighed into a test tube with 20 mL of (chloroform: methanol = 2:1, vol/vol), and homogenized with a polytroon for 5 - 10 s at high speed. The BHA dissolved in 98% ethanol added prior to homogenization. The homogenate filtered through a Whatman filter paper into a 100 mL graduated cylinder and 5 mL of 0.88% sodium chloride solution added, stopper, and mixed. After phase

separation, the volume of lipid layer recorded, and the top layer completely siphoned off. The total lipids converted to fatty acid methyl esters (FAME) using a mixture of boron-trifluoride, hexane, and methanol (35:20:45, vol/vol/vol). The FAME separated and quantified by an automated gas chromatography equipped with auto sampler and flame ionization detectors, using a 30 m, 0.25 mm inside diameter fused silica capillary column, as described. A (Model 6890N American Technologies Agilent) (U.S.A) Gas chromatography used to integrate peak areas. The calibration and identification of fatty acid peak carried out by comparison with retention times of known authentic standards. The lipid composition was determined by gas chromatography (Model 6890N American Technologies Agilent). The Pattern of fatty acids of breast samples was determined by gas chromatography (Model 6890N American Technologies Agilent). The composition of breast meat samples fatty acid of supplemented lipids is shown in Tables 3 data were statistically analyzed using one-way ANOVA, and means with significant F ratio were compared by Duncan multiple range test.

Statistical analyses

Data were analyzed in a complete randomized design using the GLM procedure of SAS version 8.2 (SAS Inst. Inc., Cary, NC).

$$y_{ij} = \mu + a_i + \varepsilon_{ij}$$

Where

y_{ij} = All dependent variable

μ = Overall mean

a_i = The fixed effect of oil levels (I = 1, 2 and 3)

ε_{ij} = The random effect of residual

Duncan multiple range tests used to compare means.

RESULTS AND DISCUSSION

The effect of canola oil on biochemical serum levels was shown in (Table 2). According to results were none significantly different on triglycerides and VLDL content in serum, while total cholesterol, HDL and LDL were significantly affected with dietary manipulation (P > 0.05). Cholesterol content has been descending rate and affected canola oil and from 148.83 mg/dl in the control group (T1) significantly reached to 114.0 mg/dl in T3 group, but compared with T2 (126.67 mg/dl) has not been significantly different. High density lipoprotein and very low density lipoprotein positively affected with CO and HDL content significantly increase in treatment contain with 5% CO (61.00 mg/dl) compared the control group, and for LDL results show that treatment with CO (T2 and T3) have lower content of LDL and significantly deferent compared with control group(P > 0.05). The present findings showed that substitution canola oil in dietary reduced the serum cholesterol concentration by 5%, whereas an addition of 2.5% decreased serum cholesterol but not significant. Canola contains 65 - 75% monoenic fatty acids and 9 - 30% polyunsaturated fatty acids (Ackman, 1990). Monounsaturated fat has also been shown to lower cholesterol (Grundy, 1988; Mensink and Katan, 1989).

Table 1. Percentage composition of experimental diets in four periods.

Ingredients[1]	4 - 8 week			8 - 12 week			12 - 16 week			16 - 20 week		
	T1	T2	T3	T1	T2	T3	T1	T2	T3	T1	T2	T3
Corn	42.50	38.00	36.00	45.60	43.00	35.00	56.64	48.50	40.00	64.41	58.00	48.00
SBM[1]	34.40	36.00	31.15	28.25	27.30	28.24	26.00	27.00	27.50	21.00	21.00	21.00
Oil	0.00	1.25	2.50	0.00	2.50	5.00	0.00	2.50	5.00	0.00	2.50	5.00
Fish	4.80	3.70	6.60	8.00	8.00	8.00	2.64	1.82	1.50	0.65	0.70	0.67
Starch	3.10	3.22	1.56	7.46	3.32	3.37	6.57	6.51	6.50	7.10	5.56	6.71
Alfalfa	3.47	5.00	6.00	3.00	5.00	6.00	1.50	4.00	6.00	1.00	3.80	6.00
DCP[2]	1.38	1.52	1.11	0.63	0.61	0.62	1.03	1.15	1.18	1.17	1.15	1.15
Met[3]	1.50	1.50	1.50	1.50	1.50	1.50	1.50	1.50	1.50	1.50	1.50	1.50
Lys[4]	1.50	1.50	1.50	1.50	1.50	1.50	1.40	1.50	1.50	1.50	1.50	1.50
Oyster	1.02	1.02	0.86	0.73	0.67	0.62	0.92	0.87	0.82	0.90	0.81	0.73
wheat bran	2.00	3.30	6.00	2.50	5.00	6.00	1.00	3.00	6.00	0.00	1.70	5.00
Vit supp[5]	0.25	0.25	0.25	0.25	0.25	0.25	0.25	0.25	0.25	0.25	0.25	0.25
Min supp[6]	0.25	0.25	0.25	0.25	0.25	0.25	0.25	0.25	0.25	0.25	0.25	0.25
Salt	0.25	0.25	0.25	0.25	0.25	0.25	0.25	0.25	0.25	0.25	0.25	0.25
Sand	3.58	3.54	4.47	0.08	0.85	3.40	0.05	0.90	1.75	0.02	1.03	1.99
	100.00	100.00	100.00	100.00	100.00	100.00	100.00	100.00	100.00	100.00	100.00	100.00
Calculated nutrient content												
ME kcal/kg	2755	2755	2755	2850	2850	2850	2945	2945	2945	3040	3040	3040
Crude protein (%)	24.7	24.7	24.7	20.9	20.9	20.9	18.1	18.2	18.1	15.7	15.7	15.7
Calcium (%)	0.95	0.95	0.95	0.81	0.81	0.81	0.71	0.71	0.71	0.62	0.62	0.62
Available P (%)	0.48	0.48	0.48	0.40	0.40	0.40	0.36	0.36	0.36	0.31	0.31	0.31
ME/CP	112	112	112	136	136	136	163	162	163	194	194	194
Ca/P	2	2	2	2	2	2	2	2	2	2	2	2

1- Soy Bean Meal 2- Di calcium phosphate 3- Methionine 4- Lysine 5 Vitamin content of diets provided per kilogram of diet: vitamin A,D, E and K.
6 Composition of mineral premix provided as follows per kilogram of premix: Mn, 120,000mg; Zn, 80,000 mg; Fe, 90,000 mg; Cu, 15,000 mg; I, 1,600 mg; Se, 500 mg; Co, 600 mg

Canola oil is an excellent source of monounsaturated fat, contains intermediate amounts of the precursor omega-6 and omega-3 polyunsaturated fatty acids linoleic acid (LA) and alfa-linoleic acid (ALA) respectively and is very low saturated fat. Canola oil as a source of phytosterols. Phytosterols (plant sterols) are structural analogs of the cholesterol found in animals and humans.

The consumption of phytosterols has been shown in numerous studies to lower blood cholesterol levels and may therefore, help reduce the risk of cardiovascular disease (Ling and Jones, 1995). Unsaturated oil could decrease the amount of harmful low-density lipoprotein (LDL) cholesterol in the serum (Mensink and Katan, 1989; Katan et al., 1995). Studies have shown that consumption

of monoenic fatty acids effectively lowers serum cholesterol concentrations (Mattson and Grundy, 1985; Sirtoni et al., 1986; Mensink and Katan, 1989; Dreon et al., 1990; Valsta et al., 1992; Grundy et al., 1988).The reduction of serum cholesterol by monoene-nich rapeseed oil agrees with earlier observations with monounsaturated fatty acids (Mattson and Grundy 1985; Sirtoni et

Table 2. Least square means for serum biochemical parameter.

	Treatment				
	T1	**T2**	**T3**	**SEM**	**P value**
TAG[1] (mg/dl)	77.75	76.66	79.50	10.769	0.9586
CHOL[2] (mg/dl)	148.83 [a]	126.67 [ab]	114.00 [b]	14.394	0.0264
VLDL[3] (mg/dl)	15.55	15.332	15.9	1.8987	0.9682
HDL[4] (mg/dl)	41.41 [b]	48.33 [ab]	61.00 [a]	7.189	0.0127
LDL[5] (mg/dl)	91.87 [a]	63.00 [b]	57.10 [c]	13.619	0.0003

[1]Triglycerides; [2]Total cholesterol; [3]Very low density lipoprotein cholesterol; [4]High density lipoprotein cholesterol; [5]Low density lipoprotein cholesterol. Values in the same row with no common superscript are significantly different.

Table 3. Least square means of fatty acid profile of breast meat turkey.

	Control	**2.5 %**	**5 %**	**P value**	**SEM**
C14:0	0.7424[a]	0.8457[a]	1.0254[a]	0.2436	0.1068
C15:0	0.2114[a]	0.2562[a]	0.2917[a]	0.8880	0.1158
C16:0	28.590[a]	19.30[b]	16.94[c]	0.0001	0.4042
C16:1 n7	7.1100[a]	5.95[b]	4.83[c]	0.0001	0.1427
C18:0	8.9800[b]	9.26[b]	10.75[a]	0.0016	0.2000
C18:1 n9	17.430[a]	15.60[b]	15.30[b]	0.0134	0.3725
C18:1 Trans t11	0.2987[a]	0.2077[a]	0.4518[a]	0.5209	0.1447
C18:2	2.5059[a]	2.8915[a]	3.1760[a]	0.2014	0.2314
C18:2 Trans t12	0.5293[a]	0.3253[a]	0.5655[a]	0.7134	0.2168
C18:2n6Cis	4.4154[c]	8.2898[b]	9.3383[a]	0.0001	0.2439
C18:3 n-3	3.5562[c]	6.7994[b]	8.2447[a]	0.0001	0.1993
C20:0	1.3194[a]	1.2867[a]	1.2688[a]	0.9898	0.2536
C20:5n-3	1.3421[b]	2.3737[a]	2.1263[a]	0.0390	0.2230
C20:1n-9	0.6001[b]	1.3501[a]	1.6164[a]	0.0141	0.1718
C22:0	0.9369[b]	2.0205[a]	2.6262[a]	0.0054	0.2291
C22: 4n-6	8.8864[a]	10.1375[a]	10.6384[a]	0.1111	0.5019
C22:5 n-3	2.7250[c]	6.7263[b]	8.3857[a]	0.0002	0.4243
C22:6 n-3	1.9138[a]	2.5467[a]	2.4275[a]	0.2282	0.2436
PUFA	25.870[c]	40.090[b]	44.8120[a]	0.0001	1.1283
MUFA	25.453[a]	23.1271[b]	22.2077[b]	0.0059	0.4539

Different superscripts in each raw indicate significant difference.

al., 1986; Mensink and Katan 1989; Dreon et al., 1990; Valsta et al., 1992). The results of a Table 1 show that used of canola oil could influence mono unsaturated fatty acids in the control group and from 25.45% significantly reduced in treatments and reached, respectively, 23.12 and 22.20% (P < 0.05). About polyunsaturated fatty acids also influence the results and from 25.87% of control with significantly increased in treatments and reached to 40.09 and 44.81% (P < 0.05).

Results show that in Table 3 breast meat saturated fatty acids include Myristic acid (C14:0) and Arachidic acid (C20:0) no significant changes compared with the control group. However, Palmitic acid (C20:0) with decline and significant rate from 28.59% in the control group, respectively reached to 19.30 and 16.94% in experimental diets containing 2.5 and 5% was canola oil

(P < 0.05). Stearic acid (C18:0) significantly from 8.97% of the control group respectively reached to 9.26 and 10.75 (P < 0.05). Also behenic acid (C22:0) increase with the use of canola oil in comparison to control group and values, respectively, reached to 2.0205 and 2.2662%. n-3 fatty acids as α-linoleic acid using canola oil had the positive effect on the values of this fatty acid and amount of this fatty acid in the control of 3.5562% reached to 6.7994 and 8.2447%, respectively in the experimental treatments (P < 0.05). Eicosapentaenoic acid (C20:5-3) also from 1.3421% in control treatment significantly reached to 2.3737 and 2.1263% in treatments and Docosohexaenoic acid (C22-n-3) in 1.9138 percent, respectively, significantly reached to 2.5467 and 2.4275% (P < 0.05). Some authors showed that the dietary polyunsaturation level of fat does not influence intramuscular lipid content of breast (Scaife et al., 1994; Crespo and Esteve-Garcı´a, 2001), but Kirchgessner et al. (1993) and Ajuyah et al. (1991) found a higher fat content in breast muscle with increasing levels of PUFA in the diet that according with this research finding. However, other authors found lower lipid content of breast of chickens fed diets enriched with polyunsaturated oils (Sanz et al., 1999). Such discrepant findings in intramuscular fat content of breast muscles may be attributed to several factors, such as the analytical procedure used to extract fat from samples.

Recent studies showed that fat content of tissues in more polyunsaturated treatments was underestimated when lipid contents were analyzed using the AOAC (1995) methodology, suggesting total FA content as an estimator of crude fat in highly polyunsaturated samples (Villaverde et al., 2003). In general, modification of FA composition of intramuscular fat seems to be more limited (Pan and Storlien, 1993; Lo´pez-Bote et al., 1997). It may be due to the fact that FA in intramuscular fat is used mainly as components of cellular membranes, and the cell has to maintain its physical characteristics to ensure fluidity and permeability of different compounds.

ACKNOWLEDGMENT

Financial support for this study was provided by Islamic Azad University, Shabestar Branch.

REFERENCES

Ackman RG (1990). Canola fatty acids-an ideal mixture for health, nutrition, and food use. In: Shahidi F, ed. Canola and rapeseed. NewYork: Van Nostran Reinhold., 2:81-98.

Ajuyah AO, Lee KH, Hardin RT, Sim JS (1991). Changes in the yield and in the fatty acid composition of whole carcass and selected meat portions of broiler chickens fed full-fat oil seeds. Poult. Sci., 70: 2304-2314.

AOAC (1995). Official Methods of Analysis. 16th rev. ed. Association

Balnave D (1970). Essential fatty acids in poultry nutrition. World's Poult. Sci. J., 26: 442-459.

Cherian G, Sim JS (1991). Effect of feeding full fat flax and canola seeds to laying hens on the fatty acids composition of eggs, embryos and newly hatched chickens. Poult. Sci., 70: 917-922.

Crespo N, Esteve-Garcı´a E (2001). Dietary fatty acid profile modifies abdominal fat deposition in broiler chickens. Poult. Sci., 80: 71-78.

Crespo N, Esteve-Garcı´a E (2002a). Dietary polyunsaturated fatty acids decrease fat deposition in separable fat depots but not in the remainder carcass. Poult. Sci., 81: 512-518.

Crespo N, Esteve-Garcı´a E (2002b). Nutrient and fatty acid deposition in broilers fed different dietary fatty acid profiles. Poult. Sci., 81: 1533-1542.

Dreon DM, Vranizan KM, Knauss RM, Austin MA, Wood PD (1990). The effects of polyunsaturated fat vs. monounsatunated fat on plasma lipoproteins. JAMA., 263: 2462-2466.

Folch J, Lees M, Sloane-Stanley GH (1957). A simple method for the isolation and purification of total lipids from animal tissues. J. Biol. Chem., 226: 497-507.

Friedewald WT, Levy RI, Fredrickson DS (1972). Estimation of the concentration of low-density lipoprotein cholesterol without the use of the preparative ultracentrifuge. Clin. Chem., 18: 499-502.

Grobas S, Mendez J, Lazaros R, Blas CD, Mateos GG (2001). Influence of source of fat added to diet on performance and fatty acid composition of egg yolks of two strains of laying hens. Poult. Sci., 80: 1171-1179.

Grundy SM (1980). Comparison of mono-unsaturated fatty acids and carbohydrates for lowering plasma cholesterol. New England J. Med., 314: 745-748.

Grundy SM, Florentin L, Nix D, Whelan MF (1988). Comparison of monounsaturated fatty acids and carbohydrates for reducing raised levels of plasma cholesterol in man. Am. J. Clin. Nutr. 47: 965-969.

Grundy SM (1980). Comparison of mono-unsaturated fatty acids and carbohydrates for lowering plasma cholesterol. New England J. Med., 314: 745-748.

Halle I (1996). Effect of dietary fat on performance and fatty acid composition of egg yolk in laying hens. Arc. Fur. Gef., 60: 65-72.

Halle I (2001). Effect of dietary fish oil and linseed oil on performance, egg component and fatty acid composition of egg yolk in laying hens. Arc. Fur. Gef., 65: 13-21.

Hargis PS (1988). Modifying egg yolk cholesterol in domestic fowl-a review. World's Poult. Sci. J., 44: 17-29.

Herod PM, Kinsella JE (1986). Fish oil consumption and decreased risk of cardiovascular disease: A comparison of findings from animal and human feeding trials. Am. J. Clin. Nutr., 43: 566-598.

Howard BV, Van Horn L, Hsia J (2006). Low-fat dietary pattern and risk of cardiovascular disease: the Women's Health Initiative Randomized Controlled Dietary Modification Trial, JAMA., 295: 655-666.

Hrdinka C, Zollitsch W, Knaus W, Lettner F (1996). Effects of dietary fatty acid pattern on melting point and composition of adipose tissues and intramuscular fat of broiler carcasses. Poult. Sci., 75: 208-215.

Kirchgessner M, Risitic M, Kreuzer M, Roth FX (1993). Einsatz von fetten mit hohen anteilen an freien fettsaren in der broilermast. 2. Wachstum sowie qualitat von schlachtkorper,

Ling WH, Jones J (1995). Dietry Phytosterols: A review of metabolism, benefits and side effects. Life Sci., 57: 195-206.

Lo´pez-Bote CJ, Rey, M. Sanz AI, Gray JI, Buckley DJ (1997). Dietary vegetable oils and -α-tocopherol reduce lipid oxidation in rabbit muscle. J. Nutr., 127: 1176-1182.

Lo´pez-Ferrer S, Baucells MD, Barroeta AC, Grashorn MA (1999a). N-3 enrichment of chickenmeat using fish oil: Alternative substitution with rapeseed and linseed oils. Poult. Sci., 78: 356-365.

Lo´pez-Ferrer S, Baucells MD, Barroeta AC, Grashorn MA (1999b). Influence of vegetable oil sources on quality parameters of broiler meat. Arch. Geflu¨ gelk, 63: 29-35.

Mattson FH, Grundy SM (1985). Comparison of effects of dietary saturated, monounsaturated, and polyunsaturated fatty acids on plasma lipids and lipoproteins in man. J. Lipid Res., 26: 194-202.

Mensink RP, Katan MB (1989). Effects of a diet enriched with monounsaturated or polyunsaturated fatty acids on levels of a low-density and high-density lipoprotein cholesterol in healthy women and men. N Engl. J. Med., 321: 436-441.

Pan DA, Storlien LH (1993). Dietary lipid profile is a determinant of tissue phospholipid fatty acid composition and rate of weight gain in rats. J. Nutr., 123: 512-519.

Salamatdoustnobar R, Nazeradl K, Aghdamshahriyar H, Ghorbani A, Fouladi P (2007). The ratio of ω6:ω3 fatty acids in broiler meat fed with canola oil and choline chloride supplement. J. Anim. Vet. Adv., 6(7): 893-898.

Sanz M, Flores A, Perez de Ayala P, Lopez-Bote CJ (1999). Higher lipid accumulation in broilers fed on saturated fats than in those fed on unsaturated fats. Br. Poult. Sci., 40: 95-101.

SAS Institute (2000). SAS Institute Inc., Cary, NC.

Scaife JR, Moyo J, Galbraith H, Michie W, Campbell V (1994). Effect of different dietary supplemental fats and oils on the tissue fatty acid composition and growth of female broilers. Br. Poult. Sci., 35: 107-118.

Sim JS, Qi GH (1995). Designing poultry products using flaxseed. Pages 315–333 in Flaxseed in Human Nutrition. SC. Cunnane and LU Thompson ed. AOCS Press, Champaign, IL.

Sirtoni CR, Tremoli E, Gatti E (1986). Controlled evaluation of fat intake in the Mediterranean diet: comparative activities of olive oil and corn oil on plasma lipids and platelets in high-risk patients. Am. J. Clin. Nutr., 44: 635-642.

Temple NJ (1996). Dietary fats and coronary heart disease. Biomed.

Pharmacother., 50: 261-268.

Valsta LM, Jauhiainen M, Aro A, Katan MB, Mutanen M (1992). Effects of a monounsatunated rapeseed oil and a polyunsaturated sunflower oil diet on lipoprotein levels in humans. Arterioscler Thromb., 12: 2-7.

Villaverde C, Baucells MD, Cortinas L, Galobart J, Barroeta AC (2003). Effects of the dietary fat unsaturation level on body fattening in female broiler chickens. Page 66 in Proceedings of Poultry Science Association Annual Meeting, Madison, WI.

Walsh RJ, Day MF, Fenner FJ, McCall M, Saint EG, Scott TW, Tracey MV, Underwood EJ (1975). Diet and Coronary Heart Disease, Report Number 18 (Canberra, Australian Academy of Science).

12

Preliminary studies on the damage symptoms and the spatial distribution of an emerging insect pest, *Mecocorynus* sp. (Coleoptera: Anthribidae) on cashew in Ghana

Dwomoh, E. A.[1]*, Ahadzie, S. K.[1], Somuah, G. A.[1] and Amenga A. D.[2]

[1]Cocoa Research Institute of Ghana, P. O. Box 8, New Tafo-Akim, Ghana.
[2]Jaman South District Directorate, Ministry of Food and Agriculture, Jaman South, Ghana.

Cashew (*Anacardium occidentale* Linn.) has become a very important non-traditional tree crop in Ghana. The crop is, however, infested by numerous insect pests at different stages of its growth. Knowledge of the insect complexity associated with the tree is essential for developing pest control strategies for the crop. Following a reported outbreak of an unknown pest by officials of Ministry of Food and Agriculture (MOFA) and farmers in March 2009 in some cashew plantations in the Brong-Ahafo Region, an investigation was initiated to collect and identify the insect species involved and assess the extent of the damage caused by the insect. The cashew bark borer, *Mecocorynus* sp. (Coleoptera: Anthribidae) was identified and found to be responsible for the damage and could kill the attacked trees within weeks after infestation. Observations at Jaman south and Jaman north of death of mature trees within few weeks of infestation were indications of emergence of new pest of cashew in Ghana. Studies on the distribution and damage characteristics of the pest was conducted from April 2009 to December 2010 in 41 major cashew growing areas within five districts of the Brong-Ahafo Region. In the Jaman South District, symptoms of infestation of the insect were found on 14.2% of the tree population and the death rate of tree population was 2.2%. The incidence of the pest in the Jaman north was also high, with death rate of 2.8% and infestation rate of 13.6%. In the Tain district death rate was 0.6%; infestation was observed only at Banda-Biema, Nsawkaw, Mendji, Donkokrom and Brodi. This paper reports the status of a new emerging pest of cashew in Ghana, which needs an urgent attention.

Key words: Damage symptoms, spatial distribution, *Mecocorynus* sp., cashew, Ghana.

INTRODUCTION

Cashew was introduced into Ghana by the government in the 1960s for afforestation in the savannah, coastal savannah and forest-savannah transition zones in Greater Accra, Eastern, Volta and Brong-Ahafo regions (Anonymous, 2005). Its cultivation was also considered essential for tree cover in eroded areas where land reclamation grammes were under way to prevent further erosion. Large scale cultivation of the crop started in 1991 and by 1997, the area under cashew cultivation nation-wide was 12,500 ha. Between 2000 and 2007, incentives were provided to farmers in the form of loans and improved planting materials to establish new and rehabilitate old plantations. Consequently, there was a marked increase in acreage from 18,000 to 75,831 ha, with a corres-ponding rise in nut yield from 3,600 to 35,915 MT (FAO, 2008).

The cashew tree is, however, infested by numerous insect pests at different stages of its growth (Eguagie, 1972; Pillai et al., 1976; Devasahayam and Nair, 1986; Malipatil and Houston, 1990; Xianli and Van Der Geest 1990;

*Corresponding author. E-mail: aedwomoh@gmail.com

Topper et al., 2001; Yidana et al., 2004). In Ghana, surveys were conducted from July 2003 to October 2005 to collect and identify the insect fauna on cashew in 13 major cashew growing areas within ten districts of the Northern, Upper West, Brong-Ahafo and Eastern Regions of Ghana and a total of 170 insect species were collected. Eighty nine species of the total collection were identified to family level, 57 of which were identified to at least the generic level (Dwomoh et al., 2008).

Insects recorded were mostly sap-suckers, defoliators, branch girdlers, stem/twig borers and fruit/nut borers. A few beneficial species were also recorded either as pollinators or predators. The most devastating species belonged to the order Hemiptera, includes: *Helopeltis schoutedeni* Reut. *Pseudotheraptus devastans* Dist., *Anoplocnemis curvipes* F., *Homoecerus pallens* F., *Clavigralla shadabi* Dolling and *Clavigralla tomentosicollis* Stal. Others were the stem and twig girdler, *Analeptes trifasciata* F. and the stem borer, *Apate telebrans* Pall (Dwomoh et al., 2008).

Following a reported outbreak of a new and unknown insect species by officials of Ministry of Food and Agriculture (MOFA) and farmers in March 2009 in some cashew plantations in the Jaman north and south districts of Ghana, a preliminary investigation was initiated to collect and identify the insect species involved and assess the nature and extent of the damage caused by the insect in the reported districts, the knowledge of which will be useful in forming the basis for developing effective control measures in the field.

MATERIALS AND METHODS

To find out and identify the insect responsible for the damaged cashew trees as reported, the bark of the trunks, branches and twigs of infested cashew trees in 10 selected plantations in each of the districts were initially examined for eggs, larvae and the adults of the insects. Furthermore, the barks of some identified infested trees from each of the plantations were carefully slashed using machete to search for eggs, larvae, pupae and adults. Samples of frass and gum exudates were also collected from branches and trunks in search of eggs, larvae, pupae and adults of the insect. Photo documentations of the different life stages of the insect were collected as well as the various parts of the infested/dead stands were made using a digital camera. Adult specimens collected were pinned and oven-dried at 80°C for 48 h and preserved with naphthalene balls for identification. The immature stages of the insect were preserved in 70% ethyl alcohol, following the techniques described by Dwomoh (2003).

Preliminary studies on the distribution and damage characteristics of the insect was conducted from April 2009 to December 2010 in 41 major cashew growing areas within five districts of the Brong-Ahafo Regions of Ghana (Table 1). To ascertain the incidence and the confines of the insect in the cashew growing communities in the region, cashew plantations in two neighbouring cashew growing districts, Tain and Wenchi were also included in the scope of assessment of the insect's distribution.

Preliminary assessment of 10 cashew plantations in 10 locations within Jaman south (Table 1) (*Kofiko, Kofidomo, Nsuansa, Kubease, Sebreni, Famkw a, Kotokware, Dwenem, Ponko* and *Gonosua*) and Jaman north (*Duadaso, Kabile, Jamara, Buko,*

Bonakira, Adabiem, Jinini, Kokoaa, Koofosua and *Buni*) as well as Tain (*Banda-Biema, Banda-Boasi, Brohani, Nsawkaw, Mendji, Sabie, Donkokrom, Bofie, Banda-Ahenkro* and *Brodi*) and Wenchi (*Awisa, Asuogya, Nkonsia, Wurompo, Nchiraa, Subinso, Branam, Perho, Droboso* and *Abotreye*) districts as well as *Mim* in the Asunafo district was carried out to determine the incidence and infestation levels of the pest by counting and recording numbers of dead stands in each plantation as well as noting symptoms associated with of the pest's feeding.

RESULTS AND DISCUSSION

The field assessment revealed the presence of only one insect species in all the infested trees found in all the 41 cashew growing communities surveyed. The insect was identified to the generic level as *Mecocorynus* sp. Earlier surveys conducted from 2003 to 2005 in the Northern, Brong-Ahafo, Central, Greater Accra, Upper West and Eastern Regions of Ghana to catalogue the insect species associated with cashew did not capture *Mecocorynus* sp. The activities of this pest, however, have been reported in Tanzania, Mozambique and other cashew growing East African countries (Topper et al., 2001). In Tanzania and Kenya, *Mecocorynus loripes* has been reported as a very serious insect pest of cashew (Millanzi, 1997; Topper et al., 2001; Anonymous, 2002). The larva, a typical weevil grub tunnelled down beneath the bark, eating the sapwood of the tree causing damage to the crop. Heavily infested trees died in a short period of time. The adult appeared to lay eggs at multiple sites of the trees and this resulted in the larvae causing extensive multiple entry/exit holes and also making irregular tunnels inside the trunk, which resulted in gum leakage and death of affected trees.

In the present study, the various life stages of the insect (egg, larva, pupa and adult) were not found in the soil samples collected from the base of the trees and also in the frass/exudates collected from the tree trunks. Larvae (Figure 1) and adults (Figure 2) were mostly found in the frass collected by chiseling the entry/exit holes observed on the branches and trunks of the infested trees.

The number of larvae collected from both branches and trunks of cashew trees was as high as 136 and 124 for Jaman north and Jaman south, respectively, while the number of adults was as low as 16 and 19 for Jaman north and Jaman south, respectively. The eggs were conspicuously absent at the time of this assessment as they might have all hatched and developed into other life stages (larva, pupa and adult).

The larva has a curled creamy-white body with wrinkled thoracic and abdominal cavities and brown head capsule (Figure 1). Larvae are typical weevil grubs which tunnelled down beneath the bark, eating the sapwood of the tree causing substantial economic damage to the crop. The adult is a large weevil with an average length of 20.5 ± 0.6 mm long. It is greyish-brown in colour. Heavily infested trees were found dead. Generally, cashew

Table 1. Fourty-one cashew-growing communities selected for *Mecocorynus* sp. incidence and distribution studies in the Brong-Ahafo Region of Ghana, April 2009 to December 2010.

Districts	Location	Latitude	Longitude	Location	Latitude	Longitude
	Duadaso	7°55'SW	2°39'NW	*Adabiem*	8°05'SW	2°33'NW
	Kabile	7°59'SW	2°42'NW	*Jinini*	8°07'SW	2°34'NW
Jaman North	*Jamara*	7°59'SW	2°40'NW	*Kokoaa*	7°52'SW	2°42'NW
	Buko	8°05'SW	2°42'NW	*Koofosua*	7°50'SW	2°41'NW
	Bonakira	7°59'SW	2°33'NW	*Buni*	7°48'SW	2°41'NW
	Kofiko	7°30'SW	2°52'NW	*Famkwa*	7°34'SW	2°52'NW
	Kofidomo	7°32'SW	2°50'NW	*Kotokware*	7°35'SW	2°50'NW
Jaman South	*Nsuansa*	7°32'SW	2°48'NW	*Dwenem*	7°42'SW	2°47'NW
	Kubease	7°35'SW	2°48'NW	*Ponko*	7°41'SW	2°47'NW
	Sebreni	7°36'SW	2°52'NW	*Gonosua*	7°37'SW	2°50'NW
	Banda-Biema	8°10'SW	2°23'NW	*Sabie*	8°04'SW	2°22'NW
	Banda-Boasi	8°11'SW	2°24'NW	*Donkokrom*	7°46'SW	2°13'NW
Tain	*Brohani*	7°57'SW	2°33'NW	*Bofie*	8°02'SW	2°26'NW
	Nsawkaw	7°53'SW	2°21'NW	*Bamda-Ahenkro*	8°10'SW	2°23'NW
	Mendji	7°56'SW	2°25'NW	*Brodi*	7°52'SW	2°37'NW
	Awisa	7°50'SW	2°07'NW	*Subinso*	7°55'SW	2°03'NW
	Asuogya	7°38'SW	2°43'NW	*Branam*	7°58'SW	2°08'NW
Wenchi	*Nkonsia*	7°43'SW	2°03'NW	*Perho*	7°55'SW	2°08'NW
	Wurompo	7°54'SW	2°07'NW	*Droboso*	7°43'SW	2°05'NW
	Nchiraa	7°55'SW	1°58'NW	*Abotreye*	7°55'SW	2°09'NW
Asunafo-North	*Mim*	6°55'SW	2°35'NW			

Figure 1. A larva (grub) of *Mecocorynus* sp.

farmers interviewed indicated that attacked or infested trees always died after a short period of time and infested trees had their leaves becoming yellowish and dropping before death (Figure 3). This observation might be due to their feeding activities which probably damage the vascular tissues, arrest the sap flow and weaken the stem thereby resulting in yellowing and shedding of leaves, drying of

Figure 2. (a) Ventral, (b) lateral and (c) dorsal views of the Adult *Mecocorynus* sp.

Figure 3. Cashew tree killed by *Mecocorynus* sp infestation at Kofiko near Drobo in the Brong-Ahafo region.

a   b

c   d

Figure 4. (a) and (b) Small entry holes in the trunks of cashew, (c) and (d) cracks and multiple entry/exit holes on cashew tree trunks caused by *Mecocorynus* sp infestation at Kofiko near Drobo in the Brong-Ahafo region.

twigs and death of the tree.

Their attack could easily be recognized by the presence of small entry holes on the trunks and branches, Figures 4a to d, gummosis, extrusion of chewed up frass and excreta at the base of infested trees as the larvae expels the frass and excreta out (Figure 5).

Figure 5. Gummosis and extrusion of chewed up cashew trunk by *Mecocorynus* sp.

The incidence of *Mecocorynus* sp on cashew plantations at Kofiko and the surrounding villages near Drobo in the Jaman South District of the Brong-Ahafo region is shown in Figure 6. Symptoms of infestation of the insect were found on 14.2% of the total tree population. The death rate of trees was 2.2%. In the Jaman South District, tree death and infestation were observed on plantations at Kofiko, Kofidomo, Kubease, Sebreni, Famkwa, Dwenem, Ponko and Gonosua, but no tree deaths were recorded at Nsuansa, and Kotokware (Figure 7).

The incidence of *Mecocorynus* sp. on cashew plantations in the Jaman North District was also high, with death rate of 2.8% and infestation rate of 13.6% (Figures 8 and 9). The insect occurred on all plantations inspected in the Jaman North District.

In the Tain District death rate was as low as 0.6% (Figure 10) while infestation was observed at only Banda-Biema, Nsawkaw, Mendji, Donkokrom and Brodi with no tree death in plantations at Banda-Boasi, Brohani, Sabie, Bofie and Banda-Ahenkro (Figure 11). Cashew plantations in the communities within Wenchi District did not record any death of trees though a few trees were found to be infested at Wurompo, Nchiraa and Droboso.

The recent pest outbreak and highest tree infestation and death rates by *Mecocorynus* sp in the Jaman North and South Districts could be possibly attributed to migration of the pest from Cote d'ivoire, a major producer of cashew in the Sub-Saharan Africa, since the two districts are closer and share territorial boundaries on the west with the former. According to Anon (2002) efficient control approach should mainly target larval stage during early stages of infestation. However, the whole control approach is basically physical confrontation, whereby adults should be collected and destroyed. Also, it is recommended that the bark around the infested area could be removed to expose the larvae, which when they are washed or blown down are never able to climb up the trunk again as they have no legs. The pupae could be destroyed by inserting a sharp object inside the chambers. Further studies will be conducted to determine the seasonal incidence, alternative host plants and probably control measures of this new emerging insect pest of cashew in Ghana.

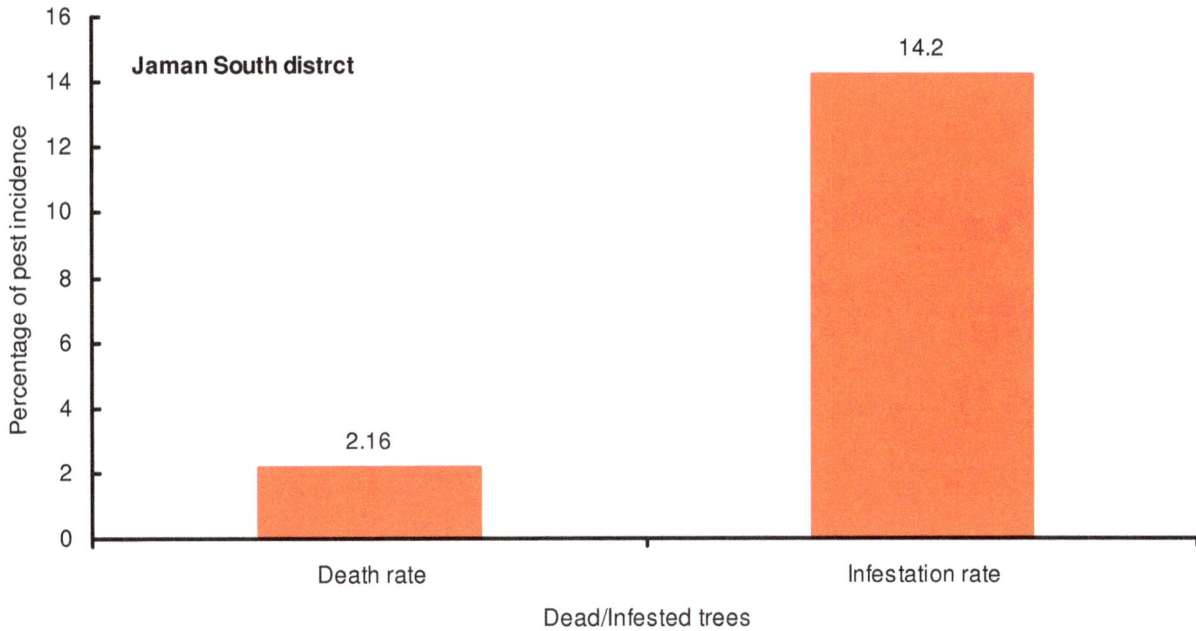

Figure 6. Mean infestation and death rates of *Mecocorynus* sp. at Jaman south district in the Brong-Ahafo region of Ghana.

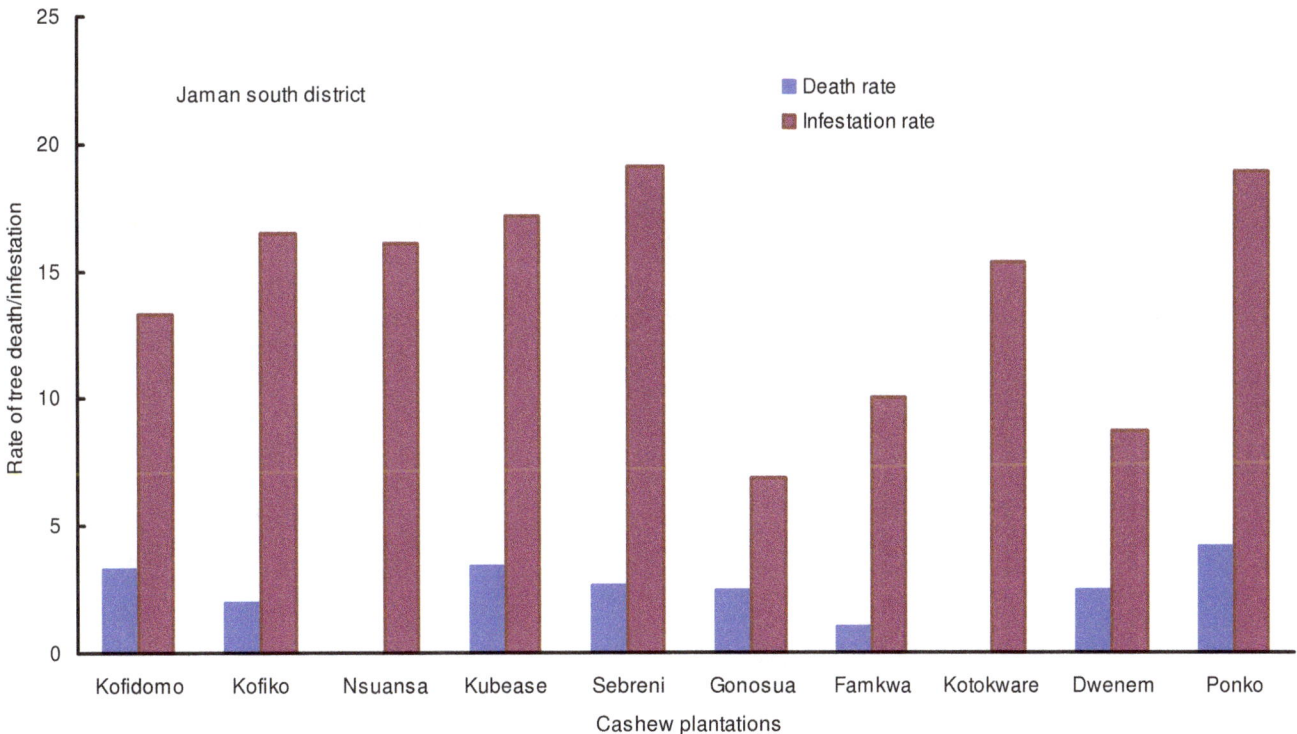

Figure 7. Incidence of *Mecocorynus* sp. in cashew plantations at Jaman south district in the Brong-Ahafo region of Ghana.

ACKNOWLEDGEMENT

Our sincere acknowledgement goes to the entire staff of Ministry of Food and Agriculture District Directorate of Jaman North and South, Wenchi and Tain. This study was funded by CRIG, supported by Cashew Development Project of Ministry of Food and Agriculture, Accra, Ghana and it is published with the kind permission

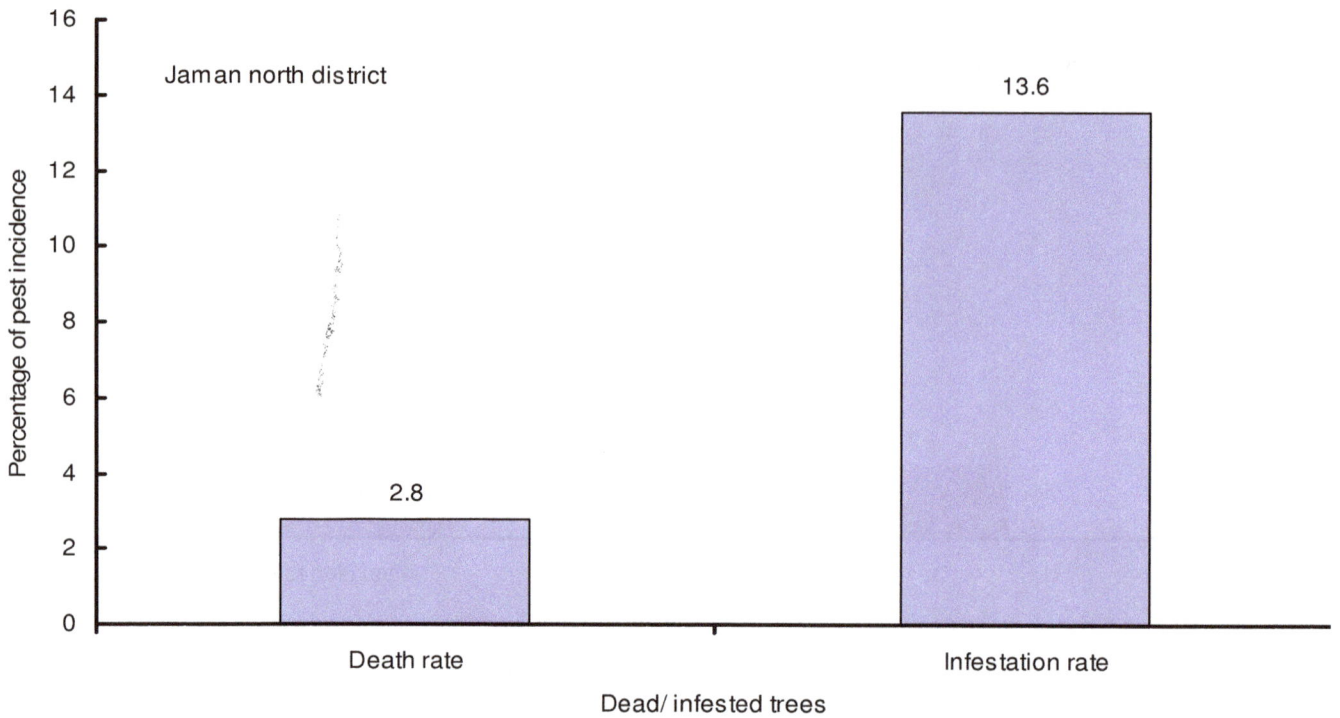

Figure 8. Mean infestation and death rates of *Mecocorynus* sp. at Jaman north district in the Brong-Ahafo region of Ghana.

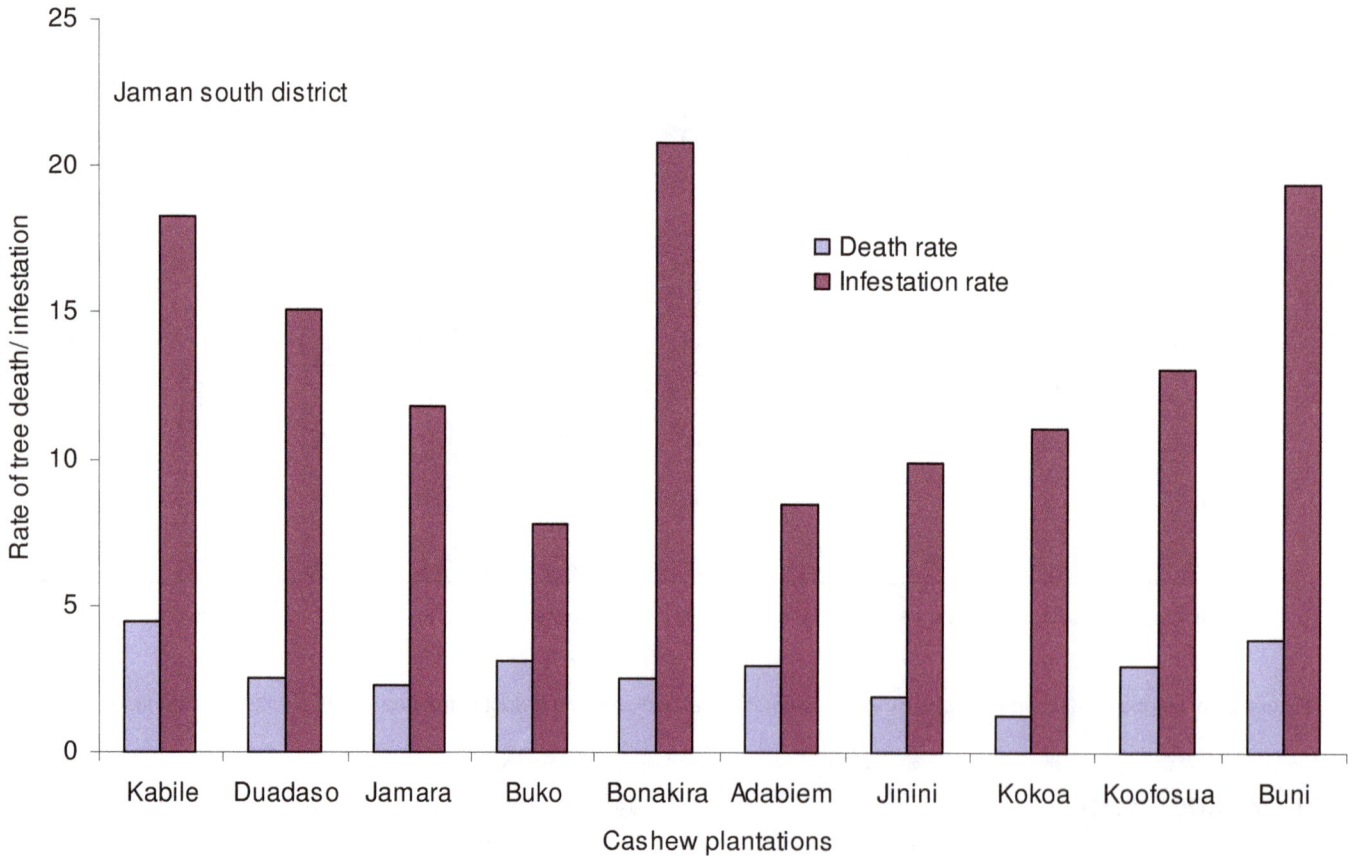

Figure 9. Incidence *Mecocorynus* sp.in cashew plantations at Jaman north district in the Brong-Ahafo region of Ghana.

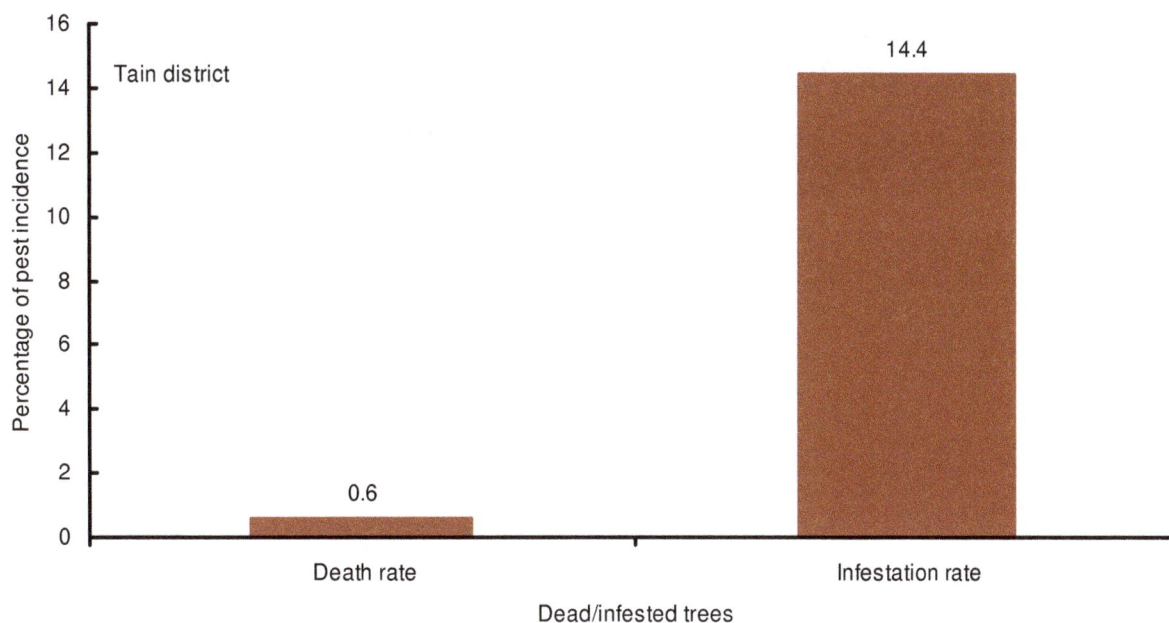

Figure 10. Mean infestation and death rates of *Mecocorynus* sp. at Tain district in the Brong-Ahafo region of Ghana.

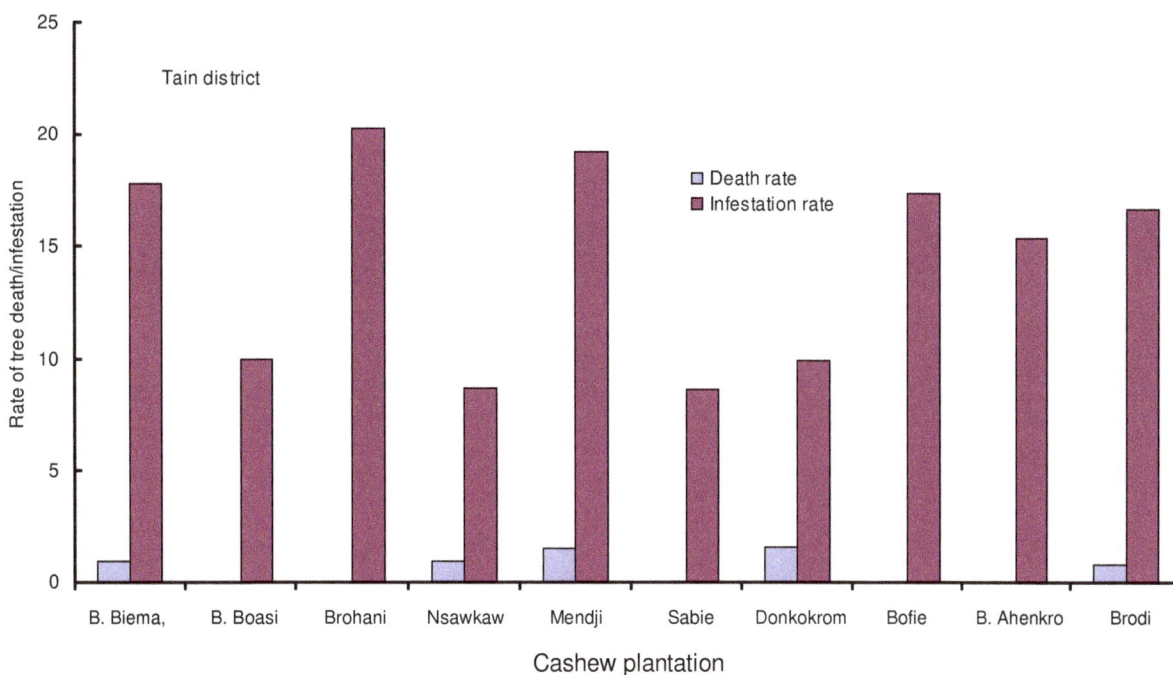

Figure 11. Incidence of *Mecocorynus* sp.in cashew plantations at Tain district in the Brong-Ahafo region of Ghana.

of the Executive Director of Cocoa Research Institute of Ghana.

REFERENCES

Anonymous (2002). Plant Pests Field Handbook: A Guide to their Management. Ministry of Agriculture and Food Security, United Republic of Tanzania, pp. 347

Anonymous (2005). Global Cashew Scenario. Cashew Development Project, Ghana Report December 2005, 13 p.

Devasahayam S, Nair CPR (1986). The tea mosquito bug *Helopeltis antonii* Signoret on cashew in India. J. Plant. Crops, 14(1): 1-10.

Dwomoh EA (2003). Insect species associated with sheanut tree (*Vitellaria paradoxa*) in Northern Ghana. Trop. Sci., 43: 70-75.

Dwomoh EA, Afun JVK, Ackonor JB (2008). Survey of insect species associated with cashew (*Anacardium occidentale* Linn.) and their distribution in Ghana. Afr. J. Agric. Res., 3(3): 205-214.

Eguagie WE (1972). Insects associated with cashew *Anacardium occidentale* in Nigeria. CRIN Ann. Rep., 1971-72: 134-137.

FAOSTATDATA (2008). Food and agriculture commodities production. www.faostat.fao.org.

Malipatil M, Houston W (1990). Bioecology of cashew insects at Wildman River, NT. Working Papers of the Third Annual Cashew Research and Development Workshop, February 21, 1990, Darwin, Northern Terr., pp. 22-25.

Millanzi KJK (1997). Screening cashew clones for reaction to *Holopeltis*. Proceed. Int. Cashew Coconut Conf., Dar Es Salaam, pp. 117-120.

Pillai GB, Dubey OP, Singh V (1976). Pests of cashew and their control in India. A review of current status. J. Plant. Crops, 4: 37-50.

Topper CP, Caligari DSP, Camara M, Diaora S, Djaha A, Coulibaly F, Asante AK, Boamah A, Ayodele EA, Adebola PO (2001). Sustainable Tree Crop Programme: West Africa Regional Cashew Survey. pp. 65.

Xianli P, Van Der Geest LPS (1990). Insect pests of cashew in Hainan, China, and their control. J. Appl. Entomol., 110: 370-377.

Yidana JA, Dwomoh EA, Yeboah J (2004). Report on survey for selection of promising cashew mother trees for development of quality planting materials. Submitted to Ministry of Food and Agriculture, Ghana, P. 27.

13

Spermatozoa quality of half blooded Ettawah goat fed with three day old green bean sprout

Djoko Winarso[1]*, Budi Purwo[2] and Y. Rina Kusuma[2]

[1]Faculty of Veterinary Medicine of Brawijaya University, Indonesia.
[2]Extension Agriculture College, Magelang, Indonesia.

The research aims to study the sperm quality of local Ettawah goat fed with three day old green bean. Sixteen males of local Ettawah goats that are one and half year old were used in this study, with one female as the libido inducer. Randomized studies were designed using four kinds of treatment groups arranged in four trial blocks. However, there were four dietary treatments: K diet without sprout was used as the control diet; PI diet contains 1 g sprout/kg body weight; PII diet contains 2 g sprout/kg; and PIII contains 3 sprout/kg body weight. Four goats were allocated to each treatment diet and all goats underwent four times sperm collection during a seven day period. Variables recorded were sperm motility, concentration, and sperm liveability percentages, while scrotum size and body weight gain were measured as the supporting data. The data obtained were analyzed statistically following analysis of variance (ANOVA). There were significant differences, and the Duncan's multiple range test was used to observe the level of their influence on the treatments. The results indicated that the use of green pea in the diets significantly increased ($P < 0.01$) motility and sperm liveability. Decreasing sperm concentration was observed on goats receiving 1 g sprout/kg body weight, but followed by inconsiderable increase on goats receiving 2 and 3 g sprout/kg body weight. It could be concluded that the use of three day old green pea sprout in the diets increased goat sperm quality, especially in the case of motility and sperm liveability, but did not increase sperm concentration, scrotum size, and body weight gain.

Key words: Local Ettawah goat, spermatozoa quality, scrotum size, body weight gain, green bean sprout.

INTRODUCTION

Artificial insemination (AI) is an entry or delivery of semen into the female genital tract by using man-made tools or AI gun (Martin et al., 2011). This technique has been introduced in Indonesia since the beginning of the fifties. Benefits expected using the AI technique include avoiding disease transmission through direct contact sex, male animal maintenance cost savings, as well as the improvement of genetic quality with the use of superior males (Djanuar, 2005). For success in marriage with the AI technique, the production of semen in the right amount or quantity/quality, if not to say high, is required. Quantity, especially the declining semen quality also reduces conception rates. There are several factors that have been commonly discussed in relation to the quantity and

quality of semen, one of which is the feed factor (Batista et al., 2009).

Food is needed quantitatively and qualitatively in male animals. However, food for reproduction need not exceed the needed amount for animal growth either in young or adult animals to sustain life in a healthy condition. In addition to forage, goats feed also requires reinforcement to meet nutritional needs. Food that is perfect is a fairlywell-balanced diet of carbohydrates, proteins, fats, minerals, water and vitamins that are essential for reproduction (Shi et al., 2010).

Green bean, as a food ingredient amplifier, besides having high enough protein is also rich in vitamin E. Many say vitamin E serves to increase fertility, as well as prevention of aging, whereas Purwanto et al. (2004) said that vitamin E is needed for reproduction in rats.

The content of alpha tocopherol in 3 days old mung bean sprouts also helps to prevent sterility and muscular dystrophy of the reproductive organs and improve the quality of spermatozoa. Active alpha-tokopherol can be regenerated by interaction with vitamin C, which inhibits free radical oxidation peroxysalts in the reproductive tract. The oxidation of free radicals is inhibited in the reproductive tract peroxysalts so that the motility and percentage can increase the spermatozoa's life. Thus, one of the two peroxysalts free radicals found in the reproductive organs becomes glucoronat conjugated when there is excretion in the kidney (Agawal, 2010). Vitamin E also strengthens the walls of blood capillaries and prevent damage to the red blood cells due to toxins in the intestine tract so that blood circulation in the manufacture of spermatozoa can run well (Rink, 2010).

Mung bean sprout as good source of vitamin E (alpha-tocopherol), has sufficient potential. Vitamin E is an antioxidant that protects cells from free radical attack. By eating the sprouts, there is the possibility that vitamin E will protect the cells of spermatozoa from damage caused by free radical attack, so that there will more motility and higher power of life (Made, 2003).

The research aims to determine the effect of 3 days old green bean sprouts on the quality of goat spermatozoa. Given that the green bean sprouts are 3 days old, they can improve the quality of spermatozoa in Ettawah goats.

MATERIALS AND METHODS

The tails of 16 PE goats (aged one and half year) with initial body weight and an average of 31.5 kg were maintained for 120 days in the stable stage, with the plasma of each tail placed separately in a box (inside a cage). The livestock feed had 10% of body weight, and it consisted of king grass and polard feed (waste grain) with a ratio of 75 versus 25%. In addition to polard feed and bean sprouts, feed supplement was given first before king grass. Green bean sprouts are manufactured by soaking the green beans in water for one night, and then placed in a container positioned in the hangar at the dark. The container is washed and drained every 3 hours for juice, after which it is ready for use. Bean sprout green was added to the polard feed/*waste grain* in accordance with treatment doses ranging from 0 g / kg bw to 3 g / kg bw for the benefit of sperm quality tests conducted for four treatments with the addition of bean sprouts (3 days old) as follows: Control (K) = 0 grams / kg bw, Treatment I (PI) = 1 gram / kg bw, Treatment II (PII) = 2 g / kg bw, Treatment III (PIII) = 3 g / kg bw. It was discovered by chemical analysis in the Chemical Analysis Laboratory of the Faculty of Pharmacy, Gadjah Mada University that every gram of the bean sprout green (3 days old) has 0.0155 mg vitamin E.

For each treatment, 4 goats were used and semen was collected from each goat in the first, second, third and fourth month of the maintenance period by using artificial vagina for microscopic examination, and the sperm motility, sperm concentration and percentage of live spermatozoa were evaluated. However, the data supporting the process of sperm production was measured by scrotal circumference and weight gain.

The motility of spermatozoa was evaluated by looking at the movement and motion of individual spermatozoa, with the end result classified into six groups of scores: special (6), very good (5), good (4), less good (3), sufficient (2) and not good (1), where the stipulation was synchronized with the percentage of the number of progressively moving spermatozoa. The score is "very good" if the mass movement of spermatozoa is + + + + and the motion of the

individual is above 90%; it is "good", if the mass movement of spermatozoa is + + + and the motion of the individual is 80 to 90%; it is "less good", if mass movement of spermatozoa is + + and the motion of the individual is 70 to 80%; it is "sufficient", if the mass movement and motion of the individual spermatozoa is + 60 to 70%; and it is "not good", if there is no mass movement and gestures of individuals less than 60%. Semen, prior to review, was previously 0.1 ml diluted with 0.9 ml Na citrate. Dilution was done to facilitate the calculation of sperm motility. A drop of semen was placed on a glass object, and was made thin. The glass was covered with glass cover, and the movement of the semen was viewed under a microscope with 45 x 10 magnification.

The concentration of spermatozoa was assessed by direct counting using a hemocytometer and was recorded in a million / cc. Erythrocyte pipette filled with semen (without dilution first) until a 0.5. 3% NaCl solution was sucked up to mark 101 on the pipette. The mixture was shaken gently to form the number 8 for 2 to 3 min.

A few drops were again discarded, and then one drop was placed under a glass cover at room count Neubeur. By using a microscope with a magnification of 40 x., the number of spermatozoa was counted in the top right, top left, middle, bottom right and bottom left of the room count in a single view. Sepermatozoa concentration was obtained from the sum total count of spermatozoa in the room Neuber multiplied by 10^6.

Measurement of the percentage of live spermatozoa was based on observations made with the microscope for the unit value of percent (%). Inspection was carried out by dripping a drop of eosin - nigrosin on the object glass, after which one drop of semen was then added and smear preparations were made. The object glass was immediately aerated to dry. Sperm was counted under a microscope with a magnification of 10 x 10. Scrotal circumference measurement was done by using a tape measurement with cm scale in the middle of the scrotum (the bulk). Body weight was measured using scales with measurements every week for four times to obtain a weight gain (kg). The data obtained from the four kinds of treatment analyzed descriptively and statistically with Random Group Draft were arranged in four trial blocks on sperm motility, sperm concentration and percentage of live spermatozoa, as the treatments were given doses of bean sprouts at 3 days with a size of K = 0 g / kg bw, PI = 1 gram / kg bw, PH = 2 g / kg body weight, and P III = 3 g / kg bw. Each treatment uses 4 goats, and the examination of each tail is repeated with semen shelter as much as 4 times the time interval of one month. The data obtained were the resource persons being statistically analyzed following analysis of variance (ANOVA). If there are significant differences, the Duncan's Multiple Range Test is used to determine the influence among the treatments.

RESULTS AND DISCUSSION

Motility spermatozoa

Based on the results of microscopic examination of semen collection during the study, the data obtained are given in Table 1 and Figure 1. Figure 1 shows the mean percentage motility of spermatozoa with 3 days old green bean sprouts at different doses, while Table 1 shows a change in sperm motility setup dose for bean sprouts at 3 different days. Here, we can see that the higher dose of bean sprouts when they are 3 days old also increased sperm motility.

Results of the microscopic examination of motility in PI, PII and PIII show that the movement of spermatozoa is fast, strong and progressive, with a mean percentage of

Table 1. Average of spermatozoa motility percentage Goat PE during 4 times semen collection at different dose giving 3 days old sprout green peanut.

Animal	Percentage of spermatozoa motility (%)			
	K	PI	PII	PIII
A1	69	75.25	79	80.5
A2	71	76	78.75	81
A3	69.25	74	78	81
A4	69.5	75.25	80.25	81
Average	69.5625[a]	75.125[b]	79[c]	80.875[c]
Total	1113	1202	1264	1294

[a,b,c] Different superscripts at the same row indicate significant differences (P < 0.05).

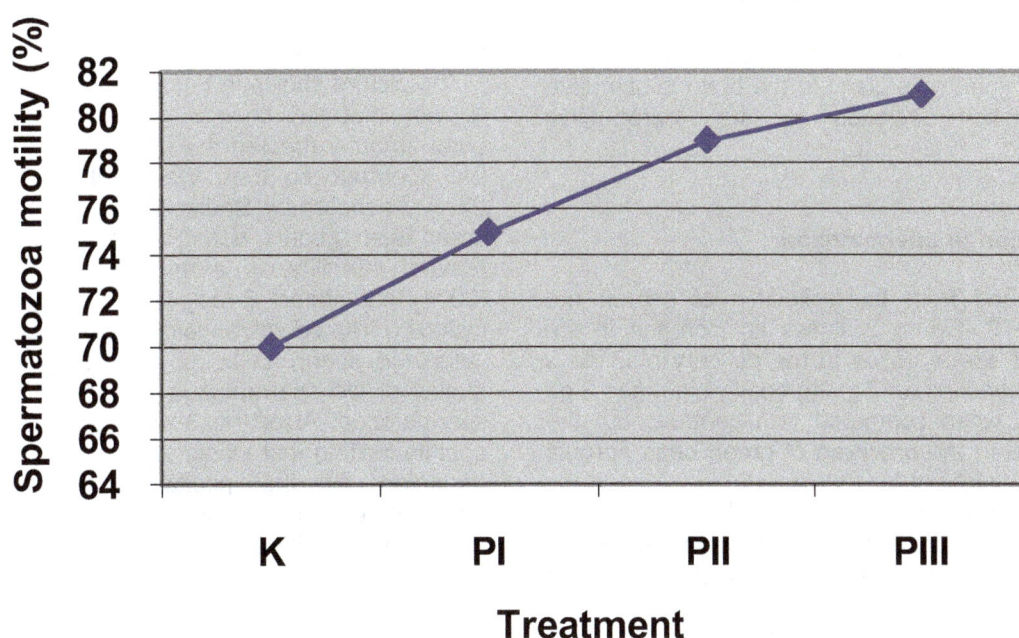

Figure 1. Sperm motility of goats given 3 days old sprout green peanut at K = 0 gram/kg bw, PI = 1 gram/kg bw, PII = 2 gram/kg bw, and PIII = 3 gram/kg bw.

motility for each treatment; although, the movement is higher than that observed by Toelihere (1998) which states that the ram has 60 to 70% motile spermatozoa.

From the results obtained by statistical tests, it was observed that different doses of 3 days old green bean sprouts (Figure 1) had an effect on the motility of spermatozoa with increased number of real or significant (P < 0.01) differences observed in the control (without giving 3 days old green beans sprouts). However, significant difference was observed between treatments II and III, but was not for treatment I.

As stated by Bearden and Naing et al. (2010), the part which provides energy for life and movement of spermatozoa is the tail of spermatozoa through the metabolic processes that take place in the mitochondrial helix. Supported by Men (2001), a reaction that produces energy in the semen only takes place in spermatozoa.

The energy itself to sperm motility by Toelihere (1998) originated from within the renovation of the sheath mitochondria ATP through its disposal reactions to ADP and AMP. He also added that most of the vital physiological activism is given by the energy source of carbohydrate or fat.

The use of 3 days old bean sprouts in the implementation of the study, other than as an energy source for the fat content of 5.997 mg / gram also has vitamin E that amounted to 1.55 mg/100 g in accordance with the results of the chemical analysis performed at the Chemistry Laboratory, Faculty of Pharmacy UGM Analyst. According to Fuquay (2002), vitamin E contains alpha-tocopherol that affects sperm trajectory where the deficiency of vitamin E will shorten the path down which affects sperm motility. Besides, it is said also by Made (2003) that deficiency of vitamin E results to loss of movement of sperm and embryos growth disruption in mice, in addition to sterility in mice and muscle damage in animals, such as: dogs, guinea pigs and rabbits, including the muscles of the reproductive tissues. The content of alpha tocopherol in 3 days old mung bean sprouts also helps to prevent muscle dystrophy sterility in reproductive organs, and it inhibits free radical oxidation peroxysalts in the reproductive tract so that the motility and percentage could increase spermatozoa life.

In addition, one of the two peroxysalts free radicals found in the reproductive organs becomes glucoronat when there is excretion in the kidney (Agawal, 2010). Vitamin E deficiency can cause reduction in the power of the body, decreased sexual activity, and abnormal fat deposits in the muscles (Epoch, 2008) so as to reduce the motility of spermatozoa. Also, Sediaoetama (2001) said that green bean sprouts contain lecithin that serves to improve the function of the hypophysis gland in the brain to stimulate hormonal secretion included in the reproductive hormones.

The concentration of spermatozoa

The data obtained from the collection of semen are shown in Table 2. Figure 2 shows an increase in the concentration of spermatozoa in the delivery of 3 days old green bean seedling for 2 g / kg body weight and 3 g / kg body weight when compared with controls, but the figure decreased in the provision of green bean sprouts at 3 days with a weight of 1 g / kg bb.

Statistical analysis showed that sperm concentration decreased the 3 days old bean sprouts at increments of 1 g / kg bw, and also showed that the numbers are neither real nor significant (P < 0.01). An increase was observed for the 3 days old green bean sprouts in control, followed by an increase in the concentration of spermatozoa with significant changes (P < 0.01). Decrease in sperm concentration occurred in the delivery of 3 days old green beans in increments of 1 g / kg bw as compared to the control, probably due to the time of the semen-making erratic weather with occasional hot climate and a sudden rain, thus affecting the concentration of spermatozoa. Toelihere (1998) observed that hot weather and climate with low humidity tends to decrease the concentration of spermatozoa. Also, he added that the concentration of spermatozoa is influenced by climate and weather, as well as by the volume of semen, age, weight, individual health, food, frequency of relocation and genetic factors.

Increasing the concentration of spermatozoa with numbers that are not significant (P > 0.01) in 3 days old bean sprouts with a dose of 2 and 3 g / kg bw as compared to the controls is likely due to differences in the dose level, which is not too flashy but demands that the feed meets its required ration in order to increase the concentration of spermatozoa, because, as expressed by Gutrie (2000), animals should be fed properly for the production of spermatozoa. Supported by Djanuar (2005),

feeding needs to spur the gonadothropine hormone secretion from the pituitary gland, so that spermatogenesis can be run perfectly.

The percentage of live spermatozoa

Examination of the percentage of live or dead spermatozoa is important, because it not only shows live die (Partodihardjo, 2000). According to Toelihere (1998), the preparations made for shaking time affect the percentage of live spermatozoa. This is reinforced by the opinions of Priyono et al. (1994) which state that the increasing time and shock will reduce the motility and percentage of live spermatozoa. Results of the data obtained by semen collection percentage for live spermatozoa are shown in Table 3 and Figure 3.

The picture shown in Figure 3, and the statistical tests, showed that any change in the dosage of 3 days old bean sprouts affected the increase in the percentage of live spermatozoa from the numbers significantly (P < 0.01). As stated by Sediaoetama (2001), the 3 days old green bean sprouts, rich in alpha-tocopherol content, can prevent infertility. It is reinforced by Purwanto et al. (2004) that vitamin E may help the regeneration of cells, including reproductive cells, especially the test organ cells and sperm cells, so that they will maintain the quality of life spermatozoa including the percentage of spermatozoa. Also, the 3 days old green bean sprouts contain lecithin that works to improve the tissue function regulated by the pituitary gland associated with expenditure mechanisms of the reproductive hormones (Sediaoetama, 2001). The content of alpha tocopherol in 3 days old mung bean sprouts also helps to prevent muscle dystrophy sterility in reproductive organs, and it inhibits the free radical oxidation peroxysalts in the reproductive tract, so that the percentage of live spermatozoa motility may increase (Agarwal, 2010).

An increasing percentage of life on the provision of bean sprouts is possible. This is supported by the opinion of Burns (2005) which states that the increased activity on the hormone testosterone metabolisms have an influence on the spermatogenesis process of improvement by increasing the trajectory of spermatozoa in the male genital tract so that the accessory gland secretion will increase and provide energy substrate for spermatozoa. However, this has more impact on increasing the vitality of spermatozoa.

Scrotal circumference

The results of scrotum circumference are shown in Table 4. It was observed that giving 3 days old green bean sprouts to Ettawah goat did not affect the size of the circumference of the scrotum. Scrotal circumference measurements were taken to support these observations.

This is because the increase in scrotal circumference

Table 2. Average of spermatozoa concentration Goat PE during 4 times semen collection at different dose giving of sprout green peanut old age 3 days.

Animal	Spermatozoa concentration (million/cc)			
	K	PI	PII	PIII
A1	225	202.50	216.25	221
A2	206	265.5	230.5	248.50
A3	264.50	211.75	305.25	250
A4	249.25	224	262.5	266.5
Average	236.1875[a]	225.9375[a]	253.5625[a]	246.625[a]
Total	3779	3615	4057	3946

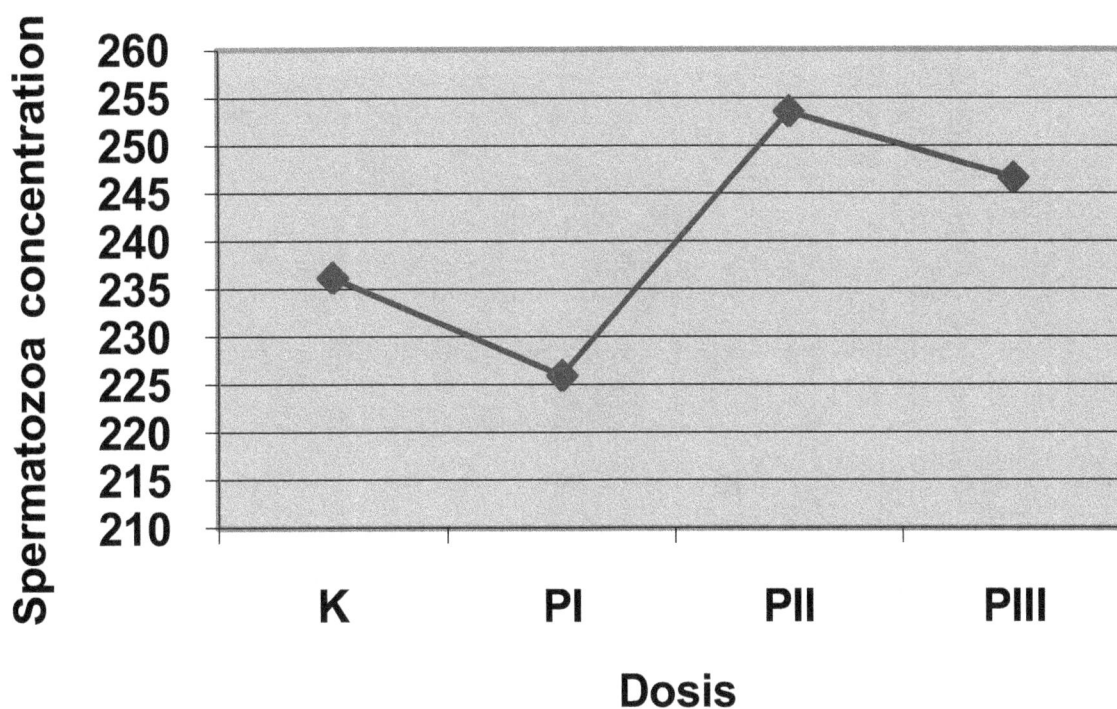

Figure 2. Spermatozoa concentration of goats given 3 days old sprout green peanut at K = 0 gram/kg bw, PI = 1 gram/kg bw, PII = 2 gram/kg bw, and PIII = 3 gram/kg bw.

size has a fair correlation with the quality and quantity of spermatozoa in young males (Gipson et al., 2001; Diwyanto et al., 1992); although, scrotal circumference size is influenced by many factors such as the nation (Coulter and Bailey, 1988), age (Vogt ct tit., 1984; Diwyanto, 1992), body weight (Diwyanto, 1992), as well as season and management (Coulter and Bailey, l988). In North America, the size of scrotal circumference is seen as one parameter used in the study of talon, such that if there is an increase in scrotal circumference sizedue to the 3 days old green bean sprouts, it can be recommended to a Shaman who can help improve thequality of male candidates as an additional information about the relationship between scrotal circumference and sperm motility, but moving spermatozoa that are about to quality spermatozoa.

Weight

The results of the analysis of weight gain of Ettawah goat (PE-) are shown in Table 5. The mean weight gain between treatments, both for control (without addition of bean sprouts at 3 days) and for other treatments (with addition of green bean sprouts), showed a slight increase in body weight on the provision of green bean sprouts (1 g / kg body weight), when compared with the control (3.92 kg: 3.82 kg). Overall, the statistical test results showed that the administration of green bean sprouts had no effect on weight gain, because the green bean sprouts provide the main influence on the reproductive tract by enhancing trajectory of the spermatozoa hormone (Bearden and Fuquay, 2002), so that the quality of spermatozoa increases. Weight gain was used in almost

Table 3. Average of spermatozoa live percentages goat PE during 4 times semen collection at different dose giving of sprout green peanut old age 3 days.

Animal	Percentage of spermatozoa live (%)			
	K	PI	PII	PIII
A1	81	83.50	87	90
A2	80.5	84.25	87.25	90.25
A3	80.25	84	85.25	90
A4	80.25	84.5	87.25	90.75
Average	80.5625[a]	84.0625[b]	86.6875[c]	90.25[d]
Total	1289	1345	1387	1444

Different superscripts at the same row indicate significant differences ($P < 0.05$).

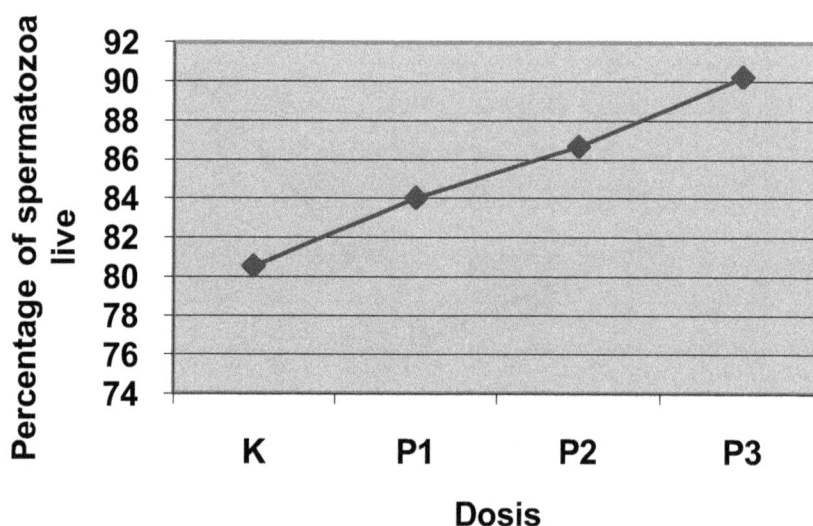

Figure 3. Percentages of spermatozoa live given 3 days old sprout green peanut at K = 0 gram/kg bw, PI = 1 gram/kg bw, PII = 2 gram/kg bw, and PIII = 3 gram/kg bw.

Table 4. Average of circular scrotum goat during 4 times semen collection at different dose giving of sprout green peanut old age 3 days.

Animal	Scrotum circular (cm)			
	K	PI	PII	PIII
A1	23.25	23.25	23.125	23.5
A2	21.25	21.5	21.75	21.875
A3	22.75	23	22.125	23.25
A4	21.25	21.375	21.375	21.625
Average	22.125[a]	22.281[a]	22.094[a]	22.562[a]
Total	254.0	356.5	353.5	361.0

all the treatments, because during the study, forage and concentrate feed that met the nutritional content of food (as seen in the results of the proximate analysis on the instrument) were used with a standard dose or dose maintenance (maintenance). According to Church and Pond (2003), the level of feed requirements for goats used in this research (in accordance with a maintenance dose) has been fulfilled by the protein concentrate, calcium, phosphorus, TDN, fat and raw fiber given to them, so the average weight gain during maintenance at the research activities is still within normal limits. Besides, green bean sprouts contain less energy, protein and fat than soy sprouts, and the germination process that occurs during the process of protein hydrolysis is lower

Table 5. Average of gain goat PE during 4 times semen collection at different dose giving of sprout green peanut old age 3 days.

Animal	Average body weight gain (kg)			
	K	PI	PII	PIII
A1	3.70	4.05	3.35	3.5
A2	3.725	3.7	3.775	3.6
A3	3.825	3.85	3.9	3.85
A4	4.075	4.075	4.025	3.975
Average	3.831[a]	3.9187[a]	3.762[a]	3.731[a]
Total	61.3	62.7	60.2	59.7

and is unable to support the weight growth (Made, 2003).

CONCLUSIONS AND SUGGESTION

From the results of this study, it is concluded that 3 days old green bean sprouts should be provided to Ettawah goat (PE) in order for it to:

1. Be able to improve sperm motility.
2. Not significantly affect the concentration of spermatozoa.
3. Be able to increase the percentage of live spermatozoa.

For study purposes, in order to improve the quality of spermatozoa, 3 days old green bean sprouts (2 g / kg bw) can be added to the feed of goat.

REFERENCES

Agarwal A, Sekhon LH (2010). Oxidative Stress and Antioxidant for Idiophatic Oligoastheno teratospermia. Indian J. Urol., 27(1).

Batista M, Nino T, Alamo D, Castro N, Santana M, Gonzalez F, Cabrera F, Gracia A (2009). Successful Artificial Insemination Using Semen Frozen and Stored by an Ultrafreezer in the Majorera Goat Breed. Theriogenology, 71(8): 1307-1315.

Bearden JH, Fuquay JW (2002). Applied Animal Reproduction. Mississipi State University.

Burns DM (2005). Produksi Kambing di Daerah Tropis. Edisl III. Penerbit ITB Bandung.

Church DC, Pond WG (2003). Basic Animal Nutrition and Feeding. Third Edition. John Willey and Sons New York.

Coulter GH, Bailey DRC (1988). Testicular Development of Sales Bullus to One Year of Age. Ca. J. Anim. Sci., 68: 961-964.

Diwyanto K (1992). The effect of Age and Body Weight Against Young Cattle's Circle scrotum. Pros. Agro-Industries in Rural Husbandry. Ciawi Livestock Research Institute, Bogor. Center for Animal Husbandry 10th to 11st August 1992.

Djanuar R (2005). Practical Artificial Insemination. Gen. Sudirman Univ. Facult. Anim. Husbandry, Purwokerto.

Epoch T (2008). Kecambah Kacang Hijau Mempunyai Nilai Gizi Tinggi. http://erabaru.net/kesehatan/34-kesehatan/160-kecambah-kacang-hijau-punya-nilai-gizi-tinggi. Up load 17 October, 2008.

Gipson E, Maxwell (2001). Salamon's Artificial Insemination of Sheep and Goat. Butterworths. Sidney.

Gutrie AH (2000). Introductory Nutritions, 11th edition. The C.V. Mosby Company, Saint Louis, hal., pp. 226-232.

Made A (2003). Mari Ramai-ramai Makan Taug : Nutrisi dan Gizinya. Indonesian Nutrition Network.Jakarta. http://www.kompas.com/kesehatan/news/0304/23/003738.htm. Diakses 5 July, 2011.

Martin JP, Policelli RK, Neuder LM, Rapohael W, Pusley JR (2011). Effect of Cloroprostenol Sodium at Final PGF2α of Ovysynch on Complete Luteolysis and Pregnancy per Artificial Insemination in Lactating Dairy Cows. J. Dairy Sci., 94(6): 2815-2814.

Men T (2001). Biochemistry of Semen and Secretion of Male Accesory Organs in Reproduction in Domestic Animals. Academic Press, New York and London.

Naing SW, Wahid H, Mohd AK, Rosnina Y, Zuki AB, Kazhal S, Bukar MM, Thein M, Kyaw T, San MM (2010). Effect of Sugars on characteristics of Boer goat semen after cryopreservation, 122(1-2): 23-28.

Partodihardjo (2000). Animal Reproduction Science. Sixth Edition. Faculty of Veterinary Medicine, Bogor Agricultural University. Pearl Source Widya. Central Jakarta.

Priyono AWS, Rachmawati B, Suwahyo AD (1994). Local Sheep and Goats in Indonesia: Influence of the time and the Tapping of Differential Sperm Motility and Local Sheep. The Report from the Center of Research and Husbandry Development, Bogor Agriculture Department BPPH.

Purwanto SL, Budiprahoto GC, Sembiring SL, Effendi, Widodo RV (2004). Animal Drug Data of Indonesia, edition 11th, PT. Grafidian Jaya, Jakarta.

Rink C, Christoforidis F, Khanna S (2010). Tocotrienol Vitamin E Protects Againts Preclinical Canine Ischemic Stroke by Inducing arteriogenesis.J. Cereb. Blood Flow Metabol., 10: 1038.

Sediaoetama AD (2001). Profoosional and Nutrition Sciences for Students, Volume IV. P.T. Dian Rakyat, Jakarta, pp. 118-121.

Shi LG, Yang RJ, Yue WB, Xun WJ, Zhang CX, Ren YS, Shi L, Lei FL (2010). Effect of Elemental Nano-Selenium on Semen Quality, Glutathione Peroxidase Activity, and Testis Ultrastructure in Male Boer Goats. Anim. Reprod. Sci., 118(2-4): 248-254.

Toelihere M (1998). Inseminasi Buatan Pada Sapi dan Kerbau.Penerbit Angkasa Bandung.

14

Histological studies of the visual relay centres in cyanide toxicity: Mode of neuronal cell death in the V1, lateral geniculate body (LGB) and superior colliculus (SC) of adult Wistar rats

Ogundele O. M.[1]*, Enuaibe B. U.[2], Igwe J.[3], Olu-Bolaji A. A.[4] and Caxton-Martins E. A.[2]

[1]Trinitron Biotech LTD, Science and Technology Complex, Sheda, Abuja, Nigeria.
[2]University of Ilorin, Department of Anatomy, Ilorin. Nigeria.
[3]National Hospital, Department of Histopathology, Central Area, Abuja, Nigeria.
[4]Department of Epidemiology and Community Health, University of Ilorin, Nigeria.

Loss of sight has been a major disorder associated with progressive neurodegenerative disorder involving motor and cognitive dysfunction. Currently, there is no effective treatment either for symptomatic relief or disease modification. This relates, in part, to a lack of knowledge of the underlying neurochemical abnormalities, including cholinergic neurons status in the diffused system of the visual relay centres. To determine histologically the possible adopted mode of cell death in the visual relay centres, previous studies have shown difference in cell death based on treatment dosage and location in brain. 50 F1 generation adult Wistar rats were treated with potassium hexacyanoferrate for a period of 30 days and the tissues were processed histologically (Hematoxylin and eosin, Gordon and sweets and crystal violet) to determine the cellular changes and characterize as apoptotic or necrotic. The predominant mode of cell death in the V1 is apoptosis while necrotic tissue sites were found to exist in the superior colliculus (SC) of the same treatment, certain cells in the lateral geniculate body (LGB) showed morphological features of both apoptosis and necrosis.

Key words: Cyanide, toxicity, visual relay centres, cell death, necrosis, apoptosis.

INTRODUCTION

Cyanide as earlier described is a potent neurotoxic substance that can initiate series of intracellular reactions leading to cell dysfunction and eventually cell death; it enhances N-Methyl-D-Aspartate (NMDA) receptor function (de Haro, 2009; de La Cruz et al., 2009). In cultured neurons, cyanide neurotoxicity is linked with NMDA receptors, mediated rise in calcium that in turn activates series of biochemical reactions leading to generation of reactive oxygen species (ROS) and nitric oxide (NO), This anti-oxidants species then mediates peroxidation of lipids (de Haro, 2009; Varone et al., 2008; Almer et al., 2006). It is concluded that oxidative stress plays an important role in cyanide induced neurodegeneration; this phenomena can cause cell death

1. Apoptosis,
2. Necrosis.

in two ways. Depending on the cell type or stimuli, a cell may die in either of the two distinct ways. Apoptosis considers the physiological form of cell death, it is an acute process with distinct morphological and biochemical features (Bathachanya and Lakshimana, 2001; Quintavalle et al., 2010). Apoptotic cells are characterized by condensed and fragmented nuclei where as necrotic cells show less plasma membrane integrity without apparent nuclei damage. This two different forms of cell death can be elicited by the same stimuli depending on the intensity of the effects (Isom et al., 1999), suggesting that an initial common event can be shared by both forms of cell death. It has been shown previously that cyanide induce cell death in a varying mode for different brain areas (Prabhakaran et al., 2007;

---

*Corresponding author. E-mail: mikealslaw@hotmail.com.

Pedro et al., 2010). Cell death occurs predominantly via apoptosis in the cortical region while necrosis was observed in the substansia nigra after the same dose of cyanide was administered in the mice. Selective vulnerability of different aspect of the brain to cyanide may be explained by the triggering of region specific toxic pathways in which oxidative stress may be a common activator (Isom et al., 1999).

In the two modes of cell death, oxidative stress was a factor but the level of ROS generated varies with the cell types, other reports that oxidative stress can be involved in apoptosis or necrosis (BathaCharrya and Tuwalsami, 2008), it is possible that when high ROS accumulated in the cell, direct and irreversible damage to cell components can lead to necrosis. Moderate levels of ROS may function as cellular messengers and regulatory molecule which mediates apoptotic cell death (Li et al., 2000), in mesencephalic cells, cyanide induced a progressive but rather small increase in necrosis at the concentrations of 100 to 300 Um and at 400 Um the data recorded shows a steep rise in necrotic index.

Organisms use physiological cell death for a variety of reasons. We need to know how many different cell death programs there are, whether and when they interact, and their precise molecular nature. In the past, cell death has been identified and studied descriptively. The term "apoptosis" is often used to describe the physiological death of mammalian cells. When individual cells die in a healthy organ, their death is not accompanied by changes that are characteristic of pathological cell deaths. Apoptosis is associated with death of isolated cells, rather than contiguous patches or areas of tissue; there is no inflammatory infiltrate; nuclear shrinkage occurs relatively early, but changes to the organelles and loss of membrane integrity are relatively late; the dying cells are phagocytosed by neighbouring cells, rather than immigrant professional phagocytes; the DNA is rapidly broken down. The degraded DNA from apoptotic cells forms a characteristic ladder when analyzed by electrophoresis because endonucleases gain access to the DNA in the internucleosomal regions (de Haro, 2009; Varone et al., 2008; Almer et al., 2006; Phrabakharan et al., 2007). The same ladder is produced from the DNA of cells killed by cytotoxic T cells (Vicente et al., 2006; Isom et al., 1999). Necrosis, on the other hand, affects many adjoining cells (Gamper et al., 2005; Pedro et al., 2010; Mathangi and Navasayami, 2000). It is characterized by cell swelling, with early loss of plasma-membrane integrity and major nucleus tends to swell. Necrosis is accompanied by inflammatory infiltrate phagocytic cells. DNA degradation, if it occurs, is a late event (Katherine et al., 2001). The term "programmed cell death" is commonly used synonymously with apoptosis. Apoptosis is more of a descriptive term, whereas programmed cell death implies that the decision to die was made by the cell-autonomously, independently of any other cells. Both terms suggest that the cell is an active participant in its own death. Unfortunately, the terms are not clearly defined,

and different investigators have used different criteria in their classification. Evidence that inhibitors of macromolecular synthesis block cell death argue that the cell is, indeed, actively involved in its own demise (Bonfoco et al., 1995), but in just as many cases macromolecular-synthesis inhibitors do not protect cells, so the cell-death machinery presumably already exists (Gilman et al., 2010). To add to the confusion, there are numerous reports of actinomycin D and cyloheximide themselves triggering apoptosis (Sharma and Chalam, 2009; Lee et al., 2010).

Thus, the effect of macromolecular- synthesis inhibitors on cell death is of no help in determining whether it is physiological. The term "programmed cell death" is also problematic. If it is used strictly, referring only to purely cell-autonomous cell deaths, most physiological cell deaths would have to be excluded, as many cell deaths are triggered by other cells, or products of other cells. The term "physiological cell death" indicates that the cell has died by a mechanism that has evolved specifically by the host to kill its own cells and probably includes most of what has been described as "programmed cell death" and "apoptosis." This term also includes cell deaths that may not exhibit some of the typical features of apoptosis, such as DNA degradation. There may be several different mechanisms that carry out physiological cell deaths. Inappropriate activation of these mechanisms, or their inhibition, may then lead to what can be recognized as pathology. The term "physiological cell death," however, distinguishes the death of the host's cells by the host's machinery from those deaths caused by external agents, such as infections, extremes of temperature, toxins, or deprivation of some vital nutrients. Nevertheless, some external agents may trigger inappropriate activation of physiological cell-death mechanisms, such as when radiation causes the interphase death of lymphocytes, or intestinal epithelial cells are exposed to anticancer agents, or when cells are exposed to extremes of temperature (Behar et al., 1999).

MATERIALS AND METHODS

Animal preparation

50 F_1 generation adult male Wistar rats (*Rattus novergicus*) each weighing on the average 250 g, were breed from parents adult Wistar rats procured from the animal facility of the National institute for pharmaceutical Research and Development, Idu, Abuja (NIPRD). The animals were weaned on 44th day after birth and were divided into five groups of 10 animals each selected at random irrespective of parental origin using the method of Svensden and Hau (1994). The weight of the animals were obtained at an interval of 3 days using a sensitive weighing balance (Jenway).

Treatment solution and mode of administration

A standard isotonic solution of 0.25 M sucrose was prepared to dissolve the potassium hexacyanoferrate III, $K_3[Fe(CN)]_6$; Mol wt = 329.25 in order to obtain a final working solution of concentration 5 mg/ml of potassium hexacyanoferrate in 0.25 M sucrose solution.

Table 1. Treatment dosage (sub-lethal dose) based on the LD_{50}.

Group 1	Group 2	Group 3	Group 4	Group 5
LD50 for oral treatment of adult Wistar rats is $K_3[Fe(CN)]_6$ 2,970 mg/kg BW (US, EPA 1990)				
20 mg/kg BW	12 mg/kg BW	6 mg/kg BW	2 mg/kg BW	0.25 M Sucrose

5 g of the CN salt was dissolved in 1000 ml of 0.25 M sucrose solution (β-D-Fructofuranosyl-α-D-Glycopyranoside; $C_{12}H_{22}O_{11}$; Mol. wt = 342.30: Sigma) (Varone et al., 2008).

Method of administration of cyanide solution

The animals were force fed orally using oral canula with a ball point at the tip. The animals were held with a glove with the left hand such that the neck region is held by the fingers to still the neck while being fed with the canula; treatment was done every morning before the animals were fed.

Grouping and treatment of animals

The treatment duration is 30 days: the total duration of the experiment is 80 days. The animals were kept under standard laboratory condition with alternating 12 h light and dark; they were fed on standard rat chow containing proteins, carbohydrate, fats, vitamins and minerals. Weight of the animals was measured at intervals of 3 days for the 30 days treatment duration Table 1 (Soler-Martin et al., 2010).

Sacrifice and specimen collection

The animals were sacrificed by cervical dislocation, the scalp was scraped and the skull immediately opened using a brain forceps, the skull was opened from the posterior part to leave the tissue intact (Van Zutphen et al., 1994). The V1 (primary visual cortex; designated as area 17 in the Brodmann's classification) was obtained from the rearmost portion of the occipital cortex, the superior colliculus was excised from the superior part of the copora quadrigermina, while the lateral geniculate body was traced to its sulcus at the base of occipital region. This was done for all the animals in each of the groups labelled Groups 1 to 5. The tissues were collected in specimen bottles containing the fixative "Formol Calcium".

Tissue processing for histology

Tissue processing for histology was done using the method of Pearse (1960) to obtain paraffin wax embedded sections; Fixation-was done using formol calcium, dehydration in ascending grades of alcohol 50, 70, 90% and absolute alcohol I and II clearing- Xylene was used as the clearing agent (2 changes). Infiltration and embedding in paraffin wax (56°C Mpt) was done in the oven at 60°C, and the tissues were blocked out in paraffin wax.

Serial sections were obtained serially and the diameter varies for each of the histological method adopted for specific structure demonstration in the brain tissues.

Crystal violet for pyknotic nuclei

Paraffin wax sections about 8 μm thick were used, the staining solution is prepared thus;

1. Crystal violet 1 g,
2. Distilled water 100 ml,
3. Acetic acid 0.25 ml.

The sections were taken to water (Pearse, 1960) and then stained in the cresyl fast violet staining solution for 25 min, they were rinsed in distilled water and allowed to stay in 96% alcohol until most of the stains has been removed (Um et al., 2010).

Gordon and sweet's method for silver impregnation of degenerating neurons

The following reagents were prepared:

1. Solution A: 2% silver nitrate,
2. Solution B: 2% potassium dichromate.

Tissue blocks were immersed in Solution B for 2 days, while dry tissue block with filter paper were immersed in Solution A for another 2 days, after which they were embedded in paraffin wax cutting sections (about 20 μm thick), and then dehydrated, cleared and mounted (Guy et al., 2010).

Photomicrography and cell count

This was done using the LCD Bresser; cell count was done stereoscopically with the use Java application (Open Office Draw) with the magnification set at X400 on the Bresser, the image size and resolution remained unchanged and a line area was drawn on specific regions of the slide image (V1: Molecular layer, LGB: magnocellular layer and the SC: stratum superficiale). The magnification is a requirement for the cell count as the cell features and appearance were required in categorizing the cells as either apoptotic, normal or necrotic. The regions of the cell obtained were approximately predetermined; the orientation of the tissues prior to processing were similar and also serial sections were made and the third sections were considered for the cell count in all the slides used for this procedure.

RESULTS

Haematoxylin and eosin

The general morphology as demonstrated in Figure 1 (Hematoxylin and Eosin) shows the appearance of the molecular layer of the primary visual cortex (V1-Brodmann's Area 17), the distribution of the neurons, glial cells as well as projections to and from this layer of the neo-cortex. In the high dose treatment (20 mg/kg BW); the V1 shows a decrease in cell size for degenerating neurons 2.5 μm Figure 1A (a) characterized by

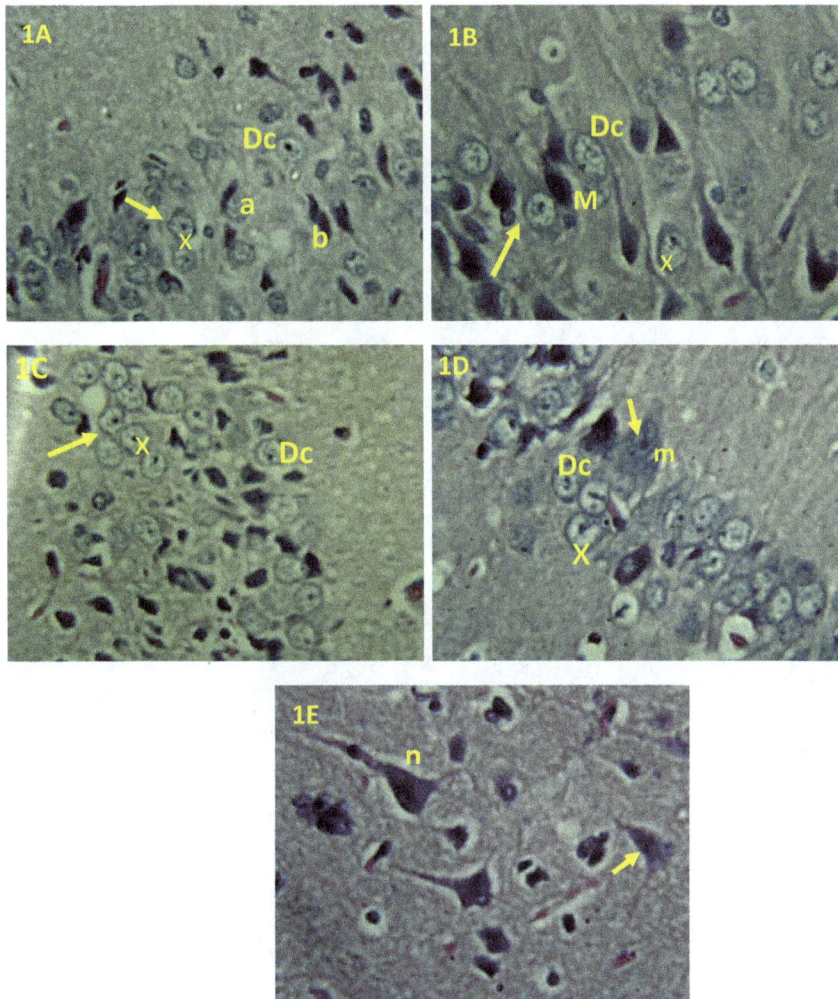

Figure 1A to E. H and E staining for general morphology of the primary visual area (V1) where A- Represents Group 1, B- 2.C- 3, D- 4 and E- 5. (M- molecular layer, X-degenerating cell, f- fibrous layer, Dc- degraded cytoplasm, N- normal cells). Arrow heads shows the relatively intact nature of the membrane and a centrally placed reduced nucleus (Magnification ×400).

fragmented cytoplasm and a centrally placed darkly stained nuclci (Figure 1A -arrow head), loosely bounded fibrous axon (arrow head Figure 1B), smaller sized spiny neurons measures about 1.25 μm in diameter. The measurement was done stereoscopically at a magnification of ×400 using a LCD Bresser microscope and analyzed on JAVA (Open office) platform. The control group consists of cells with darkly stained cell body resulting from hematoxylin staining of heterochromatin, the projections of the axons are prominent and the cytoplasm is not fragment as seen in the treatment group (Figure 1E). The cells do not show features of neuronal degeneration such as fragmented cytoplasm, loss of axons and loose fibres around a spherical cell body but rather possess a pyramidal cell body. The control group (Figure 1E) has a centrally placed nuclei measuring 0.6 μm in diameter compared with the single dot-like nuclei in the degenerating cells of thc treatment group (x) Figure 1A to D. The projections of the axons are prominent and the cell body are about 5 μm in diameter (Figure 1E) in the Group 1 (treatment group) the number of degenerating cells per/unit area (5 cm^2) is higher that the population observed in Group 2, such that in the highly dense region of group 1 (9 cells) (Figure 1A) while it is just about 3 cells over the same area in the highly dense region in Group 2 (Figure 1B), 7 in Group 3 (Figure 1C) and Group 4 (Figure 1D) has 8 cells [degenerating morphology based on parameters of Isom and Way (1999) could be described as apoptotic]. Although, the cell density per unit area is higher in Group 1, the diameter of the cell varies between the two groups with Group 2 cells measuring about 6 μm in diameter compared to the 2.5 μm observed in Group 1, Group 3 measures 5.6 μm while Group 4 measures about 5.1 μm.

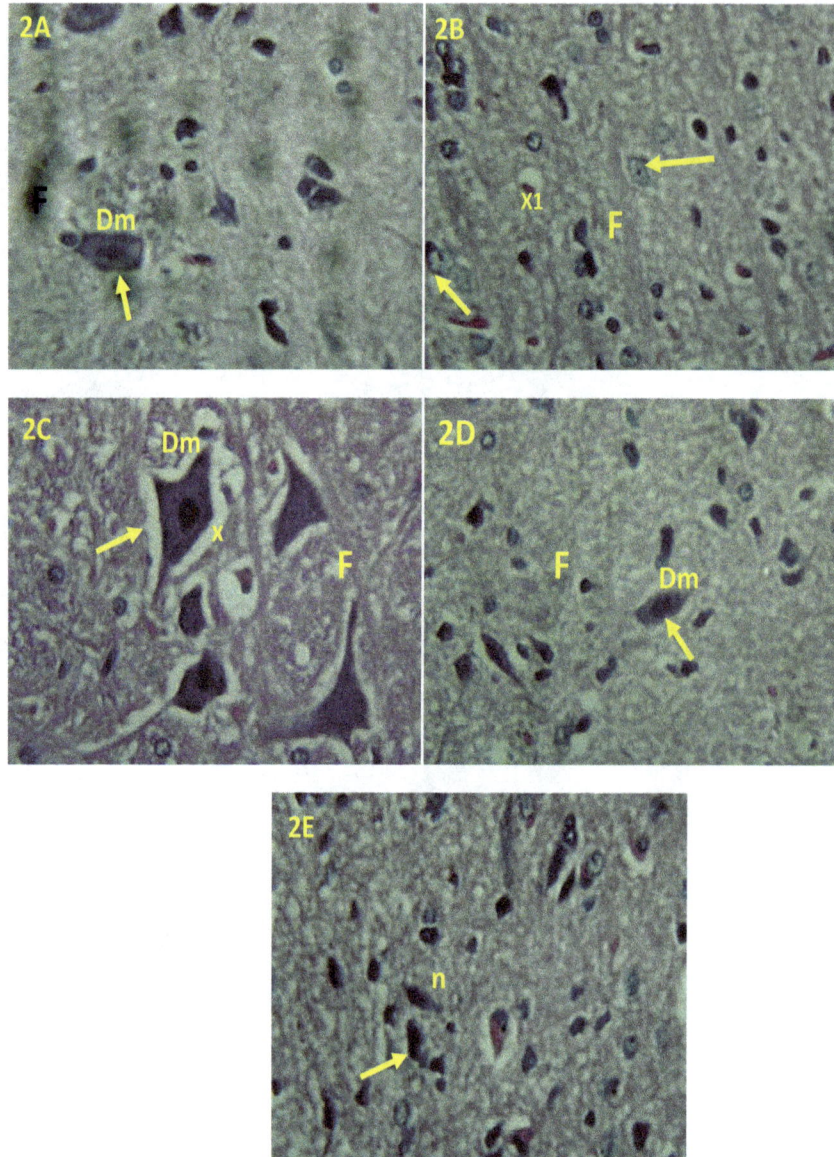

Figure 2A to E. H and E staining for general morphology of the LGB. Dm-degenerating membrane, f- fibrous layer, n- normal cells, x-vacuolar spaces) Arrow head shows thickened membrane measuring about 1.25 µm and giant nuclei of 2.5 µm in diameter (Magnification ×400).

The extreme doses; the high dose treatment (20 mg/kg BW) and the low dose treatment group (2 mg/kg BW) recorded a close range in cell count based on the number of cells characterized by fragmented cytoplasm, loose fibres around cell body and shrunken nuclei.

In the lateral geniculate body (LGB), the high dose treatment group (Figure 2A) showed an increase in cell diameter of about 5 µm compared with the control group ~1.25 µm (Figure 3E), the low dose treatment has a cell diameter close to that observed in the high dose treatment group which is about 4.5 µm, both of which are characterized by a dark stained shrunken nuclei of about 0.5 µm in diameter (Figure 2D). In the treatment Groups 1 and 4 (Figure 2A and D) loss of the fibrous layer was predominant while Groups 2 and 3 (Figure 2B and C) shows orientation of the fibrous layer with Group 2 (F) showing the highest cell density per unit area having degenerated cells (x) interspersed in between cells with normal morphology (x1) (Figure 2B). Group 3 (Figure 2C) shows the largest margin of increase in cell size with cells measuring about 6.25 µm and the nucleus is 1 µm in diameter, the thickness of the axon at the point of origin from the cell body is about 1.25 µm. In the LGB of the treatment group (Figure 2A to E) the predominant changes in the cells is characteristic cell enlargement (highest in Group 3, Figure 2C - 6.25 µm) as some of the

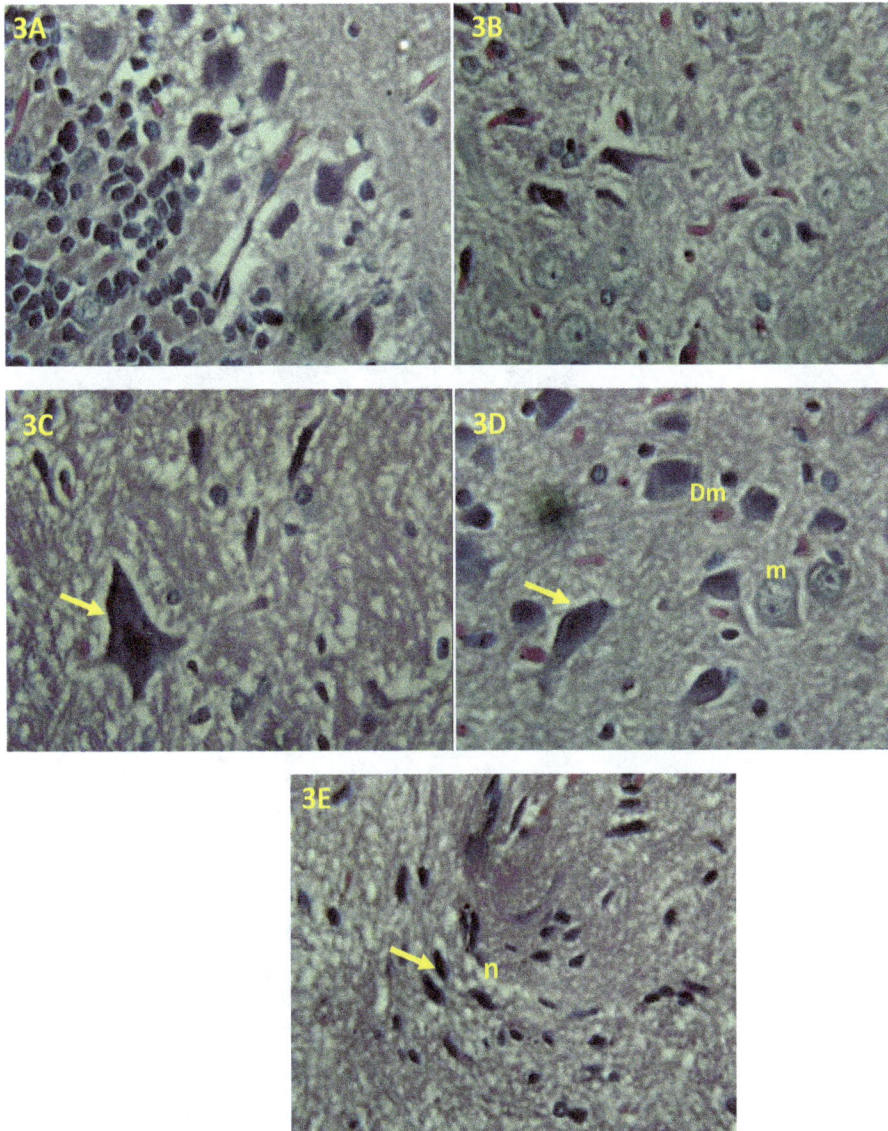

Figure 3A to E. H and E staining for general morphology of the superior colliculus. M-degenerating cells- n- normal cell, Dm- degenerating membrane, f- Fibrous layer, n-Normal cells, x- Vacuolar spaces). Arrow head shows thickened membrane measuring about 1.05 µm and giant nuclei of about 2.0 µm in diameter (Magnification ×400). Haematoxylin and Eosin staining to demonstrate the general morphology of the treatment and control animals after 30 days of treatment (V1- H and E micrograph for the primary cortex, L1- lateral geniculate body and S1-superior colliculus).

cells shows loss of nuclei and thickening of the membrane or rather pronounced increase in cell diameter (Features of necrosis).

The cell distribution per unit area in the superior colliculus (SC) Group 1 (3 cells), 5 cells in Group 4, 6 cells in Group 4 and 12 cells in the control group, with the cells differing in morphology post-treatment, the cells shows fragmented cytoplasm and a pyknotic nuclei, in Group 1 LGB the membrane is highly distorted (y); characterized by thickening and retraction of the axonal projections from adjacent neuronal networks, some points on the membrane appears invaginated. Group 3 (Figure 3C) arrow head shows enlarged cells measuring 6 µm also similar to the enlarged cell size observed in the LGB, also larger than the average cell enlargement observed in normal morphology cells of between 3.5 to 4 µm observed in (Groups 2, 3 and 4; Figure 3B, 3C and 3D), the nuclei are centrally placed and the number of degenerating cells is higher in Group 1, 2 and 4 which is about 3 cells (per 5 cm sq at a magnification of X400). This is a silver iron-method for demonstration of cells with loss of cytoplasmic materials or rather fragmented

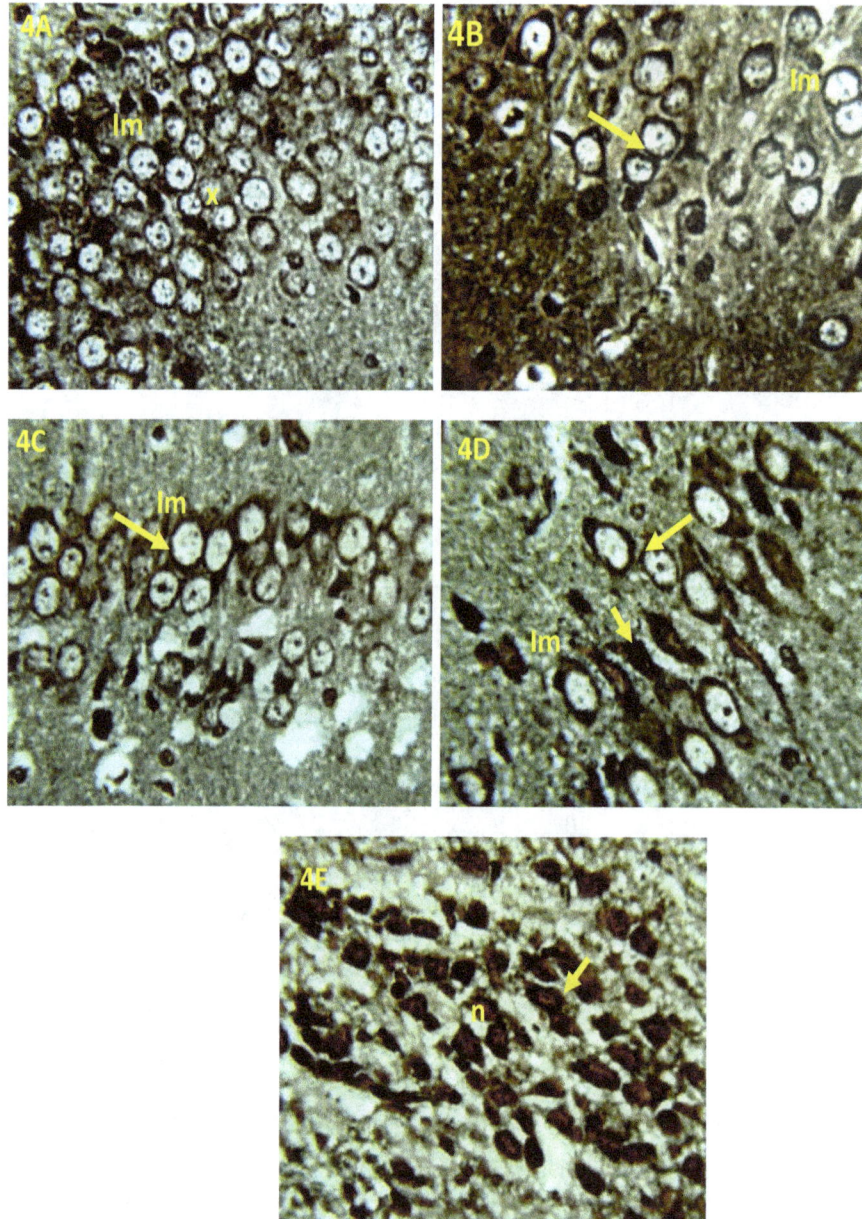

Figure 4A to E. Gordon and sweet's silver-Iron staining of the V1. (Im- intact membrane, x- cytoplasmic degeneration and vacuolar spaces. This is an indication of apoptosis in the V1 of the treatment groups (Figure 4A to D) Magnification ×400.

cytoplasm (x) and degenerating axonal projections from the cells (a), this technique was used to support the observations from H and E for cells with degenerating cytoplasm as well as the distribution of these cells per unit area. The silver-iron method shows membrane integrity of the neurons and presence of reduced nuclei distinguishing it from cytoplasmic fragments observed in H and E and crystal violet, thus confirming the presence of a reduced nuclei in the treatment groups of the V1, in Figure 4A (High dose treatment group), the distribution of cells with unstained (Silver-iron negative), centrally or eccentric nuclei (Silver-iron positive) as well as intact

membrane suggests the cell are apoptotic. The neuronal projections are well defined in the 4th group (Figure 4D); on comparison with the high dose treatment Group 1 (Figure 4A) shows that although both have close cell diameter other structural differences exist between these two categories, which include presence of axonal projections in the low dose treatment group (arrow head Figure 4D) and total absence of such projections on Group 1, also the cell cluster observed in Group 1 was as a result of degenerated axonal projections thus causing the close approximation.

The degenerating fibrous layer of the SC and LGB

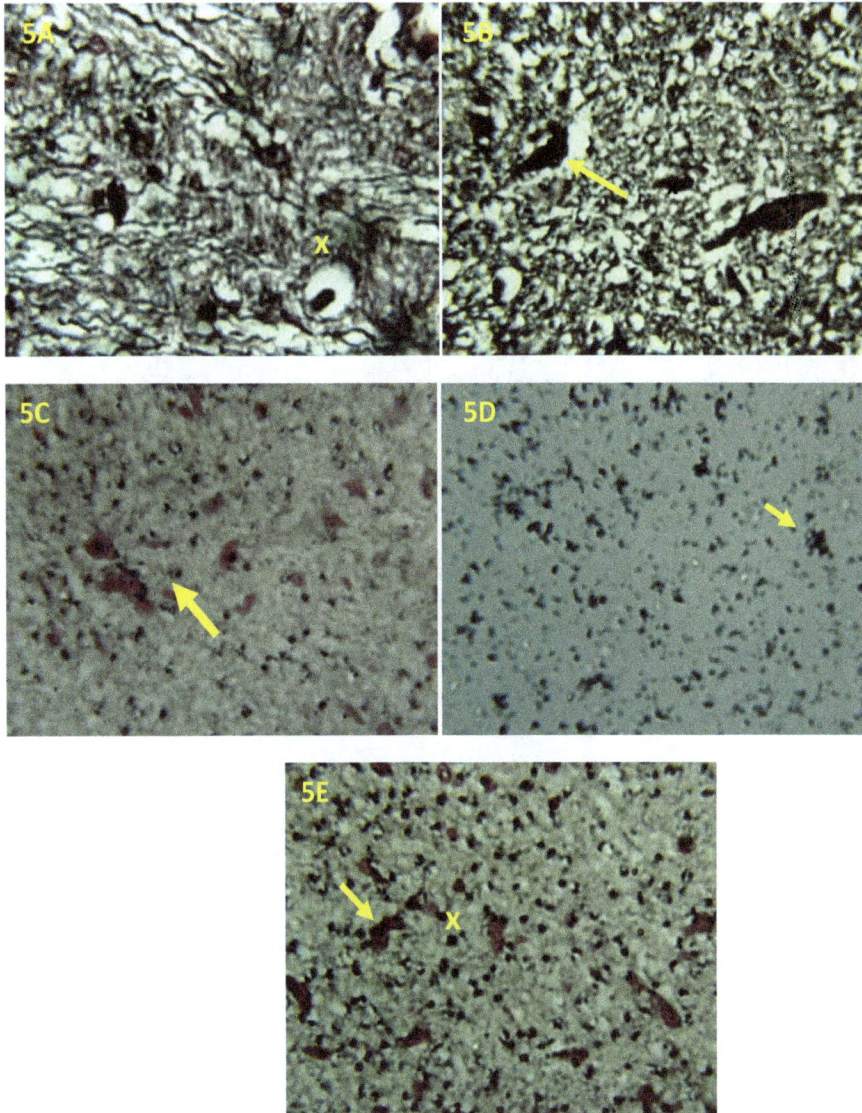

Figure 5A to E. Gordon and sweet's silver-Iron staining of the LGB. Arrow head shows presence of degenerating membrane and no clearly bounded membrane structure, which suggests necrosis as the predominant cell death mode in this LGB (Magnification ×400).

blobbed out to differentiate the fibrous layer (Arrow head) in the SC (Figure 5A to E). The degenerative features observed in the treatment group was absent in the control as the cytoplasmic content was stained rather than blank as seen in the treatment group (Figure 4A to D) compared with Figure 5E. Moreover, the Gordon and sweet's method for demonstration of neurodegeneration (×400) is shown in Figures 4 to 6.

Apoptosis; characterized by shrunken nuclei or pyknotic nuclei, fragmented cytoplasm, intact cell membrane is a predominant feature of cells in the V1 treatment group. Other parameters such as axonal projections, cell diameter and cell count per unit area vary with the treatment dose. Enlarged cell bodies features such as enlarged nuclei (prominent feature of

late stage necrosis), thickening of membrane, prolapsed membrane were the major observation in the LGB and SC also cell loss gave structures with predominant fibrous layers rather than an alternating cellular and fibrous layer in the laminae.

The Sc and LGB are laminated structures with alternating cellular and fibrous layers, it is observed that the fibrous layer is predominates in the treatment group as much as there is a reduction of the cellular layer, although certain number of cells are interspersed in between the fibrous layers some of these cells are characterized by enlarged cell body and nuclei, thickened cell membrane and loss of parts of the cell membrane (x) Figure 5A and B while necrotic sites with degenerating cells were observed in cells distributed in the fibrous

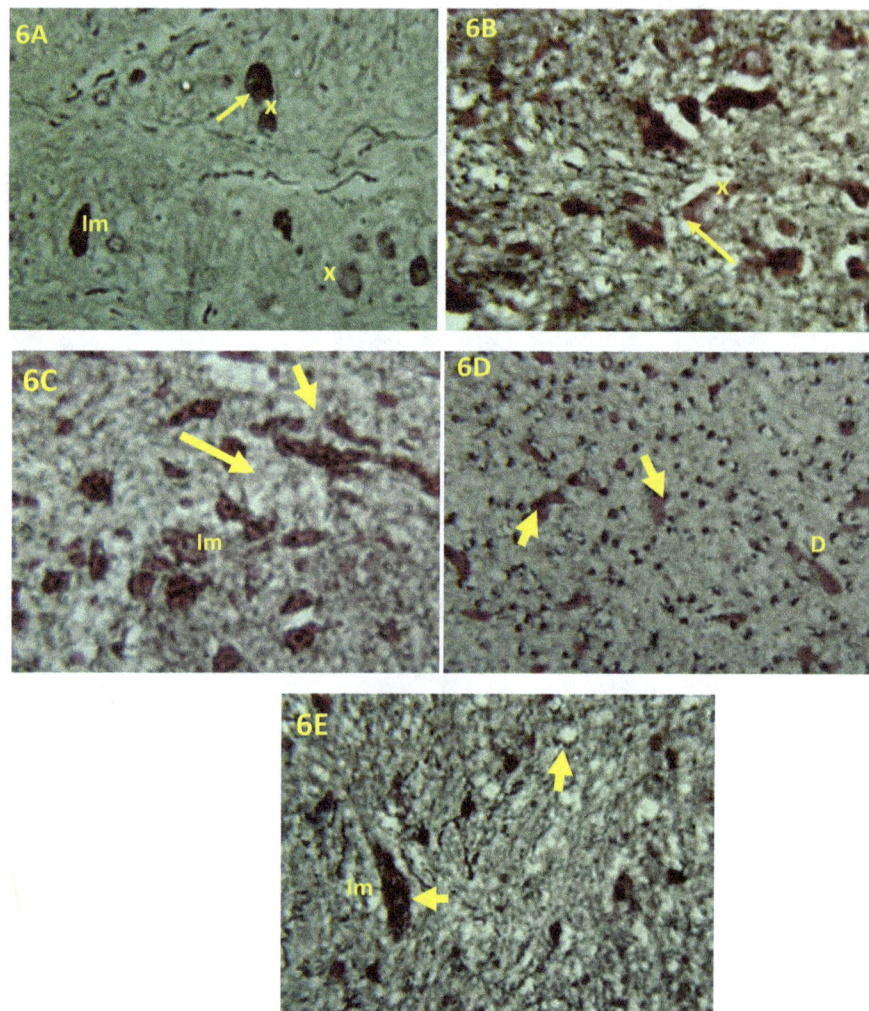

Figure 6 A to E. Gordon and Sweet's silver-Iron staining of the SC. Dm- degenerating membrane, Im- intact membrane. Arrow head shows presence of degenerating membrane and no clearly bounded membrane structure, which suggests necrosis as the predominant cell death mode in this SC although apoptotic features was also observed (Magnification ×400).

layer of the Groups 3 and 4 Sc (Figure 6C and D) arrow head.

Crystal violet

The cells in Group 3 (6 mg/kg BW/day) were observed in all the staining techniques employed to have undergone cell enlargement (Figures 7C, 8C and 9C). The treatment group shows loss of nissl substance in their cytoplasm (Arrow head), loss of nissl substance in the LGB was prominent in Groups 1 and 2 with loss of nissl substance in the cells of the V1 and relatively no loss in the SC (Figure 9A to E) arrow heads shows intense staining for nissl, although cell size is higher in cells with defective nissl substance (3.5 μm) compared to the 2.5 μm in the nissl stained cells as well as in the control group.

LGB: It was observed that nissl substance for enlarged cells were intensely stained with crystal violet (Figure 8A to E). Also, the fibrous layer is more prominent (f) with cells interspersed in the fibrous layer (arrow head Figure 8B and C). Some cells were also found to have apoptotic features such as fragmented cytoplasm, shrinked nuclei (y) in Figure 8C. The co-occurrence of apoptosis and necrosis in certain regions of the LGB and SC was demonstrated in the staining for nissl substance; Figure 9C (High dose treatment Group 1)- presence of necrotic tissue site (x) with prolapsed cell membrane and large eccentric and sometimes centrally placed nuclei with extracavations around degenerating cells. The cells in Group 3 (6 mg/kg BW/day) were observed in all the staining techniques. Crystal violet staining of the sections for V1, LGB and SC (Mag X400) are shown in Figures 7 to 9.

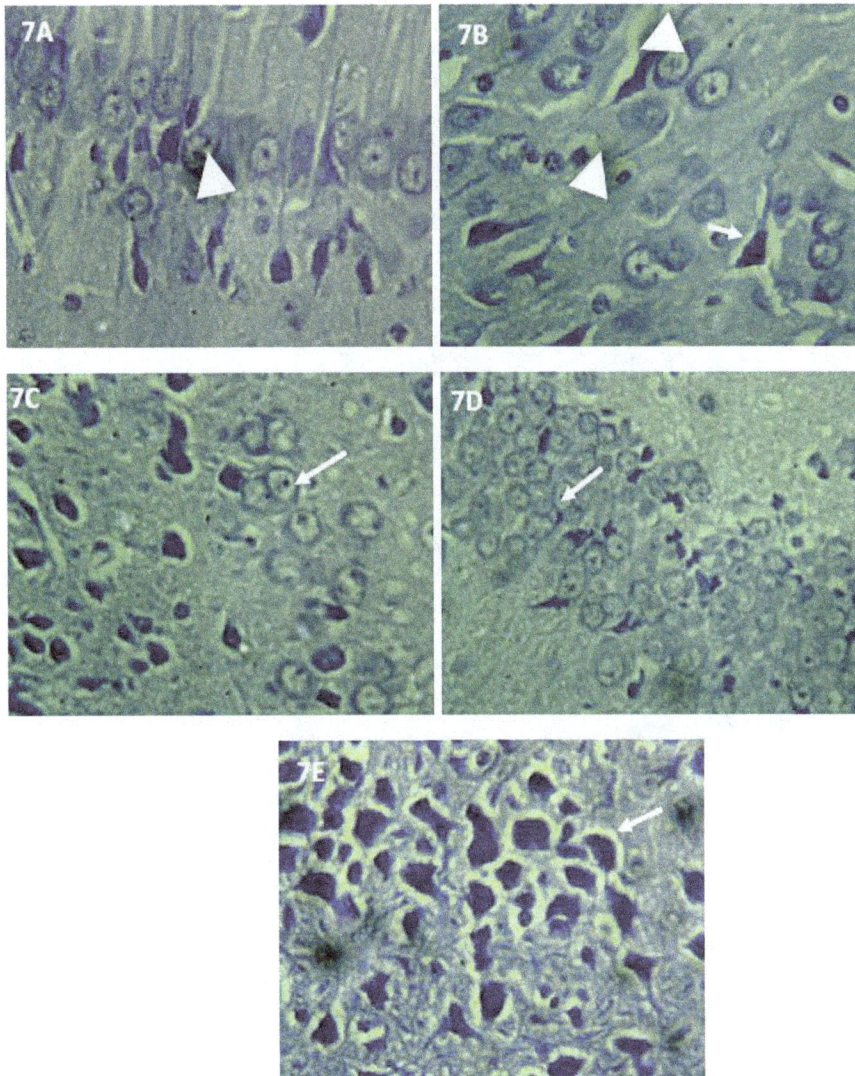

Figure 7A to E. Crystal violet (modified for nissl) for the V1 (Magnification ×400).

In 14 N, arrow heads shows 3 adjacent cells a, b and c, where (a) is characterized by cell enlargement, knobbed axonal projection, (b) shows fragmented cytoplasm, pyknotic nuclei, intact cell membrane and (c) is a degenerated cell presumably in the advanced stage of necrosis characterized by small cell body and large space of about 3 μm around the cell.

We determined the One-sample T test value for the normal cells and abnormal cells with either apoptotic or necrotic feature (Tables 3 and 4).

From graphical illustration the distribution of Φ in the Graph 1 shows that the number of normal looking cells was highest at the 6 mg/kg BW treatment group (Φ-$_{V1/Group\ 3}$) with an equilibrium almost attained in the rate of cellular degeneration at this treatment concentration (Ψ-$_{V1/Group\ 3}$) in Graph 2. The highest normal cell distribution was observed in the control ($\Phi5_{-V1}$) as the least cell degeneration was found in $\Psi5_{-V1}$ also the control group.

Cell death was highest in $\Psi4_{-V1}$ which is the 2 mg/kg BW treatment group with most of the resultant cell death found to be mainly apoptotic on comparing the cell number per unit area. The distribution pattern in Φ_{LGB} was found to be similar to those observed in Φ_{-V1} and Φ_{-SC} with treatment, we found a closely related rate of cell survival in the LGB and SC, and this can also be correlated with the confidence interval values Table 3. Cell survival is highest at $\Phi5_{-SC}$ (control) while $\Phi2_{-SC}$ Table 4 has the highest survival rate among the treatment groups, $\Psi1_{-SC}$ recorded the highest number of degenerating cells per unit area with most of the degeneration found to be regional and characterized with vacuolar spaces around the cells; in the micrographs cell number per unit area was close to those of the control, special training techniques shows that most of this cells have characteristic loss of metachromasia and vacuolar spaces around these degenerating cells (necrosis). In the

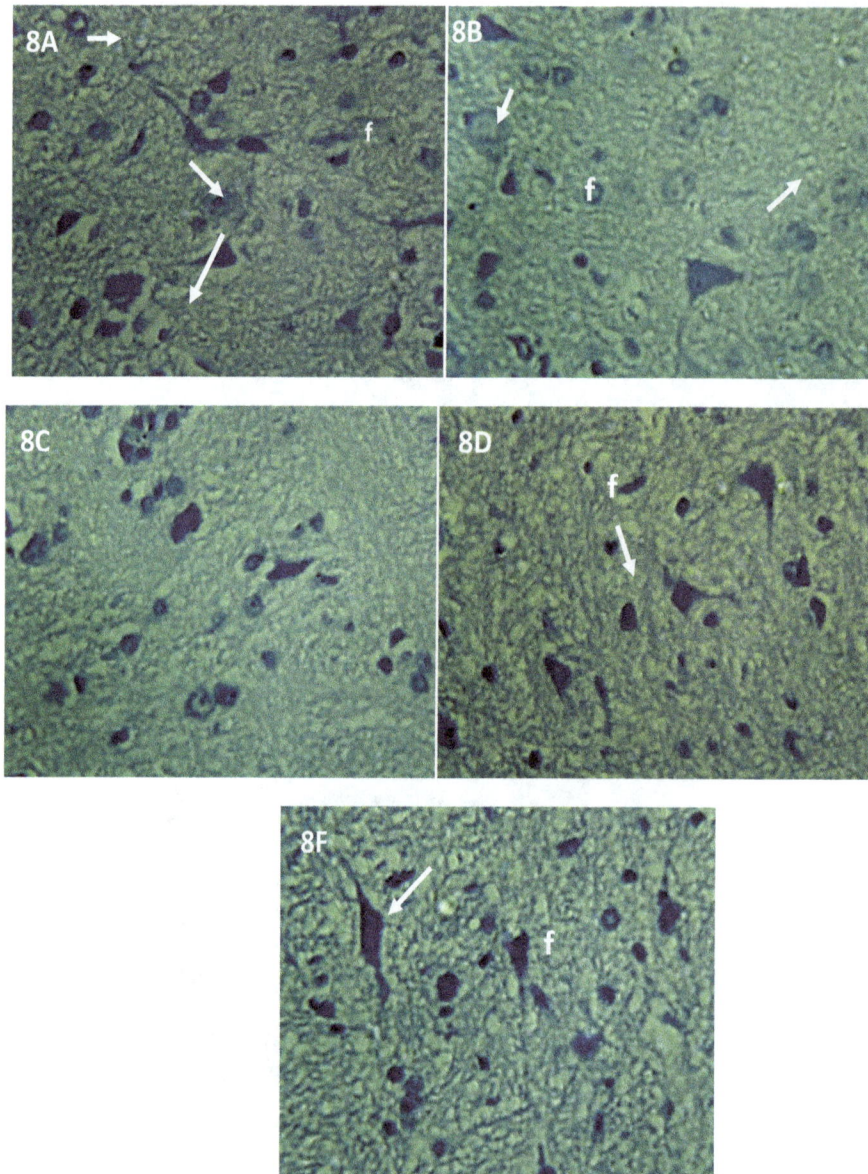

Figure 8A to E. Crystal violet (modified for nissl) for the LGB. f- Fibrous layer. Arrow head shows predomination of the fibrous layer of the LGB (Magnification ×400).

V1 cells, degeneration was conserved to the cytoplasm while the cell sizes were increased considerably; similar changes were observed for the entire treatment group even though the number of cells per unit area differs considerably, with cell number greatly reduced in the high dose treatment group Table 2.

DISCUSSION

The present study uses chemical induction of cell death, to examine the morphology of neurons in the cerebral cortex (molecular layer) of rats, this was aimed at understanding the earlier findings involving the wide distribution of movement disorder and loss of vision in cassava endemic regions of Tanzania, Mozambique, Niger and Nigeria (Osuntokun et al., 1981), for cell death analysis we choose crystal violet staining for fragmented nuclei and nissl (Sharma and Chalam, 2009), Gordon and sweet silver-iron staining for degenerating neurons (Guy et al., 2010), Hematoxylin and eosin to demonstrate cell morphology and immuno staining to demonstrate T-cell infiltration of cell death sites (apoptosis or necrosis). These complimentary and independent methods all demonstrated a large increase in cell death in the V1, LGB and SC of rat models treated with potassium hexacyanoferrate to mimic movement disorders and loss of vision associated cassava endemicity and

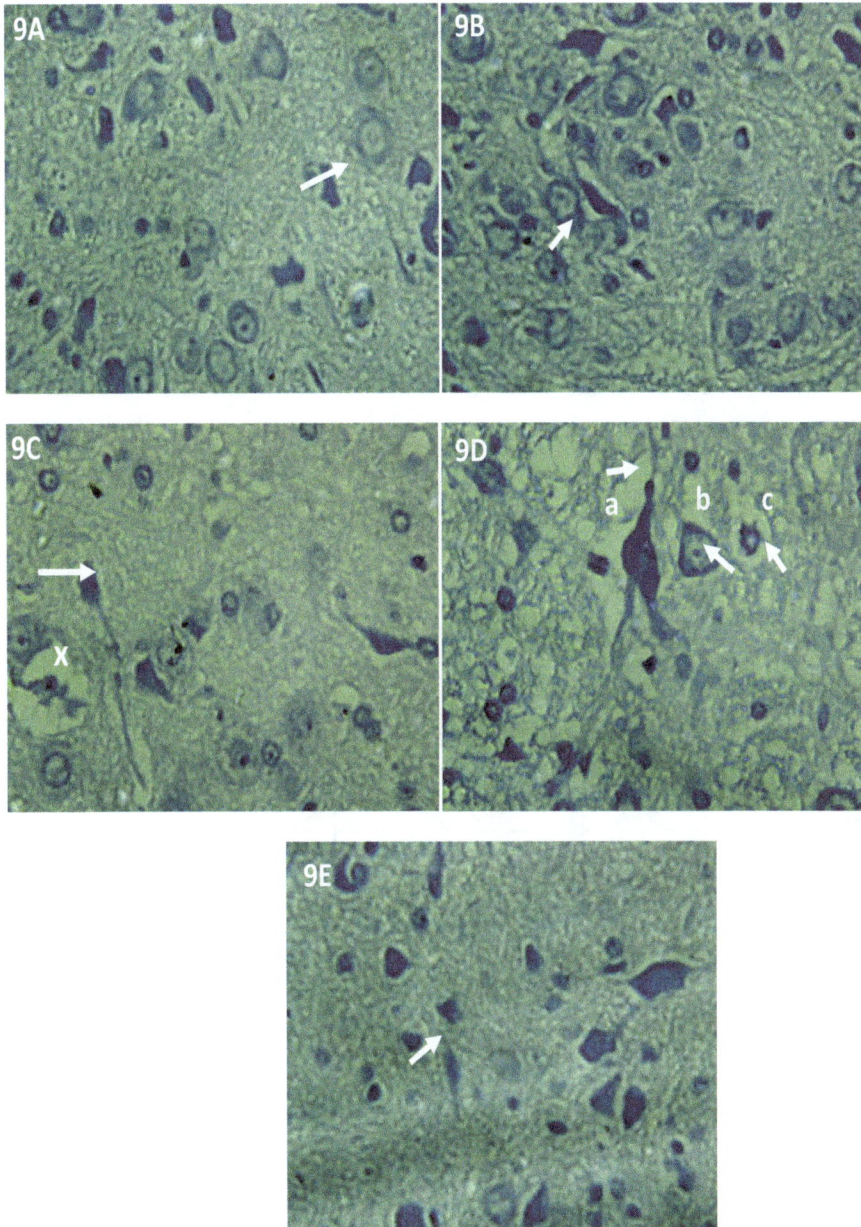

Figure 9A to E. Crystal violet (modified for nissl) for the SC. X- Degenerated cell, a-enlarged cell with prominent projections, b- characteristic apoptosis, c- characteristic necrosis (Magnification ×400).

consumption (note that cassava also contains scopoletin and afflatoxin which are known neurotoxins). Although none of this method is specific for any particular neuron type, the advantages of using these five methods together are as follows: (1) H and E staining reveals cell morphology to include nuclei, nissl and membrane as well as extents of neuronal connections and necrotic sites at higher magnifications; (2) Analysis of pyknotic nuclei using nissl staining allows on the same section both a quantitative measurement of cell death and a direct evaluation of anatomical parameters for example, (the area of the V1); (3) Silver-iron staining method

demonstrates agyrophyllic extracellular materials, degenerating neuronal membrane and loss of cytoplasmic materials.

Cellular composition has impact on how the cell moderates its demise or rather the pattern of cell death adopted by the cells in excitotoxicity (like those observed in cyanide intoxication), Previous studies we conducted on the phosphatase system showed that increased lysosomal activity by a measure of acid phosphatase (ACP) was associated with rapidly degenerating cytoplasm at high magnifications while necrotic sites and disruption of membrane structure was found to correlate

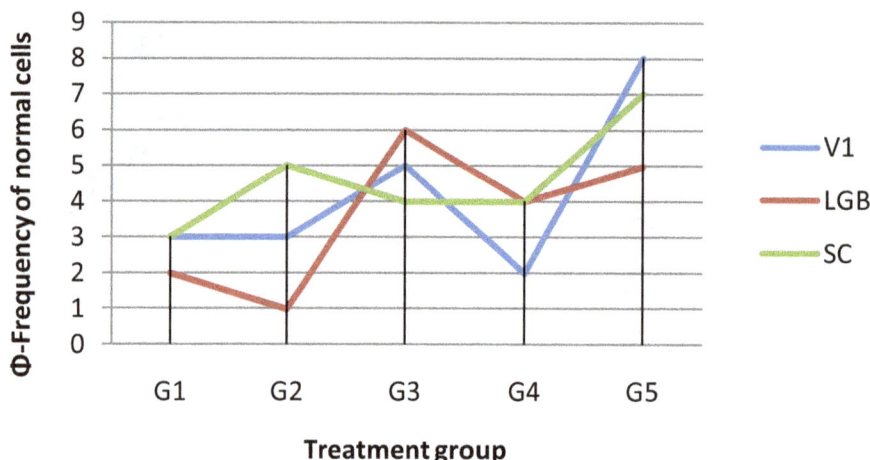

Graph 1. Φ-distribution of normal cells in the treatment groups and the control.

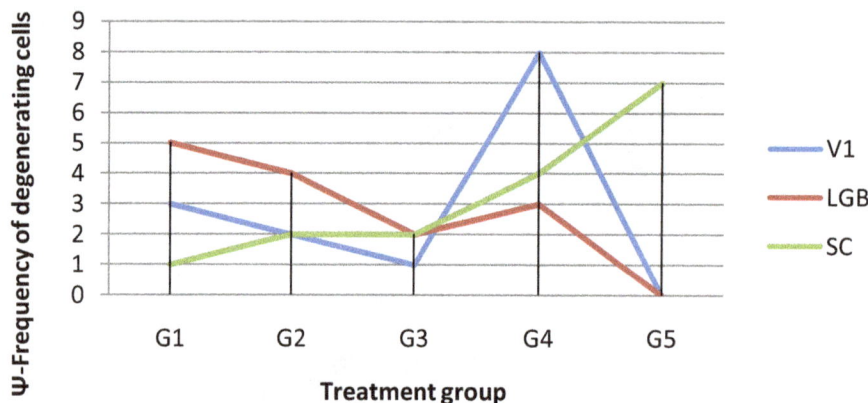

Graph 2. Ψ-Distribution of degenerating cells in the treatment and the control group.

with drastic rise of alkaline phosphatase activity (ALP) at first followed by a steep drop. Other experiments by Jensen et al. (1999) has shown that deletion of NMDA R1 receptors in mice rescues developing cerebellar granule cells, presumably due to excitotoxicity, and a link has been found between cyanide and NMDA receptors; cyanide is capable of potentiating NMDA receptors thus allowing the cell to be prone to excitotoxicity, this could also be attributed to the rather large sizes of the neurons although the neuronal size did not follow a regular pattern. Aside increase in cell death, there was also a reduction in cell number per unit area in the V1, LGB and SC with the highest effect found in V1 for the treatment groups when they were compared with the control, an early increase in size followed a great size reduction in certain cells may be explained by prolonged use "Excitability" followed by peaked oxidative stress, then a loss of cytoskeleton and membrane which might account for the reduced cell size and the presence of necrotic sites around this regions and were observed to have lost

neuronal connections. However, overlap of cell death with synapse loss in necrosis raises important issues as this differ from the observations in apoptosis where neuronal connection was relatively unchanged it is possible that defective membrane observed in necrosis causes defective synaptic connection then increased cell death it will be important to investigate actin nucleation assembly in dendritic branching of wild type (WT) and WT models treated with K-hexacyanoferrate Table 5.

Necrosis is accompanied by inflammatory infiltrate phagocytic cells. DNA degradation, if it occurs, is a late event (Katherine et al., 2001), In the LGB Figure 2C and the SC (Figure 3C) were found to have enlarged cells of about 6.1 µm in diameter and giant nucleus reaching about 1 µm in length, It was also observed that the adopted mode of cell death is dose dependent such as that which was observed in the treatment Group 3 (6 mg/kg BW) was found to have characteristics of necrotic cells in the LGB and SC (Figure 2C and 3C), the high dose treatment (20 mg/kg BW) and the low dose

Table 2. Cell count.

Group	V1		LGB		SC	
	Φ v	Ψ v	Φ L	Ψ L	Φ s	Ψ s
1	3	3	2	5	3	1
2	3	2	1	4	5	2
3	5	1	6	2	4	2
4	2	8	4	3	4	4
5	8	0	5	0	7	0

Ψ represents the number of cells having a normal appearance, Φ represents the number of cells having necrotic appearance, which includes vacuolar spaces around the cell, loss of neuronal connection and damaged membrane while features of apoptosis (Pyknotic nuclear, reduced Metachromasia of the membrane and relatively intact cell membrane on comparison with the normal.

Table 3. Descriptive statistics for the number of normal cells observed.

Parameter	N	Minimum	Maximum	Mean	Std. Deviation
V1	5	2.00	8.00	4.2000	2.38747
LGB	5	1.00	6.00	3.6000	2.07364
SC	5	3.00	7.00	4.6000	1.51658
Valid N (listwise)	5				

Table 4. Descriptive statistics for number of cells found undergoing either apoptosis or necrosis

Parameter	N	Minimum	Maximum	Mean	Std. Deviation
V2	5	0.00	8.00	2.8000	3.11448
LG2	5	0.00	5.00	2.8000	1.92354
SC2	5	0.00	4.00	1.8000	1.48324
Valid N (listwise)	5				

V1- normal cells in the visual cortex, while V2 represents damaged cells found to have one or more features of neurodegeneration, LG2- cells undergoing degeneration in the lateral geniculate body and SC2- degenerating neurons of the superior colliculus.

Table 5. Table showing distribution of mode of cell death in the V1, LGB and SC.

Parameter	Group 1	Group 2	Group 3	Group 4	Group 5
V1	Apoptosis	Apoptosis with partial necrosis	Apoptosis	Apoptosis	Normal
LGB	Necrosis	Necrosis/Apoptosis	Necrosis	Necrosis	Normal
SC	Necrosis	Necrosis/Apoptosis	Necrosis	Apoptosis/Necrosis	Normal

treatment (2 mg/kg BW) were seen at certain instances to elicit similar changes in cell morphology, in the V1 (1A and 1D) cell diameter was 5.0 and 5.1 µm, respectively with both having features like fragmented cytoplasm, loose fibres around the cell body, centrally placed shrunken nuclei and intact cell membrane (apoptosis) also the cell count per unit area was 8 and 9 respectively Table 2, another finding to support this hypothesis is in the LGB where the cell diameter in Group 1 is 4.4 µm in Group 1 and 4.5 µm in Group 4 with both having enlarged nuclei measuring 0.5 µm, as a matter of general observation the fibrous layer was predominant in LGB of both groups. The question about the similarity in features induced at these two treatment dosages is that, is the rate of progression of these features at the same rate?, the answer to this was indirectly invested by the DAB immunostaining for light chains immunoglobin turned out to show that even though generally morphology suggests similar changes, IgG studies show that both are in different stages of the cell change, while immuno studies revealed it. The loose fibres counterstained around the cell bodies of the Group 4 treatment group are entirely absent in Group 1, thus implying loss of counter stain of axonal projections. Although, these projections were observed in H and E, crystal violet (CV) and Gordon and sweet's method, it is observed that protein turnover required to detect this projections were relatively absent in Group 1 at this stage but still present in Group 4, likewise the intensity of the DAB-PAP silvers intensification was found to be of higher intensity in the Group 1 with most of the activity covering the membrane and cytoplasmic area and lower intensity in the Group 4 with most of the activity restricted to the membrane area, the predominating fibrous layers in the LGB and SC of Groups 3 and 4 was shown by intense brown deposits of fibers interlaced around the cells. Aside dose dependence in adopted mode of cell death it also varies for different parts of the brain with apoptosis predominant in the visual cortex and necrosis in the LGB and SC although certain level of occurrence of apoptosis may be found in the LGB and SC. This could explain the variation in region specific cytotoxic pathways observed in the different parts of the brain.

Conclusions

Cyanogenic neurotoxicity involves chemical induction of cell death; the adopted mode of cell death was found to be dose dependent in a particular region of the visual relay centre, and also will vary for the various V1, LGB and SC for the same treatment dose, this however, supports the fact that the different regions of the visual cortex has different cytotoxic pathway and the mode of cell death will depend on how the region respond to the assault. The predominant mode of cell death in the V1 is apoptosis while necrotic tissue sites were found to exist in the SC of the same treatment, certain cells in the LGB showed morphological features of both apoptosis and necrosis. The extreme doses gave similar events in terms of cell morphology and count per unit area.

ACKNOWLEDGEMENTS

The authors are grateful to the West African Health Organization; Trinitron Biotech Ltd; University of Ilorin; Departments of Anatomy and Histopathology, National Hospital, Abuja, Nigeria; and to Pharm. Mike Omotosho.

REFERENCES

Almer M, Van der S, Nelson L (2006). Acute cyanide toxicity: mechanisms and manifestations. J. Emerg. Nurs., Aug; 32(4 Suppl): S8-11.

Behar TN, Scott CA, Greene CL, Wen X, Smith SY, Maric D, Liu QY, Colton CA, Barker JL (1999). Glutamate acting at NMDA receptors stimulates embryonic cortical neuronal migration. J. Neurosci., 19: 4449-4461.

Bhattacharya R, Tulsawani R (2008). In vitro and in vivo evaluation of various carbonyl compounds against cyanide toxicity with particular reference to alpha-ketoglutaric acid. Drug Chem. Toxicol., 31(1): 149-61.

Bhattacharya R, Lakshmana RPV (2001). Pharmacological interventions of cyanide-induced cytotoxicity and DNA damage in isolated rat thymocytes and their protective efficacy in vivo. Toxicol. Lett., 119: 59-70.

Bonfoco E, Krainc D, Ankarcrona M, Nicotera P, Lipton SA (1995). Apoptosis and necrosis: Two distinct events induced respectively by mild and intense insults with NMDA or nitric oxide/superoxide in control cell cultures. Proc. Natl. Acad. Sci. USA., 92: 7162-7166.

de Haro L (2009). Disulfiram-like syndrome after hydrogen cyanamide professional skin exposure: two case reports in France. J. Agromedicine, 14(3): 382-384.

de la Cruz Cosme C, Medialdea Natera P, Romero Acebal M (2009). Amyotrophic lateral sclerosis syndrome-plus and consumption of cassava (Manihot). Is this a new presentation of the neurotoxic motor-neuron syndrome? Neurologia. 2009 Jun, 24(5): 342-343.

Gamper N, Li Y, Shapiro MS (May 18 Epub 2005). Structural requirements for differential sensitivity of KCNQ K$^+$ channels to modulation by Ca^{2+}/calmodulin. Mol. Biol. Cell, 8: 3538-51.

Gilman S, Koeppe RA, Nan B, Wang CN, Wang X, Junck L, Chervin RD, Consens F, Bhaumik A (2010). Cerebral cortical and subcortical cholinergic deficits in Parkinsonian syndromes. Neurology. 2010 May 4, 74(18): 1416-1423.

Guy S, Catherine T, Xiao-Qing T, Sara W, Feng Z, Xiaodong L, Poonit K, Goran P, Tanya D, Antoniette J, Paul B, Nicole B, Grant ED, Mark Z, Donald SK (2010). Positive allosteric modulators of the human sweet taste receptor enhance sweet taste PNAS 2010, 107(10): 4746-4751.

Guy S, Catherine T, Xiao-Qing T, Sara W, Feng Z, Xiaodong L, Poonit K, Goran P, Tanya D, Antoniette J, Paul B, Nicole B, Grant ED, Mark Z, Donald SK (2010). Positive allosteric modulators of the human sweet taste receptor enhance sweet taste PNAS 2010, 107(10): 4746-4751.

Isom GE, Gunasekar PG, Borowitz JL (1999). Cyanide and neurode-generative disease. In Chemicals and Neurodegenerative Disease (Bondy SC, Ed.), Prominent Press, Scottsdale, AZ, pp. 101-129.

Jensen P, Surmeier DJ, Goldowitz D (1999). Rescue of cerebellar granule cells from death in weaver NR1 double mutants. J. Neurosci., 19: 7991-7998.

Katherine L, Barrett JM, Willingham A, Julian G, Mark CW (2001). Advances in cytochemical apoptosis. J. Histochem. Cytochem., 49: 821-832.

Lee B, Park J, Kwon S, Park MW, Oh SM, Yeom MJ, Shim I, Lee HJ, Hahm DH (2010). Effect of wild ginseng on scopolamine-induced acetylcholine depletion in the rat hippocampus. J. Pharm. Pharmacol., 62(2): 263-271.

Li R, Sonik A, Stindl, R, Rasnick D, Duesberg P (2000). Aneuploidy vs. gene mutation hypothesis of cancer: Recent study claims mutation but is found to support aneuploidy. Proc. Natl. Acad. Sci. USA, 97: 3236-3241.

Mathangi DC, Namasivayam A (2000). Neurochemical and behavioural correlates in cassava-induced neurotoxicity in rats. Neurotox. Res., 2(1): 29-35.

Osuntokun BO (1981). Cassava diet, chronic cyanide intoxication and neuropathy in Nigerian Africans. World Rev. Nutr. Diet, 36: 141-173.

Pearse AGE (1960). Histochemistry. Little, Brown and Company. Boston 1960.

Pedro RN, Thekke-Adiyat T, Goel R, Shenoi M, Slaton J, Schmechel S, Bischof J, Anderson JK (2010). Use of Tumor Necrosis Factor-Alpha-coated Gold Nanoparticles to Enhance Radiofrequency Ablation in a Translational Model of Renal Tumors. Urology. 2010 May 6 (Ahead of Print).

Prabhakaran K, Li L, Zhang L, Borowitz JL, Isom GE (2007). Upregula-tion of BNIP3 and translocation to mitochondria mediates cyanide-induced apoptosis in cortical cells. Neuroscience, 150(1): 159-167.

Quintavalle C, Incoronato M, Puca L, Acunzo M, Zanca C, Romano G, Garofalo M, Iaboni M, Croce CM, Condorelli (2010). G. activity. Cell Death Differ., 2010 May 28.

Sharma RK, Chalam KV (2009). In vitro evaluation of bevacizumab toxicity on a retinal ganglion cell line. Acta Ophthalmol., 87(6): 618-22.

Soler-Martín C, Riera J, Seoane A, Cutillas B, Ambrosio S, Boadas-Vaello P, Llorens J (2010). The targets of acetone cyanohydrin neurotoxicity in the rat are not the ones expected in an animal model of konzo. Neurotoxicol. Teratol., 32(2): 289-294.

Svendsen P, Hau J (eds.) (1994). Handbook of Laboratory Animal Science, volume I, Selection and handling of animals in biomedicalresearch. CRC Press.

Um YJ, Jung UW, Kim CS, Bak EJ, Cha JH, Yoo YJ, Choi SH (2010). The influence of diabetes mellitus on periodontal tissues: A pilot study. J. Periodontal Implant Sci., 40(2): 49-55.

Van Zutphen LFM, Baumans V, Beynen AC (1994). Principles of Laboratory Animal Science. Elsevier.

Varone JC, Warren TN, Jutras K, Molis J, Dorsey J (2008). Report of the investigation committee into the cyanide poisonings of Providence firefighters. New Solut., 18(1): 87-101.

Varone JC, Warren TN, Jutras K, Molis J, Dorsey J (2008). Report of the investigation committee into the cyanide poisonings of Providence firefighters. New Solut., 18(1): 87-101.

Vicente T, Eva S, Margaret MM, Robbert HC, Afshin S, Luis S, Wim JQ (2006). Designed tumor necrosis factor-related apoptosis-inducing ligand variants initiating apoptosis exclusively via the DR5 receptor. Proc. Natl. Acad. Sci., 103(23): 8634-8639.

15

Raising community awareness about Zoonotic diseases with special reference to rift valley fever, the roles of professionals and media

May M. EL Rehima[1], Atif E. Abdelgadir[2]* and Khitmat H. ELMalik[2]

[1]Ministry of Agriculture and Animal Resources, Khartoum State, Sudan.
[2]Department of Preventive Medicine and Veterinary Public Health, Faculty of Veterinary Medicine, University of Khartoum, Sudan.

This study was carried out to through light on the factors that affect circulation of fact statement reported and listed in media. Media plays an important role in spread of information, which formulate knowledge and attitude of communities. The study also focuses light on factors that determine the roles and efficiency of veterinarian and physicians as professionals in handling problems, and study targeted people that have relationship with animal owners and methods used to deal with health problems. The recent outbreak of Rift valley fever (RVF) was taken as a case study. A questionnaire was designed and distributed in different locations in Khartoum, White Nile, Sennar states. Questions were mainly focus in how the disease is reported, controlled and prevented. Knowledge sources and development of professional carrier were examined. Another questionnaire was distributed to farmers. Both urban and rural communities were included to estimate the main factors that affect disease control and prevention. A comprehensive review of articles that were published in the daily newspapers about the issue was made. The analysis of the questionnaire was done using simple statistically methods of percentage and frequency. Also, a similar analysis was made for newspapers materials. Most veterinarians have good idea about zoonosis especially RVF reported from the field and mentioned widely in media. This reflected adequate knowledge in all diseases. However, veterinarians know much about zoonosis, but contribute little to the field of raisy awareness about the disease. Focusing light on roles and responsibilities of professionals of Ministry of Health, Ministry of Animal Resources and Press to deal with problems. Showed that: The linkage between them, responsibilities for extension, and disease reporting were not clear. The role of individual who have relationship with animals is necessary to avoid epidemics and zoonotic diseases need to have a defined body , with media representative as partners to reduce the hazard of dissemination of wrong information which leads to a state of un necessary horror among the communities. Observing professional ethics by both professionals and journalists should be a safeguard against spread of wrong in formations scare among the communities.

Key words: Zoonotic disease, professionals, media.

INTRODUCTION

Rift valley fever (RVF) is peracuty or acute zoonotic disease of domestic ruminant in Africa. It is caused by a single serotype of a mosquito–borne bunyavirus of the genus phelovirus. The disease occurs in climatic conditions favoring the breeding of mosquito vectors and is characterized by liver damage. The disease is most sever in sheep, goats and cattle, in which it produces abortion in pregnant animals and a high mortality rate in the new born. Older non pregnant animals, although, susceptible to infection, are more resistant to clinical

*Corresponding author. E-mail: atifvet@yahoo.com.

disease (Esia and Obied, 1977a; Esia et al., 1980). Humans are susceptible to infection by handling infected materials and through transmission by mosquito vectors. It Infect humans by handling infected material and through transmission by mosquito vectors. An epizootic of the disease was also reported in Kosti district, the white Nile province of the Sudan in 1973, it covered almost every single locality in the district and spread to Blue Nile province, sheep, goat, cattle and human were involved (Esia and Obied, 1977a). Mortality rates were highest in lambs followed by goat kids and calves (Eisa and obied, 1977a). The virus isolated and identified as RVF using serum neutralization test (SNT) (Esia and obied, 1977b). Esia (1984) surveyed domestic animals of the Sudan for precipitating IgG antibodies to RVF antigens .The prevalence of the infection was 34.3% in sheep, 33.2% in cattle, 22% in goat, 7.9% in camels and 4% in donkeys. Antibodies to RVF were not detected in sera from horses.

A recent serological survey was conducted in-patient admitted to Hag Elsafi Hospital, Khartoum, Sudan for detection of antibodies to RVF. Approximately 3% of the patients were suspected to be infected with RVF, as determined by detection of IgM antibodies (Kambal, 1997).

Infection of human by vectors is a striking feature in countries with a relatively small population of animal hosts. In such areas, RVF may be recognized first in humans. It has caused serious disease in laboratory workers and must be handled with high level biosecurity (FAO, 2003).

Media particularly daily newspapers, are wide spread source of information to the public. The degree of authentic correct information depends on the source information and authority of this information. This study was planned with the general objectives to investigate the degree of precision in spreading information in newspaper and the possible factors that affect this precision from the professional sources. The reflection and implications were measured in the livestock owner's knowledge, attitude and practice. The specific objectives of the study are:

1. To review the history, geographic distribution and socio-economic impact RVF,
2. To analyze the information about the rift valley fever reported and listed in the news paper and media with the intention to determine the rate of true facts statement,
3. To assess the scientific knowledge of the authors that write this information,
4. To assess the professional knowledge of veterinarian and physicians handling the problem,
5. To analyze the knowledge of livestock farmers with regards to animals disease identification and prevention
6. To suggest a model of defining the roles of the different partners in delivery of information.

MATERIALS AND METHODS

The study was designed to identify the deficiency in epidemic disease Rift Valley fever disease reporting and information dissemination systems, taking RVF as a case study. The risk that can be encountered from wrong diagnosis, management, and control because of lack of skills. Shortage in knowledge and co-ordination of efforts to compact outbreaks were also examined. The opportunities for identification of gaps in practice are to be identified through a survey of available knowledge.

Study design

Two methods were applied:
1. An active survey of knowledge attitudes and practice by all concerned parties including veterinary and medical professionals and the animal owners.
2. Secondary data collection and analysis from published materials in newspapers about the issue under consideration. This is detailed as follows:

A questionnaire was designed to cover the following categories

1. Veterinarians
2. Livestock farmers
3. Physicians

These categories of people are more subject to the risk of RVF. The foregoing-mentioned three questionnaires were used to collect data to be analyzed. One was administered to veterinarian, who help animals during birth or perform inspection at slaughter or postmortem examination .The focus of the study was on the knowledge, how to diagnose prevent, and control the disease, and acquire and keep enough updated information. A convential sample composed of 40 veterinarians was included in the questionnaire; the sample was composed of both males, and females. The second questionnaire was for physicians on how they contact cases in hospitals, and their awareness of are health care to be taken to estimate the knowledge of doctors about the zoonotic disease , and their risk, diagnosis and treatment .the second questionnaire included (30) study samples. The final questionnaire was designed for individuals who are owners animals and are at risk of RVF, due to their contact frequent with animals. Important habits were included in the questionnaire e.g. drinking milk and the practice of heat treatment, handling raw meat. The third questionnaire includes (50) random sample.

Retrospective study of published information in daily newspapers

All publish material was reviewed and categorized by:

1. Daily paper
2. Authors
3. Information: Included source of information and validity.

To assess the scientific validity of information disseminated to the public.

Data analysis

To assess the contribution of each of the mentioned stakeholders, it is necessary to assess the degree of knowledge and ability to transfer it to communities. Microsoft Excel 2003 was used for data analysis. Descriptive statistic (frequency and percentage analysis system) was used for variables. Results are presented in tables, while Bar and Pie charts were used for some variable.

RESULTS

The veterinarian's general knowledge

Analysis of questionnaires administered to veterinarians showed that their knowledge about disease is good as they were able to list and rank the important infectious disease affecting live stock (Table 1). They were able to differentiate zoonotic diseases from others (Table 2). Theirs knowledge was also confirmed by listing diseases of young animals or disease affecting more than one species (Table 3). The responsibilities to deal with RVF, the disease under study ,was not clear as it varied from different ministries, departments and responsible persons (Table 4). The responsibilities in disease reporting and handling sample collection (Table 5) showed that most vets (27.78%) are involved in disease reporting, the other responsibilities are less dealt with the role of vets in awareness raising, is however, very limited (5.56).

Figures 1 shows the knowledge of veterinarians about disease that cause abortion, brucellosis and RVF were the highly mentioned diseases. However, some diseases mentioned were not among those which affect mammals primary (AI), others may incidentally cause abortions but it is not pathognomonic (Rabies, TB, Anthrax, Hydatosis). The sources of information used by used by vets were mainly the internet and scientific journal, while media was the last used (Figure 2). The constrains faced were lack of libraries and internet connections, economic reasons ranks last.

Constrains facing veterinarians accessing information source

Source of knowledge

About 30.56% of veterinarians get their information from meeting and conferences, and 55.56% did not go to conference. While 27.78% went to internal conference, and 11.11% went to external. On the other hand, it was found that about 50% read material related to veterinary problems and 38.89% did not. 100% of veterinarians have knowledge related to zoonosis.

The owners

The owners who answered the questions have health problem in their herd, which was classified as: 15.55% having some disease, 8.88% have inflammations, and

31.11% have no health problem (and 44.46% not respondent). They reported that veterinarians are responsible of 26.66% of these problems and the government is responsible for about 11.11% of this problem, and about 6.66% have no body to care for their health problem of their herds (and 5.57% not respondent). The ways of meat inspection is 26.66% depended on veterinarians and 8.88% depended on slaughterhouse-inspected meat, and about 31.11% do not know. In addition, 31.11% of owners have a role in slaughtering, and 51.11% have no role. 80% of those owners eat meat, and 17.77% of them did not. Ways of drinking milk include: about 70.33% of them drink boiled milk, and 17.44% drink raw milk, and 10.55% drink fermented milk and 2% use milk in other ways (cheese – cooked). 15.55% of herd Gynecology care is the responsibility of owners and trader, 11.11% of pastoralist, and 20% of veterinarians (53.34% not respondent). The help in birth of animals depend on (31.11%) owner and trader, (40%) on pastoralist, and 13.33% on veterinarians. About 84.44% of owners add new animal to stock, and 17.77% did not. About 28.88% have mixed species herds, and 66.66% did not. There are no wild animal in 95 .55% of the stock.

The livestock owners' knowledge

Table 6 shows the knowledge of owners about the source of disease spread; it is a reflection of their understanding of extension services available to them. The owner's perception for control of vectors was seen as use of insecticides in the first place, then the role of vets as being responsible to limit the spread of vector came second. The responsibilities of milking was taken by milkers (30%), 20% did not practice milking; some owners and herders boys may practice milking (Figure 3). Where milk is not an important product, the role is not clearly assigned to certain people.

Vaccination

93.33% vaccinated the animals, and 8.88% did not. Component of the flocks was: about 62.22% had one species, and 40% had different species .The pasture was 60% dry and 2.22% rainy. The stock is 5.55% staying in one place, and 11.11% did not. The death in animals: 24.44% say yes, and 62.22% say no. How they diagnose the disease: About 84.44% of owners diagnose by clinical signs, and 35.55% consult to veterinarians.

The physicians

80% of these physicians are journalist, while 15% of them are specialist. How they report the zoonotic disease:

Table 1. Veterinarians ranking of the most important infectious animals disease.

Disease	Percentage of respondent (%)	No of respondents
CBPP	77.78	28
FMD	69.44	25
Brucellosis	58.33	21
HS	52.78	19
Rift valley fever	50	18
RP	44.44	16
Anthrax	27.78	10
CCPP	25	9
Rabies	22.22	8
Sheep pox	22 .22	8
PPR	16.67	6
BQ	16.67	6
LSD	8.33	3
TB	2.78	1
Blood parasite	2.78	1
Internal parasite	2.78	1

Table 2. Veterinarians list for the most important zoonotic disease.

Disease	Percentage (%)	No of respondents
Rabies	72.22	26
Brucellosis	94.44	34
RFV	86.11	31
Avian influenza	52.78	19
T B	38.89	14
Anthrax	25	9
Toxoplasmosis	16.67	6
Hydatosis	8.33	3
Tape worm	5.56	2
Yellow fever	2.78	1

About 55% said there are institutions responsible for zoonotic disease, and 45% said no. 5% said the responsibility is shared between institutions, 45% said the Ministry of Health is responsible and 56% did not.

The physician's knowledge

Table 7 shows medical doctors knowledge about zoonosis to be reported, however, medical doctor knowledge about important zoonosis is listed in Table 8. Increase of awareness and treatment of infected animals is the most important methods of control (Table 9). As they mentioned by physicians. Physician's suspension and confirmation of zoonosis is not a direct process .The only suspect in 10% of attempt when contact with animals was mentioned. History is not considered seriously (2%). (Table 10). Table 11 also confirms this finding where physicians gave consumption of raw food as the main method of zoonosis transmission, followed by insect bite and discharge from animals (20% each). Contact with animals and raw milk were the least suspected (10% each).

Knowledge related to zoonosis

90% say yes, no one say no. Relationship with veterinary: about 40% say yes, and 45% say no and 65% say there is no share role contained in the disease, and 40% say yes. 60% of doctors have knowledge, and 40% have not. Source of this knowledge: About 25% read scientific books only, and 70% junoral. 75% have challenge to get information, and 15% have not, this challenges are: No library (15%), no internet (5%), economic reasons (15%) and no sufficient time (55%).

Table 3. Veterinarians knowledge on diseases that affect young animals and or/more than one species.

Disease	Effect on young animal	Effect on multiple species
RVF	+	+
PPR	+	-
FMD	+	+
Diarrhea	+	+
Rabies	-	+
Brucellosis	-	+
Anthrax	-	+
Pox	-	+

+ positive answer.- Not applicable.

Table 4. Veterinarian awareness about reporting and Institution responsible for control of RVF disease.

Responsible department	Percentage of positive respondents (%)	No of positive respondents
Animal health and epizootic control Dept	22.22	10
Veterinary unit /hospital	16.67	6
Direct superior	11.11	4
Organization NGOS	16.67	6
Central veterinary research laboratory	5.56	2
Ministry of health	36.11	13
Owner / pastoralists / others	30.56	11

Table 5. Role of veterinarians in reporting and / or sample collection.

Role	Percentage of positive role (%)	No of positive role
Take a sample and sent to lab	16.67	15
Increase of a awareness	5.56	5
Disease reporting	27.78	25
Vaccination	5.56	5
Quarantine	13.89	12

They do this in space time (55%), any time (30%) and some read when need information (35%). The percentage of scientific correct information was very low (13.13%), when compare with general information (90.90%) this because there was no scientific background, and most authors are not specialized to deal with health problem.

DISCUSSION

This study was planned to probe and assess the degree of professional information that is made to available to the community. The recent suspected outbreak of Rift Valley fever in Sudan was taken as a case study.

Because of the uniqueness of this study, comparing findings with others was not easy, if not impossible. Yet it could be taken as a baseline study. Media play an important role in dissemination of information to communities, because it is very fast in reaching a wide zone. For these reasons, information accuracy are vital to avoid wrong facts and ideas to be spread.

In this study, information in the last epidemic of RVF was evaluated, focusing light on this information as correct or wrong details listed. Also, the role of professional related to zoonosis disease problem, because zoonosis affect all level of community economically particularly in trade and exporting of animal. By analysis of data from daily news paper articles about RVF epidemiology, it was found that about 99 news paper wrote about the topic, 90% of them wrote genera ideas about RVF 13% of those mentioned correct scientific information and 12% wrote incorrect scientific information. 74% of information was mixed of correct and wrong, although there is a great role of the press to provide correct information (the code of press release

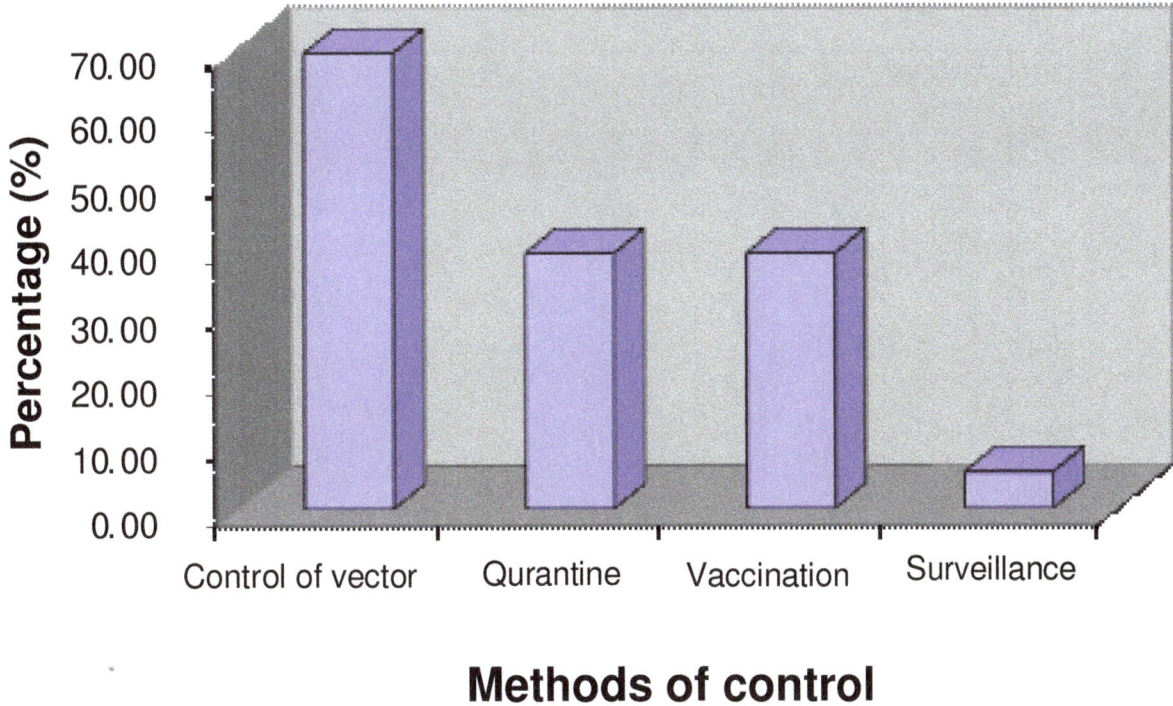

Figure 1. Methods of control Rift Valley Fever.

Constrians regarding recieving informaton

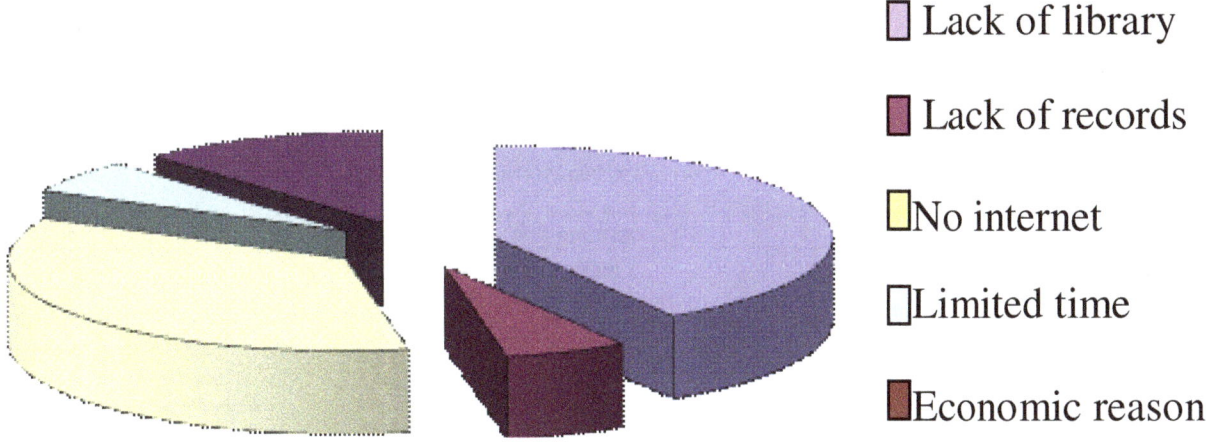

☐ Lack of library

■ Lack of records

☐ No internet

☐ Limited time

■ Economic reason

Figure 2. Source of veterinary information used by veterinarians.

and publication for the year 2004).

It was clear that the writers were not from science qualifying institutes, this was reflected in writings more towards fiction and propaganda than quiet narration of facts. The outcome is mostly a lot of debate by non-qualified audience. In contradiction, in developed countries matters which are very specialized are handled by specialized authors. The law in Sudan clearly shows the limits, especially ethical conduct, but abiding by the rules is sometimes not observed. Other media portals, as television, radio and internet contributed, although not included in the study. The general trends was as short

Table 6. Knowledge of owners about disease spread.

Way of spread	Percentage of respondent (%)
Labour	33.13
Air	26.66
Water	11.11
Insect	4.44
Do not know	11.11

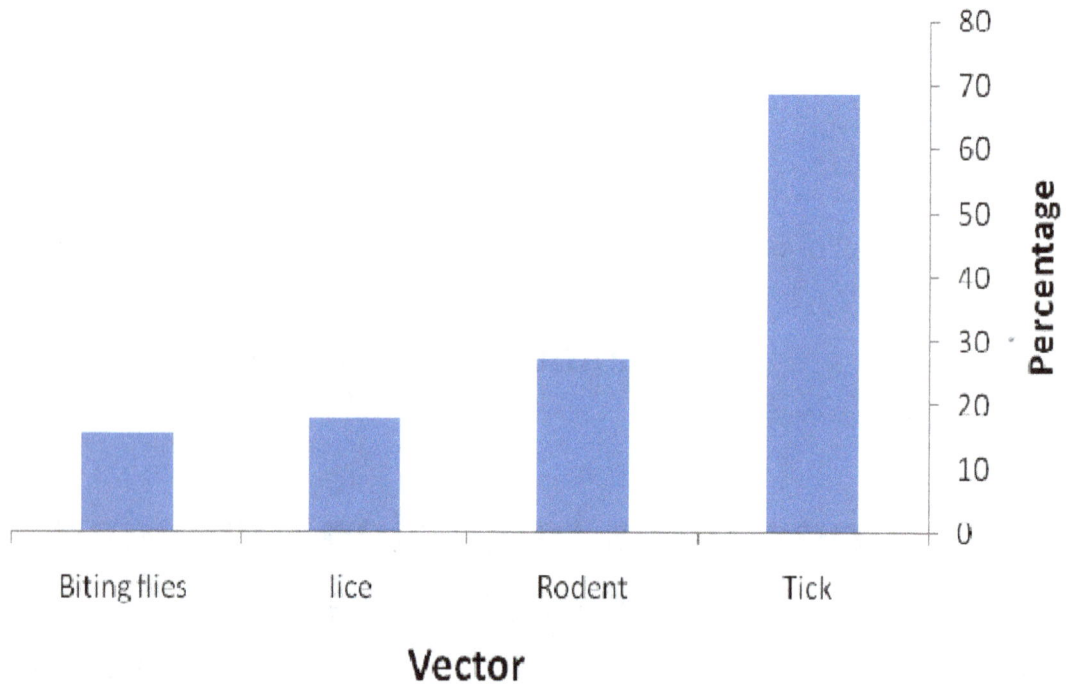

Figure 3. Role in milking by different individuals of the community.

Table 7. Medical doctor's knowledge about zoonosis to be reported.

Disease	Percentage (%)	No of respondents
Rift valley fever	65	13
Rabies	30	6
Avian influenza	25	5
Brucellosis	20	4
Anthrax	20	4
Yellow fever	5	1
T.B.	5	1

clinical articles, release by officials or interviews. In most of these, the source of information was not mentioned.

Continuing professional development (CPD) is the responsibility of veterinarians, especially in disease diagnosis. Veterinarians know contagious diseases which occur repeatedly in the field. Knowledge was however variable ranging between 2.78 to 77.78%.

Awareness about disease is also affected by media as in the case of RVF. Other means of knowledge improvement such as attendance of conferences was also variable, 55% never attended, while 11.44% attended international conferences, 45% attended national

Table 8. Medical doctors knowledge about important zoonosis.

Disease	Percentage (%)	No of respondents
Brucellosis	80	16
Rabies	40	8
Rift valley fever	40	8
Tape worm	25	5
Avian influenza	25	5
Yellow fever	5	1
T. B.	10	2
Toxoplasmosis	5	1
Mad cow	15	3
Anthrax	25	5

Table 9. Method of control of zoonosis as mentioned by physicians.

Method of control	Percentage (%)
Avoidance contact with animals	10
Avoidance of raw food	10
Vaccination	5
Treatment of infected animals	30
Early diagnosis	1
Increase of awareness	35

Table 10. Physicians suspension and confirmation of zoonosis.

Suspicion / confirmation	Percentage of respondents (%)
Laboratory diagnosis	25
Clinical signs	50
Contact with animal	10
History of disease	2

Table 11. Methods of transmission of zoonosis as mentioned by physicians.

Methods	Percentage (%)	No of respondents
Contact with animal	10	2
Contaminated milk	10	2
Bite	20	4
Disgorge from animal	20	4
Raw food	75	15

conferences. Lack of libraries was mentioned as a constraint. Most veterinarians saw that the internet is the most important source of information (33.33%). About 2% just take the information from media, as they may not have access to veterinary published material; it was noticeable that pamphlets, bulletins and books on veterinary subject are not readily available. Also, the language barrier constitute a big problem for young

professionals to cope with and understand recent developments in knowledge. The responsibilities of the Sudan Veterinary council, the Sudan veterinary association and the relevant ministries in CPD are not well taken. Concerned statuary bodies however are the focal points for professional development. It is their mandate to assure the standards of practicing veterinarians (Sudan Veterinary Council Act, 2004). 86%

of vets mention RVF as an example of zoonotic disease, they mentioned correctly the role of mosquitoes and other means in transmission and correct diagnosis because media talked about RVF for many months, although most veterinarian (61.11%) never diagnosed RVF in the field, and the percentage was even higher in physicians (90%). This reflects the strong theoretical knowledge as compared to practical experience, practice leads to improving the skills, and practical training depends on good theoretical background. However, dissemination of knowledge through inter-net based learning, using cases as demonstration, could be a solution to lack of cases in the field. Inter – disciplinary approaches are important in control of epidemics, particularly zoonosis as it plays a major role to contain the disease. A high percentage of physicians (65%) saw that there is no complementary role in containing disease, 45% of physicians responded that there is no relationship between the two specializations, and high percentage (45%) thought that the ministry of health is responsible of zoonosis. In the current situation, there is a clear dichotomy between medical and veterinary practices in developing countries, Sudan are no exception. This is probably the cause of this response from physicians. In developing countries public health issues, including zoonosis are handled in an integrated manner by both professions. 65% of physicians thought that RVF is the most reported disease because of the effect of the media, and (95%) answered that most people susceptible to disease are those who have contact with animals, and raw food is the most important way of transmission (75%).

Raising awareness is the responsibility of veterinarians. Media is a forum of dissemination of extension messages.26.66% of livestock owners saw that veterinarians are responsible of meat inspection, while 31.11% said there is no role, this support the opinion about extension, as the role of vets in meat inspection is not clear to all community members. 84.44% acquire new animals to be added to their herds, which could be one of important ways of spread of diseases. 84.44% traditionally diagnose disease from signs known to them, and (35.55%) consult veterinarians. This is also the responsibility of veterinarians and media particularly use of radio because it spread widely to livestock owners. Extension messages needed to be formulated to suite different strata of livestock communities, with varying levels of knowledge. Also, messages should reach target communities everywhere. As mentioned in the foregoing, this study although not comparable to other studies, yet it reveled gaps between academic knowledge and practice.

The root cause could be embedded in the classical curricula taught at different institutions which do not include community development as a strong component.

From and overall look at the results of this analytical study, it is evident that the knowledge available to veterinarians and medical doctors at the academic training level are good. They were easily retrieved and given in response to questionnaires or interviews. However, self-learning was not found to be at the same level, particularly in practical application, practice is limited to tentative diagnoses and treatment. Also, integration with other professions is vague. The media does not employ experienced professionals to review and correct published materials .the situation of compartmentalization of professions leads to creating an environment of suspicion and loss of trust in the information released. It is thus recommended to:

1. Form a national body of all concerned stakeholders to deal with public health issue particularly zoonosis.
2. Encourage experienced professionals to have periodicals that address issues of concern to the communities in simplified ways.
3. To make available to the different sectors of the community guidelines for the handling and management of suspected cases.

REFERENCES

Eisa M, Obied HMA (1977a). Rift valley fever in the Sudan .1.Result on field investigations of the first epizootic in Kosti District , 1973 .Bull. Anim. Health Prod. Afr., 24: 343-347.

Esia M, Obied HMA (1977b). Rift valley fever in the Sudan 11.Isolation and identification of the virus from a recent epizootic in Kosti District , 1973. Bull. Anim. Health Prod. Afr., 24: 349-355.

Esia M, Kheirelseed ED, Shommein AM, Meegan GM (1980). An outbreak of Rift valley fever in Sudan-1976.Transaction of the Royal society of tropical medicine and hygiene, 47(3): 417-419.

Esia M (1984). Preliminary survey of domestic animals of the Sudan for precipitating antibodies to rift valley fever virus. J. Hyg., 93: 629-637.

Food and Agriculture Organization (FAO) (2003). www.fao.org.

Sudan Veterinary Council Act (2004). Report, Ministry of Animal Resources and Fisheries, Khartoum, Sudan.

16

Erythrocyte osmotic fragility of Wistar rats administered ascorbic acid during the hot-dry season

A. W. Alhassan[1]*, A. Y. Adenkola[2], A. Yusuf[1], Z. M. Bauchi[3], M. I. Saleh[1] and V. I. Ochigbo[2]

[1]Department of Human Physiology, Faculty of Medicine, Ahmadu Bello University Zaria, Kaduna State, Nigeria.
[2]Department of Physiology and Pharmacology, College of Veterinary Medicine, University of Agriculture, Makurdi, Benue State, Nigeria.
[3]Department of Human Anatomy, Faculty of Medicine, Ahmadu Bello University, Zaria, Kaduna State, Nigeria.

The experiment was carried out with the aim of investigating the effect of an antioxidant ascorbic acid (AA) on erythrocyte osmotic fragility (EOF) of Wistar rats during the hot-dry season. Fifteen adult Wistar rats administered with AA at the dose of 100 mg/kg *per os* and individually served as experimental animals, and 15 others administered orally with sterile water were used as control animals. The animals were kept in the laboratory and the meteorological parameters within the period of study were determined using wet and dry bulb thermometer while the blood samples for EOF determination was obtained at the end of the experiment which lasted 8 weeks and this was done using standard procedure. EOF decrease significantly ($P < 0.05$) in experimental rats compared to the control group. The results indicated that AA protected the integrity of the erythrocyte membrane in experimental rats administered with AA as demonstrated by lower percentage haemolysis, and thus may alleviate the risk of increase in haemolysis due to heat stress in rats during the hot-dry season.

Key words: Wistar rats, ascorbic acid, hot-dry season, erythrocyte osmotic fragility.

INTRODUCTION

Environmental heat which animals are unavoidably subjected to produces heat stress by causing discomfort, irritation and some degree of psychomotor disturbances (Rakesh and Amit, 2004). The changes in thermal environment caused by fluctuations in ambient temperature (AT) and relative humidity (RH) have been demonstrated to induce a variety of physiological responses, which may adversely affect productivity and health in livestock (Ayo et al., 1998a and b; Adenkola and Ayo, 2009a). Heat stress is one of the most important stressors in the hot regions of the world (Altan, 2003) resulting in the generation of enormous free radicals and other reactive oxygen species (ROS).

The ravaging effects of oxidative free radicals are quenched by antioxidants (Akinwande and Adebule, 2003). Ascorbic Acid (AA) is an outstanding antioxidant found in the human blood plasma (Frei et al., 1989) and it stabilizes free radicals and terminates free radical induced lipoperoxidation of cytochromes, thereby maintaining the structural integrity of cells (Chews, 1995; Candan et al., 2002; Adenkola and Ayo 2009b). It has also been established that AA ameliorates heat stress and the adverse effects of environmental conditions (Tauler et al., 2003; Adenkola and Ayo, 2006a; Adenkola and Ayo, 2009a). Currently AA is the most widely used vitamin supplement throughout the world (Naidu, 2003).

Therefore, AA supplementation may attenuate the negative responses of Wistar rats to heat stress during hot-dry season, described as one of the most thermally stressful season in the Northern Guinea Savannah zone of Nigeria (Igono et al., 1982).

The aim of the present study was to determine the modulatory role of AA on the erythrocytes osmotic fragility in Wistar rats during the hot-dry season prevailing in the Northern Guinea Savannah zone of Nigeria.

MATERIALS AND METHODS

Experimental site

The experiment was carried out during the hot-dry season at the

*Corresponding author. E-mail: abdulwhb2002@yahoo.co.uk.

Table 1. Composition of pelletised growers feed.

	Percentage
Crude protein	14.5
Fat	7.0
Crude fibre	7.2
Calcium	0.8
Available phosphorus	0.4
Metabolisable energy	2.500 kcal/kg

Data were obtained from the manufacturer (Vital Feeds, Jos, Nigeria).

Department of Human Physiology, Faculty of Medicine, Ahmadu Bello University, Zaria (11° 10′ N, 07° 38′ E), at elevation of 650 m above sea level located in the Northern Guinea Savannah zone of Nigeria (Akpa et al., 2002). The area has three climatic seasons which consists of the cold dry season (November – February), hot-dry season (March – May) and the wet season (June – October) (Igono et al., 1982) with an annual rainfall of 1107 mm (Rekwot, 2002).

Meteorological data

During the study period, wet and dry-bulb temperatures (DBT) were recorded at 06:00, 13:00 and 19:00 h using dry- and wet-bulb thermometers (Brannan, England), and relative humidity (RH) was calculated using the manufacturer's standard manual attached.

Experimental animals and management

A total of 30 adult Wistar rats comprising of 14 males and 16 females weighing between 140 – 160 g. The animals were fed with pellets made from grower's mash (Table 1), maize bran and groundnut cake in the ratio 4:2:1 with wheat flour as binder . The rats were allowed to acclimatize to the environment for two weeks before the experiment commenced and they were divided into two groups. Group 1 made up of 15 rats (7 males and 8 females) which served as the control were given tap water orally for two weeks, while the second group were made up of (7 males and 8 females) and this served as the experimental animals. The experimental animals were administered with ascorbic acid (AA) at a dose of 100 mg/kg (Chervyakov et al., 1977) daily for a period of eight weeks. Each tablet was dissolved in 1 ml of distilled water to obtain 100 mg/ml suspension, just prior to its administration. At the end of the experiment, the animals were euthanized and 4 ml of blood sample was collected into bijou bottles containing the anticoagulant, disodium salt of ethylene diaminetetra-acetic acid at the rate of 2 mg/ml of blood (Oyewale, 1992). After collection, the samples were transferred to Physiology Research Laboratory, Department of Physiology and Pharmacology, Faculty of Veterinary Medicine, Ahmadu Bello University, Zaria, where erythrocyte osmotic fragility (EOF) test was carried out as described by Faulkner and King (1970).

Erythrocyte osmotic fragility determination

Sodium chloride (NaCl) solution was prepared according to Faulkner and King (1970) in volume of 500 ml for each of the samples in concentrations ranging from 0.05 to 0.85% at pH 7.4. A set of 10 test tubes, each containing 10 ml of NaCl solution of concentrations, ranging from 0.05 to 0.85%, were arranged serially in a test tube rack. One set was used to analyse each sample. The test tubes were labeled with corresponding NaCl concentration. One millilitre pipette was used to transfer exactly 0.02 ml of blood sample into each of the ten test tubes. Mixing was performed by gently inverting the test tubes for about 5 times. The test tubes were allowed to stand at room temperature (26 - 27°C) for 30 min. The contents of the test tubes were maintained at pH 7.4. Thereafter, the contents of the test tubes were re-mixed and centrifuged at 1,500 x g for 15 min. The supernatant of each test tube was transferred into a glass cuvette. The concentration of haemoglobin in the supernatant solution was measured using a spectrophotometer (Spectronic-20, Philip Harris Limited, Shenstone, UK) at 540 nm by reading the absorbance. The same procedure was repeated for every blood sample of each pig used for the study. The percent haemolysis was calculated using the formula (Faulkner and King, 1970):

$$\frac{\text{Optical density of test}}{\text{Optical density of distilled water}} \times 100 = \text{Percent haemolysis}$$

Erythrocyte osmotic fragility curve was obtained by plotting percent haemolysis against the saline concentrations.

Statistical analysis

All data obtained were subjected to statistical analysis using Student's t-test using Graph Pad Prism version 4.00 for Windows (www.graphpadprism.com). Data were expressed as mean ± standard error of mean. Values of $P < 0.05$ were considered significant.

RESULTS

The dry bulb temperature recorded during the study period was highest at 13:00 h with a value of 30.00 ± 2.25°C and the lowest dry bulb (26.00 ± 2.65°C) reading was obtained at 17: 00 h which was significantly ($P < 0.05$) different. The overall reading of relative humidity was 15.00 ± 1.73% (Table 2). There was a statistically ($P < 0.05$) different in the value of male Wistar rats at NaCl concentration of 0.5, 0.6, 0.7, 0.8 and 0.9% (Figure 1). Also the erythrocyte osmotic fragility was significantly ($P < 0.05$) increased at NaCl concentration of 0.4, 0.5, 0.6, 0.7 and 0.8% in the female Wistar rats (Figure 2). At 0.5% NaCl concentration a significant increase ($P < 0.05$) exist between a male and female Wistar rats. However, the erythrocyte osmotic fragility was not significantly ($P > 0.05$) different between the two group of rats at other concentration.

DISCUSSION

The meteorological results obtained in the present study showed that the microclimatic condition prevailing during the study period were thermally stressful to the rats. Although mortality was not recorded during the study period, it was demonstrated that temperature exceeding 28.5°C could lead to thermo tolerance being overcome

Table 2. Meteorological parameters during the study period.

Hour of the day	Meteorological parameters	
	Dry bulb (°C)	Relative humidity (%)
06:00	28.00 ± 2.02	12.00 ± 1.35
13:00	30.00 ± 2.25	15.00 ± 2.00
19.00	26.00 ± 2.69	18.00 ± 2.30
Overall mean ± SEM	28.00 ± 1.56	15.00 ± 1.73

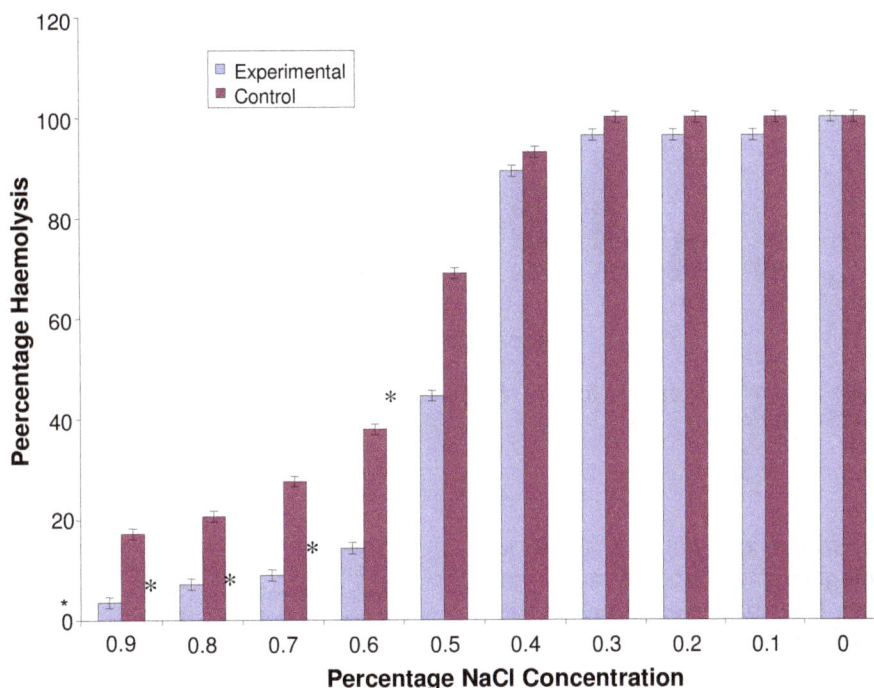

Figure 1. Erythrocyte osmotic fragility of experimental and control male Wistar rats during the hot-dry period.
* = "P < 0.05".

which could result in marked mortality (Zapata and Gernet, 1995). The meteorological results obtained in this study thus indicated a need to ameliorate the risk due to the adverse effects of hot-dry season thermal stress. It has been demonstrated that heat stress is one of the most important stressors in the hot regions of the world (Altan, 2003) resulting in the generation of enormous free radicals and other reactive oxygen species (ROS) (Chihuailaf et al., 2002) which play a vital role in cellular and tissue damage (Sudha et al., 2001; Tkaczyk and Vizek, 2007) and they have been demonstrated to have adverse effects on erythrocytes (Sumikawa et al., 1993; Avellini et al., 1995; Adenkola and Ayo, 2009b).

The significant difference observed between the experimental and control rats with a lower haemolysis in experimental rats could be attributed to the enormous ROS produced in the control animals as a result of thermal stress which play a vital role in tissue damage

and have deleterious effects on erythrocyte cytomembrane (Avellini et al., 1995; Adenkola and Ayo, 2009b). Although free radicals were not measured directly in this study, it has been shown that they are generated in animals subjected to stress (Halliwell, 1996; Senturk et al., 2001; Chihuailaf et al., 2002). The observed increase in EOF in control rats further supports this fact. Drooge (2002) demonstrated that ROS and other free radicals generated contribute to protein degeneration, lipid peroxidation and DNA oxidation. The membrane of erythrocyte is rich in polyunsaturated fatty acids which is susceptible to lipid peroxidation and this result in the loss of membrane fluidity and cellular lysis (Brzezinska-Slebodziiska, 2003) hence higher haemolysis in the control rats.

However in the experimental rats that was administered AA the lower percentage of haemolysis recorded was in agreement with observations of Senturk et al. (2001),

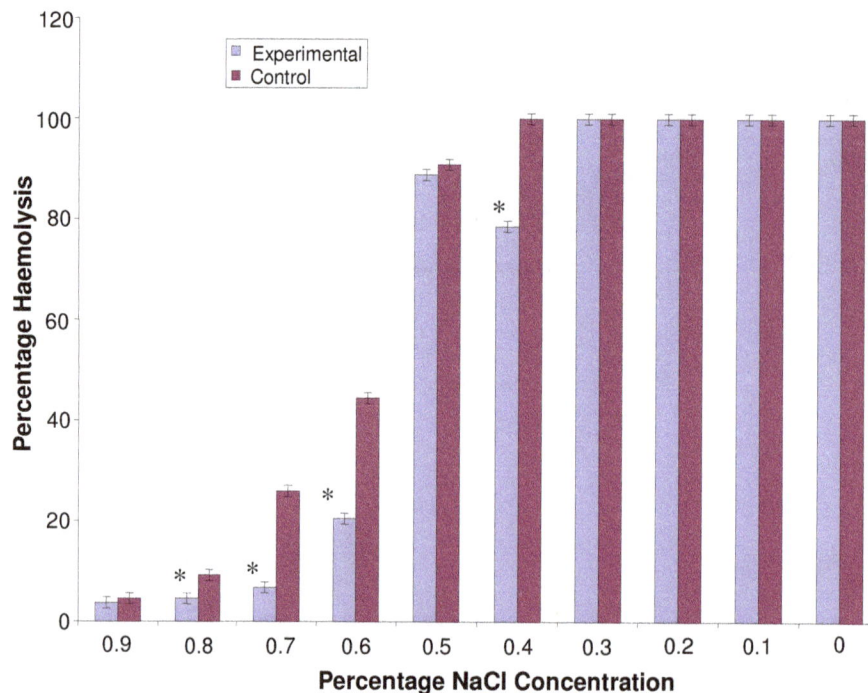

Figure 2. Erythrocyte osmotic fragility of experimental and control female Wistar rats during the hot-dry period.
* = "P < 0.05".

Candan et al. (2002) and Adenkola and Ayo (2009) that AA consolidates the integrity of erythrocyte membranes of and, therefore reduces their oxidative damage. Oxidative stress occurs when the antioxidant defence systems in the body are overwhelmed by free radicals (Williams et al., 2008). AA administration to experimental rats apparently, reduced the intensity of oxidant stress by enhancing the antioxidant defense mechanisms and suppressing the thermal stress which greatly minimized the destruction of erythrocyte. The result of this study agrees with those of Chen et al. (2000), Frei (2004) Adenkola et al. (2009a) and Adenkola and Ayo (2009b) that AA is an effective antioxidant in various biological systems. Antioxidant AA could thus be administered to animals during the hot-dry season in the region in order to prevent the unavoidable heat stress which is one of the most important stressors that animals are subjected to in the hot regions of the world.

REFERENCES

Adenkola AY, Ayo JO (2009a). Effect of ascorbic acid on diurnal variations in rectal temperature of indigenous turkeys during the hot-dry season. Int. J. Poult. Sci. 8(5): 457- 461.

Adenkola AY, Ayo JO (2009b). Effect of eight hours road transportation stress on erythrocyte osmotic fragility of pigs administered ascorbic acid during the harmattan season. J. Cell Anim. Biol 3(1): 004-008.

Akinwande AI, Adebule AOA (2003). Ascorbic acid and beta-carotene alleviate oxidative effect of London Kingsize® cigarette smoke on tissue lipids, Nig. J. Health Biochem. Sci. 2(1): 12-15.

Altan O, Pabuccuoglu A, Konyalioglu S, Bayracktar H (2003). Effect of

heat stress on oxidative stress, lipid peroxidation and some stress parameters in broilers. Br. Poult. Sci. 44: 54 - 55.

Avellini L, Silvestrelli M, Gaiti A (1995). Training-induced modification in some biochemical defences against free radicals in equine erythrocytes. Vet. Res. Comm.19: 179 - 184.

Ayo J O, Oladele SB, Ngam S, Fayomi A, Afolayan SB (1998a). Diurnal fluctuations in rectal temperature of the Red Sokoto goat during the harmattan season. Res. Vet. Sci. 66: 7 - 9.

Ayo JO, Oladele SB, Fayomi A, Jumbo SD, Hambolu JO (1998b). Body temperature, respiration and heart rate in the Red Sokoto goats during the harmattan season. Bull. Anim. Health Prod. Afr. 46: 161 - 166.

Akpa GN, Asiribo OO, Alawa JP, Dim NI, Osinowo OA, Abubakar BY (2002). Milk production by agropastoral Red Sokoto goats in Nigeria. Trop. Anim. Prod. 34: 526 – 533.

Brzezinska-Slebodzinska E (2003). Species differences in the susceptibllity of erythrocytes exposed to free radicals in vivo. Vet. Res. Comm. 27: 211 - 217

Candan F, Gultekin F, Candan F (2002). Effect of vitamin C and zinc on fragility and lipid peroxidation in zinc-deficient haemodialysis patient. Cell Biochem. Function 20: 95- 98.

Chen K, Suh J, Carr AC, Morrow JD, Zeind J, Frei B (2000). Vitamin C suppresses oxidative lipid damage in vivo, even in the presence of iron overload. Am. J. Physiologic. Endocrinol. Metaboilsm, 279(6): 1406 – 1412.

Chew BP (1995). Antioxidant vitamin effect food animal immunity and health. J. Nutr. 125: 18045.

Chervyakov DK, Yevdokimov PD, Vishker AS (1977). Drugs in Veterinary Medicine. Kolos Publishing House, Moscow, (in Russian) 496p.

Chihuailaf RH, Conteras PA, Wittwer FG (2002). Pathogenesis of oxidative stress: Consequences and evaluation in animal health. Vet. Mex. 33(3): 265- 283.

Drooge W (2002). Free radicals in the physiological control of cells. Physiol. Rev. 82: 47 – 95.

Faulkner WR, King JW (1970). Manual of Clinical Laboratory Procedures. Published by the Chemical Rubber Company, Cleveland,

Ohio p. 354.

Frei B (2004). Efficacy of dietary antioxidants to prevent oxidative damage and inhibit chronic disease. J. Nutr., 134(Suppl.): 3196 – 3198.

Halliwell B (1996). Vitamin C. An antioxidant or pro oxidant *in vivo*. Free Rad. Res. 25: 439– 454.

Igono MO, Molokwu ECI, Aliu YO (1982). Body temperature responses of savanna brown goat to the harmattan and hot-dry season. Int. J. Biometetrol. 26: 225 -230.

Naidu KA (2003). Vitamin C in human health and disease is still a mystery ? An overview. Nutr. J. 2: 1 - 10.

Oyewale JO (1992). Changes in osmotic resistance of erythrocytes of cattle, pigs rats, and rabbits during variations in temperature and pH. J. Vet. Med., A39: 98 - 104.

Rakesh KS, Amit, KR (2004). An assessment of changes in open field and elevated plus-maze behaviour following heat stress in rats. Iraninan Biomed. J., 8(3): 127 – 133.

Rekwot PI (2000). The influence of bull biostimlation and season on puberty and postpartum ovarian function in cattle. Ph. D Dissertation, Ahmadu Bello University, Zaria.

Senturk UM, Filz G, Oktay K, Mehmet R, Aktekin DP, Ozlem Y, Melek B, Akin Y, Oguz KB (2001). Exercise-induced oxidative stress affects erythrocytes in sedentary rats but not exercise-trained rats. J. Appl. Physiol. 91: 1999- 2004.

Sudha K, Rao AV, Rao A (2007). Oxidative stress and antioxidants in epilepsy. Clin. Chem. Acta. 303: 19 - 24.

Sumikawa K, Mu Z, Inoute T, Okochi T, Yoshida T, Adachi K (1993). Changes in erythrocyte membrane phospholipids composition induced by physical training and physical exercise. Eur. J. Appl. Physiol. 67: 132- 137.

Tauler P, Aquilo A, Giimelo I, Fuentaespina E, Tur JA, Pons A (2003). Influence of vitamin C diet supplementation on endogenous antioxidant defences during exhaustive exercise. Eur. J. Physiol., 446: 658- 664.

Tkaczyk J, Vizek M (2007). Oxidative stress in the lung tissue-sources of reactive oxygen species and antioxidant defence. Prague Med. Rept. 107(2): 105- 114.

Williams CA, Gordon ME, Betros CC, Mckeever KH (2008). Apoptosis and oxidant status are influenced by age and exercise training I horses. J. Anim. Sci. 86: 576 – 883.

Zapata LF, Gernat T (1995). The effect of four levels of ascorbic acid and two levels of calcium on egg shell quality of forced-molted White Leghorn hens. Poult Sci. 74: 1049 - 1052.

17

Histological and histomorphometric study of gametogenesis in breeders and helpers of sub-tropical, co-operative breeder jungle babbler, *Turdoides striatus*

Bharucha Bhavna* and Padate Geeta

Division of Avian Biology, Department of Zoology, Faculty of Science, M. S. University of Baroda, Vadodara, Gujarat, India.

In majority of birds, the reproductive commotion is restricted to a favorable but short period of time. This results in crowding of several physiological and histological changes during its breeding state. Such changes take place in both sexes for successful breeding. Jungle Babblers are termed social/co-operative breeders because of the fact that the breeding pair is assisted by other individuals of the flock (termed as "helpers"). These helpers forgo their own breeding and show allo-parental behavior and help the parents in taking care of all the reproductive chores except egg formation and laying. To understand the breeding cycle, histological and histomorphometric studies of gonadal tissues were carried out in these birds. The testicular cycle of Jungle Babbler could be roughly divided into seven stages showing different types and numbers of germ cells. As per histomorphometric studies, the seminiferous tubule diameter and germ layer thickness increases in breeders. Interstitium reduces in diameter in breeders but is densely packed with leydig cells resulting in high steriodogenesis. Ovaries also show varying degree of follicular maturation in breeders, non-breeders and helpers. The breeding ovaries had maximum number of large follicles with largest diameter. The granulosa cells of the mature follicles are responsible for the production of progesterone the fact which is also supported by the progesterone titers. The non-breeding ovaries consists of maximum number of small follicles in cortical region whereas helpers show both mature as well as small follicles along with large number of atretic follicles. Helpers show intermediate number of mature follicles which results in subdued production of progesterone. Progesterone is responsible for parental behavior as well as oviduct development. Helpers lack a fully functional and active oviduct due to subdued progesterone levels but the titers are enough to evoke the allo-parental behavior in them. To support the findings, in this paper hormonal titers (viz. testosterone, progesterone (Bharucha and Padate; 2009), cholesterol and ascorbic acid concentration in gonads in different individuals are also represented. Therefore, considering the lack of knowledge of reproductive biology about *Turdoides striatus,* the purpose of this paper consists of reporting a study on testicular and ovarian cycle in breeders and helpers. From this study it could be inferred that gonads of jungle babblers show cyclicity of development and regression during breeding and non-breeding states.

Key words: Jungle babbler, testicular cycle, mature follicle, atretic follicle.

INTRODUCTION

Reproduction is a biological phenomenon/ which exhibits a regular recurrence of pattern of activities in a cyclic manner (Lofts and Murton, 1973). Birds have an intense reproductive rhythm which is synchronized by seasonal changes of the environment and by the external factors, because of seasonality and rhythmicity, the reproductive functions exhibit meticulous regulation of initiation and development of gonadal functions which have to be a balanced interaction between environmental and physiological conditions (Phillips et al., 1985). Birds are

---

*Corresponding author. E-mail: cyprea_bb@yahoo.com.

adaptively diversified group of vertebrates and detailed information is available regarding correlation of environmental factors and their reproductive patterns. Voluminous literature exists on the annual breeding cycles of birds and greater attention has been given to the seasonal breeding cycles of the temperate seasonal breeders with more emphasis on the migratory birds. However, information regarding subtropical co-operative species is inadequate. Cooperative" or "communal" breeding occurs when more than two birds of the same species provide care in rearing the young from one nest. About 3% (approximately 300 species) of bird species worldwide are cooperative breeders. There are two types of cooperative arrangements: those in which mature non breeders ("helpers-at-the-nest" or "auxiliaries") help protect and rear the young, but are not parents of any of them, and those where there is some degree of shared parentage of offspring. Cooperative breeders may exhibit shared maternity, shared paternity, or both. Females generally disperse and pair after one or two years of helping. Helpers participate in all nonsexual activities except nest construction, egg laying, and incubation. Pairs with helpers are more successful; they fledge one and a half times more than pairs without helpers. In the most common form of cooperative breeding, such as occurs in the Jungle Babbler, most helpers are non breeding individuals that have remained on their natal territory. Jungle Babbler (*Turdoides striatus*) occur in groups that consist of a single breeding female, a dominant male breeder, and up to 4 subordinate, non breeding helpers, most of which are females. Helpers participate in territory defense, construction, and maintenance of nest cavities, incubation, brooding, feeding nestlings, and tending fledglings. Helpers have been shown to significantly improve the reproductive success of the breeding pair.

However, the helpers may be incapable of reproducing because they are sexually immature (delayed maturation) (Reyer et al., 1986) or helpers may be physiologically incapable of reproducing because of poor body condition and /or high levels of stress (physiological suppression) (Wingfield et al., 1991).

Gonadal architecture of seasonally breeding birds, poultry birds and migratory birds is well known, but the information regarding the anatomy of gonads of co-operative birds is less known and unfamiliar. One intriguing question about the co-operative breeders is that whether there is any disparity in the structure of gonads in helpers as compared to breeders and non-breeders and if so the possible physiological cause behind that dissimilarity.

Thus, the study pertaining to the histology and histomorphometry of gonads of Jungle Babbler was undertaken as it is an endemic bird to the Indian sub-continent. Jungle Babbler is also an integral part of the agro-ecosystem where the standing crops can be benefited by their insectivorous activity. Thus, understanding their reproductive cycle through the histological, histometric and physiological changes was taken up.

The breeding season of Jungle Babblers (*T. striatus*); seem to differ slightly in different parts of India (Ali, 1993; Whistler, 1949; Andrews, 1968). Jungle Babbler is a very common feral bird around, which has kept up its originality of reproductive rhythms without undergoing a modifying influence of urbanization. It is a social co-operative bird which lives in a flock of 7-8 birds and during breeding season it breaks up into a group of 3-4 birds. In this group of 3-4 birds, apart from the breeding pair, other individual members termed "helpers" forgo their breeding in order to assist the breeding pair. Hence, the aim of the present study was to inspect the histological and histomorphometric changes occurring in the gonads of breeders and especially in helpers in Jungle Babbler, a bird from semi arid subtropical region of India. This paper also presents the results of the investigations carried out to establish the correlation between changes in gonadal cholesterol and ascorbic acid concentration in different individuals along with the hormonal titers, with the histological, histomorphometric studies of the gonads (Bharucha and Padate, 2003).

The relationship of cholesterol concentration in the gonadal tissue to maturation and steriodogenesis as well as gametogenesis is been studied in birds (Marsa and Aoki, 1976; Ikegwuonu and Aire, 1977; Chand et al., 1978; Kanwar et al., 1977), while the reproductive effects of ascorbic acid have been researched extensively in mammals. The antioxidant property of ascorbic acid is essential to maintain membrane and genetic integrity of sperm cells by preventing oxidative damage to sperm DNA. Attempt has also been made to study the gonadal ascorbic acid depletion in phases of increased steriodogenesis activity during maturation in cockerels (Chand et al., 1978), because of their key role in the phenomenon of steriodogenesis (Kitabachi, 1967). The antioxidant property of ascorbic acid may delay formation of degenerative cells, MGC: multinucleated giant cells (Neuman et al., 2002). Reports were estimates of gonadal cholesterol or ascorbic acid concentrations in relation to gonadal maturation in this species were not found in literature.

Spermatogenesis is a complex process involving mitotic cell division, meiosis and the process of spermiogenesis. The regulation of spermatogenesis involves both endocrine and paracrine mechanism. The endocrine stimulation of spermatogenesis involves both FSH and LH, the later acting through the intermediary testosterone produced by the leydig cells of testes (de Kretser et al., 1998). Porter et al. (1989) proposed a "3-cell theory" for avian ovarian steriodogenesis which explained the different steroid production of the follicular tissues (granulosa: progesterone, theca interna: androgen, theca externa: estrogen). There is an increased progesterone concentration in granulosa and decrease in estrogen production of theca during follicular maturation. Ovarian hormones affect the reproductive state, the secondary sexual characters, metabolism and behaviour in birds (Sturkie, 1965). Silver et al. (1974) observed a significant correlation between follicle development and progesterone. The dominating synthesis of androgen was localized in the theca interna of immature follicles (Hackl et al., 2003). Therefore, in the present study the entire the aspect of the breeding biology of *T. striatus* is evaluated.

MATERIALS AND METHODS

The study was carried out according to the guidelines of the Committee for the Purpose of Control and Supervision of Experiments on Animals, India and approved by the Animal Ethical Committee of Department of Zoology, The M.S University of Baroda, Vadodara, India. The study carried out has been funded by ICAR (Indian Council of Agricultural Research) due to the importance of the bird in IPM (Integrated Pest management) approach. The objective of the project was to assess the relationship of carbohydrate, protein and lipid metabolism along with the histology, histochemistry and reproductive/breeding physiology including hormonal interaction during different seasons. So sufficient amount of tissue (especially in non-breeders where gonad size is too small) was required to carry out all the above said parameters, that is why the number of birds used was 8, even though for statistical analysis 6 are sufficient.

The birds were procured locally by means of a professional net to seize the flock. Thus the birds of the same flock were used for the above said studies. The number of birds used for the study for breeding and non-breeding males was 8, respectively. In case of females, the number of individuals for breeders, non-breeders and helpers was 8, respectively. They were kept in an open aviary with food and water *ad libitum*. After laparotomy, and the uncovering of the sternal breast plate and evisceration of the alimentary tract, the gonads were removed and blotted free of the tissue fluid. Sections were made, immersed in Bouin's fluid and later dehydrated in 70% alcohol, according to the technique of histological routine. The testis/ovary was embedded in paraffin and histological sections (5 μ thickness) were obtained to be later stained using Hematoxylin-Eosin technique. 20 cross-sections of testis/ovary were chosen at random to obtain the morphometric measures.

The following measurements were done with the aid of occulomicrometer.

For male birds

1. Diameter of at least 20 seminiferous tubules (in T.S.) of both breeding and non-breeding males.
2. Thickness of the germ layers of at least 20 seminiferous tubules (in T.S.) of both breeding and non-breeding males.
3. Diameter of interstitium of at least 20 seminiferous tubules (in T.S.) of both breeding and non-breeding males.

For female birds

1. The increase and/or decrease in the number of different types of follicular cells (*viz.* small, medium, large and atretic follicles).
2. Diameter of the largest follicle.

For hormonal studies

Blood was collected in the heparinised test tubes from the ventricles of the anesthetized birds and later centrifuged for 60 min at 3000 rpm. After centrifugation plasma was collected in Eppendorf tubes and stored at -4°C. The separated plasma fraction was used for quantitative measurements of progesterone and testosterone (by EIAgen kit, Biochem; Italia) respectively in both male and female Jungle Babblers.

EIAgen testosterone/progesterone kit

To evaluate testosterone/ progesterone content in blood plasma, a micro plate solid phase enzyme immunoassay kit was used. The EIAgen testosterone/ progesterone kit contains: a testosterone/ progesterone micro plate, testosterone/progesterone calibrators, testosterone/progesterone conjugates, washing solution, TMB H_2O_2 HS, stop solution (H_2SO_4). The solid phase enzyme immunoassay for testosterone/progesterone is a competitive type immunoassay wherein HRP labelled testosterone/progesterone competes with the testosterone/progesterone present in the sample (10 μl) for a fixed and limited number of antibody sites immobilised on the wells of the microstrips.

Once the competitive immunoassay reaction has occurred, the wells are washed and the HRP- testosterone/ progesterone fraction bound to the antibody in the solid phase is measured by adding the chromogen/substrate solution which is converted to a blue compound. After 15 min of incubation, the enzyme reaction is stopped with H_2SO_4, which also changes the solution to a yellow colour. The absorbance of the solution is measured photometrically at 450 nm and is inversely related to the concentration of the testosterone/progesterone present in the sample (10 μl). Calculations of the testosterone/ progesterone content in the sample are made by reference to a calibration curve.

Calculations of results for testosterone/ progesterone

To calculate the mean absorbance of calibrators and samples (A), the absorbance of the chromogen blank (Ac) is subtracted from the absorbance of all the samples. This is considered as the corrected value. Corrected values of the sample are divided by the corrected absorbance of the zero calibrator (Ao) and multiplied by 100 (A - Ac / Ao - Ac × 100). The respective testosterone/progesterone values are plotted on the logit log or semi log graph paper and the concentration of testosterone/progesterone in the samples are determined by the interpolation from the calibration curves.

Total cholesterol

The total cholesterol estimation was carried out by the method described by Crawford (1958) and is expressed as mg cholesterol/ 100 mg of tissue. Cholesterol was extracted in 3: 1 chloroform-methanol mixture. 2 ml of the extract was taken and dried completely in air oven. After drying of the tubes, 3 ml of $FeCl_3$ was added and boiled for 5 min. After cooling, 2 ml of conc. H_2SO_4 was added and mixed thoroughly. The brown color developed was measured colorimetrically after 30 min at 540 nm.

Ascorbic acid

Ascorbic acid estimation was carried out by the method described by Roe (1954) and is expressed as mg ascorbic acid/100 mg tissue. Tissue was homogenized in prechilled mortar-pestle with 6% TCA. Norit was added which acts both as oxidizing and clarifying agents. The solution was allowed to stand for 15 min and filtered. 2 ml aliquote was taken in a test tube, 2 drops of thiourea and 0.5 ml of 2, 4 DNPH were then added and left for incubation in boiling

Table 1. The histological changes observed in male Jungle babbler.

	Body weight (gms)	Testes weight (mgs)	GSI	Avg. diameter of seminiferous tubule (µ)	Avg. thickness of germinal layer (µ)	Avg. diameter of interstitium (µ)
Breeding	61.4 ± 2.27	205.0 ± 14	0.326 ± 0.032	55.25 ± 0.11	16.25 ± 0.09	4.72 ± 0.03
Non-breeding	56.7 ± 2.24	13 ± 1	0.025 ± 0.004	24.37 ± 0.1	8.12 ± 0.088	10.72 ± 0.036

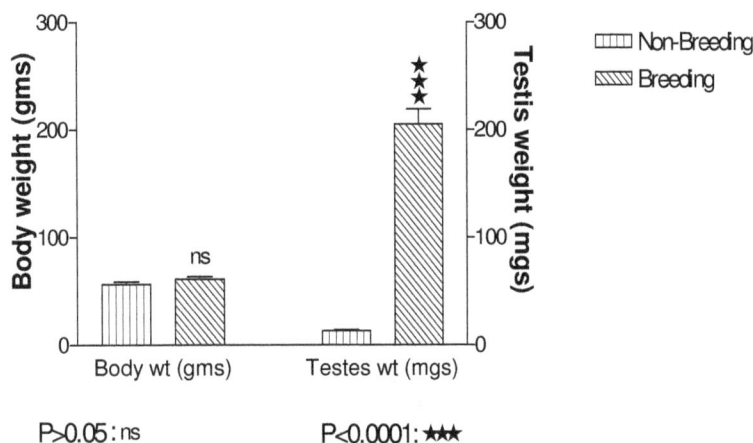

Figure 1. Body and testis weight in breeding and non-breeding male jungle babbler (*Turdoides striatus*).

water bath for 15 min. At the end of incubation, tubes were transferred to ice bath and 2.5 ml of 85% H_2SO_4 was added and allowed to stand for 30 min. The color developed was read at 540 nm.

Statistics

Statistical evaluation for males was done by T-test (non-parametric test), while the data for the females was done by one way ANOVA (non-parametric) followed by the Bonferroni multiple comparison post test and results are expressed as mean ± S. E (P<0.0001: ***, P<0.001: **/ ♦♦, P<0.01: *, P>0.05: non-significant), using Graph Pad Prism version 3.0 for Windows, Graph Pad Software, San Diego California, USA.

The student's t-test was done to analyse the male data, because t-test assesses only two groups (in this case breeder and non-breeder males) and help identify whether the means of two groups are statistically different from each other. This analysis is appropriate when comparing the means of two groups, and especially appropriate as the analysis for the post test-unpaired, non-parametric and would yield identical results. While one way ANOVA was done to analyse the female data because it enables the difference between more than two samples means (in this case breeders, non-breeders and helper females), achieved by subdividing the total sum of squares. It enables all classes to be compared with each other simultaneously rather than individually; it assumes that the samples are normally distributed.

RESULTS

The histomorphometric measures which includes the

average thickness of diameter of the seminiferous tubule, average thickness of the germinative layer and the average diameter of the interstitium along with the body weight, testes weight and GSI (Gono somatic index) for breeding and non-breeding males are given in Table 1, Figures 1 and 2 whereas body weight, weight of the ovary, weight of oviduct with GSI and different follicular sizes are given in Table 2, Figures 3 and 4. The histological features of testes are given in Plate I and of ovary in Plate II. Variations in hormones and certain metabolites involved in steriodogenesis in breeding, non-breeding and helper birds are represented in Table 3.

TESTES

The right testis of Jungle Babbler was always found to be superior in size and higher in weight than the left one. These differences were more pronounced during the breeding season. A regular well defined spermatogenic cycle as reported by (Andrews, 1968) was also found in the present study. The spermatogenic cycle of the Jungle Babbler is divided into the following stages:

Stage I. Resting spermatogonia only
Stage II. Dividing spermatogonia with a few spermatocytes.
Stage III. Many spermatocytes.
Stage IV. Spermatids and spermatozoa.

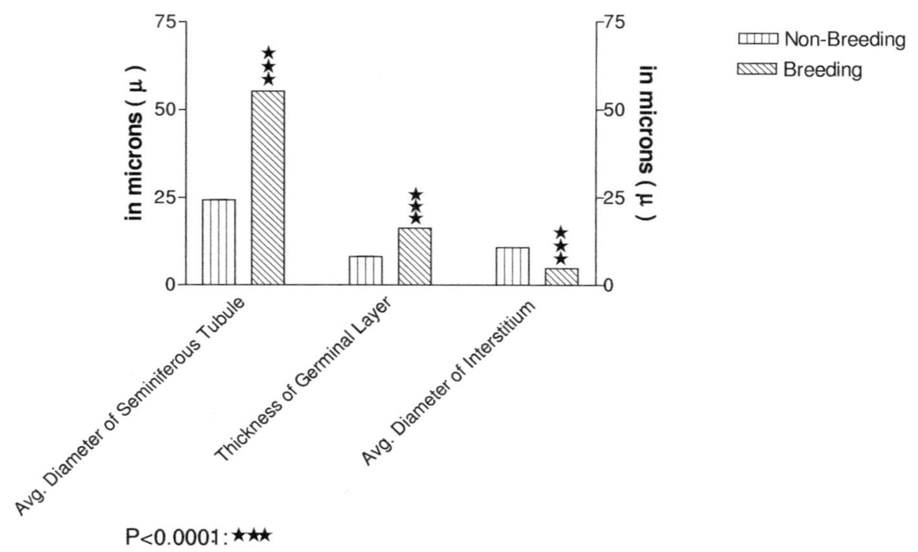

Figure 2. Average diameter and thickness of seminiferous tubules, interstitum and germinal layer of breeding and non-breeding male jungle babblers (*Turdoides striatus*).

Table 2. Variation in gono-somatic ratio and percentile of the follicles in female Jungle babblers.

	Body weight (gms)	Ovary weight (mgs)	Oviducal weight (mgs)	GSI	Small follicles (%)	Medium follicles (%)	Large follicles (%)	Atretic follicles (%)	Mean diameter of largest follicle (mm)
Breeding	58.66 ± 4.5	72.66 ± 9.39	181.66 ± 30.17	0.129 ± .024	4	12	10	5	8.90
Non-breeding	54.0 ± 2.47	9.0 ± 1.27	8.71 ± 2.87	0.028± 0.011	17	5	2	6	1.98
Helper	58.14 ± 1.86	65.71 ± 6.03	47.57 ± 8.30	0.113± 0.010	8	10	4	8	5.60

Stage V. Many spermatozoa.
Stage VI. Full spermatogenic activity with many spermatozoa still attached to the tubular wall.
Stage VII. Regressing testis.

Most of the breeding testes with spermatids and spermatozoa (Plates I, 4) were found from May to November indicating that Jungle Babbler are in readiness to breed for long period from May to November. However, non-breeding testes with resting spermatozoa were also found all throughout the year (Plate I, 1). In breeding testes the diameter of the seminiferous tubule was 55.25 ± 0.11 μm and in non-breeding testis it was 24.37 ± 0.1 μm ($F = 1.210$, P value = 0.3410, $t = 207.7$). When spermatids and spermatozoa were present (Plate I, 4) the seminiferous tubule diameter was larger than stage in which spermatozoa were still attached to germ layer (Plate I, 6). Maxi-mum diameter was observed in seminiferous tubule with primary and secondary spermatocytes.

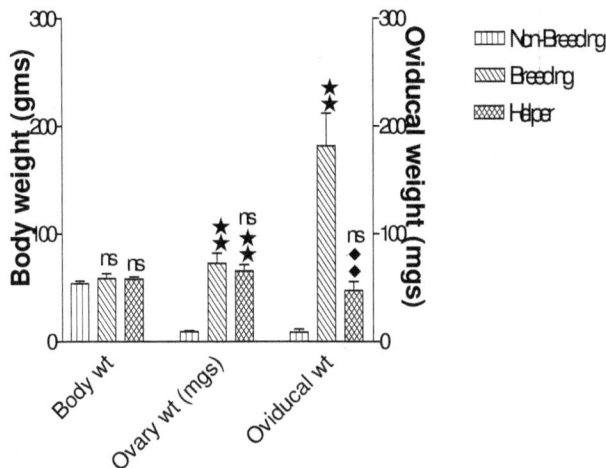

Figure 3. Body, ovary and oviducal weights in non-breeding, breeding and helper female jungle babblers (*Turdoides striatus*).

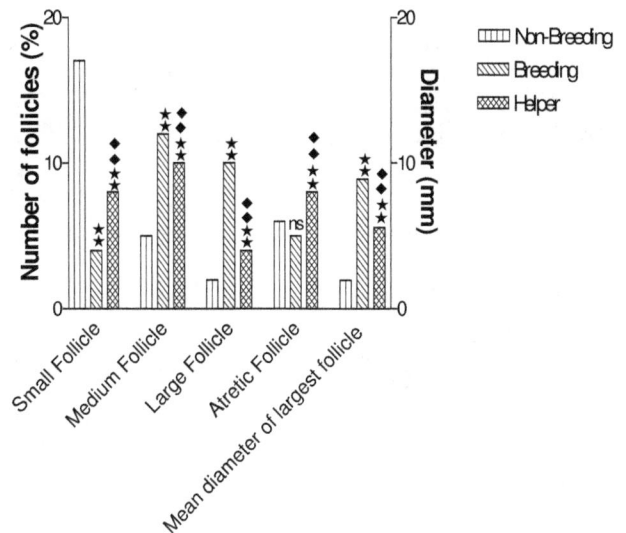

Figure 4. Percentile of various follicles and mean diameter of largest follicle in non-breeding, breeding and helper females of jungle babblers (*Turdoides striatus*).

The average thickness of germinal cell layer in breeding and non-breeding testes were 16.25 ± 0.09 and 8.12 ± 0.088 μm (F = 1.046, P value = 0.4615, t = 64.59), respectively. The average interstitium diameter in breeding and non-breeding males was 4.72 ± 0.03 and 10.72 ± 0.031 μm (F = 1.440, P value = 0.2170, t= 128), respectively (Table 1). In the regressing testes the rows of spermatogonia and occasionally a few spermatocytes were seen.

The Jungle Babbler combines characteristic of the interstitium found variably in mammals (Aire, 1997) with centrally located blood vessels and Leydig cells (Plate I, 3). The basal lamina resting on a closely associated homogenous microfibrillar layer can be seen in (Plate I; 2, 3, 4 and 6). The basal lamina in resting testes was not distinct. In Jungle Babbler, the Leydig cells were distinct in all stages except the stage when spermatozoa are present still attached to the germ layer (Plate I; 6).

The average body weight, testis weight, GSI and hormonal titers along with the metabolites involved in steriodogenesis in breeding and non breeding males are given in Tables 1, 3 and Figure 1.

OVARY

The ovaries were covered by single columnar epithelium, containing many follicles in the cortical zone. The type and degree of development varied between the stages. The inner medullar stroma was composed of well vascularised and innervated connective tissue, with spaces (lacunae) whose covering epithelium changed in each period. The various follicular stages observed in the

ovary of Jungle Babbler in present study were categorized as small follicles, medium follicle, large follicles and atretic follicles. These stages were noted for all the three females' *viz.* breeding and non-breeding females along with the helper females. The percentile of these follicular stages in the above mentioned females are given in Table 2. The percentiles of small follicles present in breeding, non-breeding and helper females were 4, 17 and 8 {F = 860.3, P value = 0.9589, t (B vs. Nb) = 40.49, t (B vs. H) = 12.46, t (H vs. Nb) = 28.03}, respectively, whereas the percentage of medium sized follicles were 12, 5 and 10 {F = 407.7, P value = 0.9773, t (B vs. Nb) = 10.16, t (B vs. H) = 28.26, t (H vs. Nb) = 17.66}, respectively. The large sized follicles were seen maximum in breeding females at 10% followed by helper females at 4% and least were observed in the non-breeding females at 2% {F = 346.9, P value = 0.9755, t (B vs. Nb) = 25.31, t (B vs. H) = 18.98, t (H vs Nb) = 6.327}. The atretic follicles were observed maximum in helper females at 8% followed by non-breeding and breeding females at 6 and 5% {F = 44.87, P value = 0.8481, t (B vs. Nb) = 3.10, t (B vs H) = 9.30, t (H vs. Nb) = 6.20}, respectively. The mean diameter of the largest follicle present in the breeding females was 8.90 mm while that in helper females was 5.60 mm and least follicular diameter was observed in non-breeding females at 1.98 mm {F = 58.64, P value = 0.9600, t (B vs. Nb) = 10.83, t (B vs. H) = 5.17, t (H vs. Nb) = 5.64}. The active ovary consisting of number of large follicles which are passing through the final stages of yolk accumulation are shown in Plate II; 2. The follicle consists of number of layers, the theca externa, the

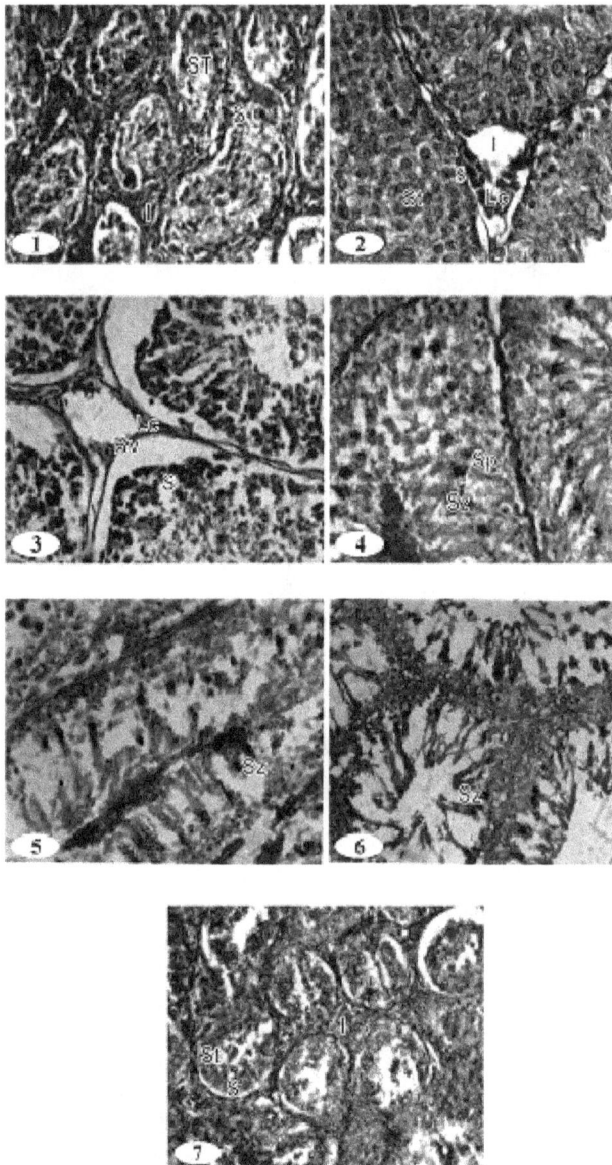

Plate I. Spermatogenic cycle in male jungle babbler *(Turdoides striatus)*. Stage 1. Resting spermatogonia (800 X). Stage 2. Dividing spermatogonia with a few spermatocytes (800 X). Stage 3. Many spermatocytes (800 X). Stage 4. Spermatids and spermatozoa (800 X). Stage 5. Many spermatozoa (800 X). Stage 6. Full spermatogenic activity with many spermatozoa still attached to the tubular wall (500 X). Stage 7. Regressing testis (800 X). S, spermatogonia; St, spermatocytes; Lc, Leydig cells; BV, Blood vessel; Sp, Spermatids; Sz, spermatozoa; L, Lumen; ST, Seminiferous tubule; I, Interstitium.

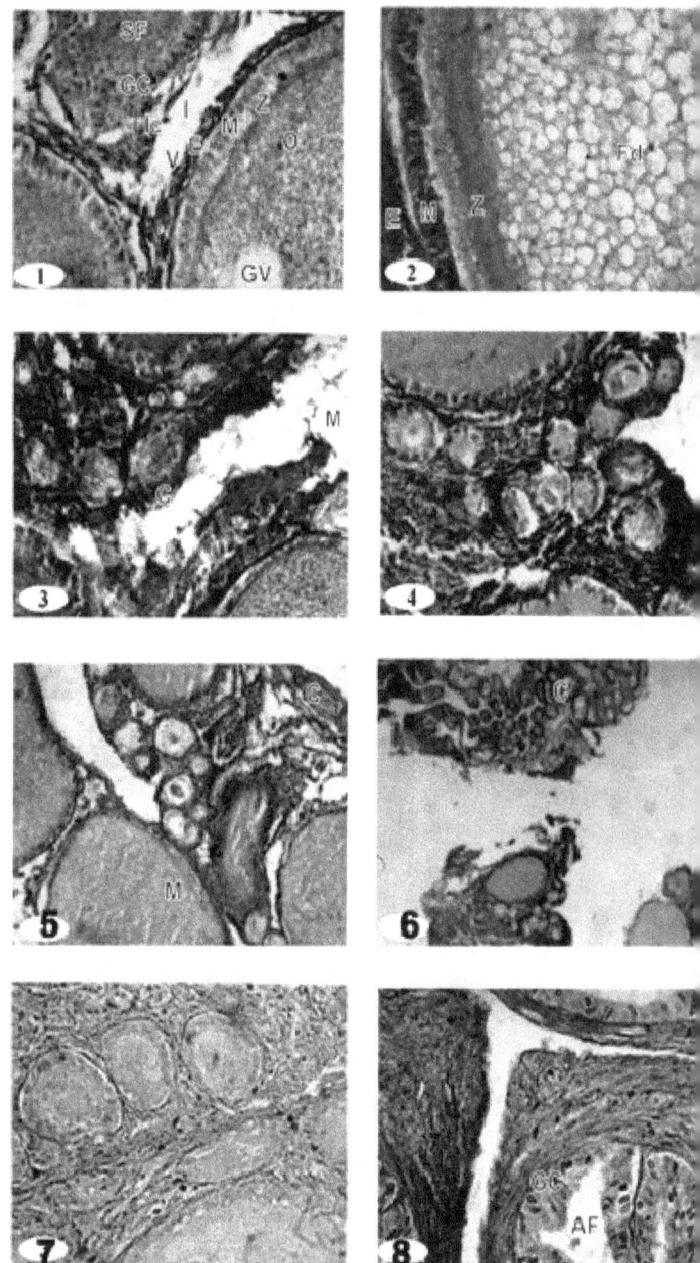

Plate II. Ovary of breeding, non-breeding and helper female jungle babblers. Figure 1. Ovarian follicles of large and medium size (800 X). Figure 2. Enlarged (large) follicle, within ova fat droplets (Fd) (800 X). Figures 3 and 4. Cortex with many small sized follicle (800 X). Figure 5. Many small and medium sized follicles (500 X). Figure 6. Cortical region with many small follicles (200 X). Figure 7. Non-breeding ovary with large number of small follicles (800X). Figure 8. Atretic follicle (AF) with cellular mass invading follicular cavity (C) (800 X). SF, Secondary follicle; O, Oocyte; GC, Granulosa cells; GV, Germinal vesicle; Ic, Interstitial cells; Z, Zona radiate; M, Membrana granulosa; V, Theca externa; M, Medulla; C, Cortex; AF, Atretic follicle; GC, Granulosa cells.

theca interna, the basement membrane, the membrana granulosa and the perivitelline membrane as shown in Plate II, 2.

The theca externa is known to comprise the greater part of the thickness of the follicular wall and contains muscle fibers. In birds the theca interna is a much thinner layer (Plate II, 2) which consists of collagen fibers. Basement membrane is a layer of cells separating the theca interna from the membrana granulosa. The follicular epithelium (membrana granulosa) is a layer immediately adjacent to the ovum (Plate II, 2). Zona

Table 3. Variations in hormones and certain metabolites involved in steriodogenesis in breeding, non-breeding and helper birds.

Parameters	Male (Testes)		Female (Ovary)		
	Non-breeding	Breeding	Non-breeding	Breeding	Helpers
Testosterone	0.21 ± 0.040	0.76 ± 0.027***	0.18 ± 0.049	0.65 ± 0.050***	0.63 ± 0.033**/ns
Progesterone	0.15 ± 0.054	0.3 ± 0.025*	0.1 6 ± 0.042	1.13 ± 0.066**	0.7 ± 0.00**/♦♦
Cholesterol	0.041 ± 0.001	0.029 ± 0.001***	0.04 ± 0.001	0.012 ± 0.003**	0.011 ± 0.004**/ns
Ascorbic acid	3.46 ± 0.53	2.93 ± 0.67 ns	2.55 ± 0.63 ns	1.85 ± 0.41 ns	2.4 ± 0.25 ns

(P< 0.0001: *** P< 0.001: **/ ** P< 0.01: * P >0.05: non-significant).

radiata is the most peripheral region of the oocyte. Atretic follicles are known to occur normally and regularly in active ovaries. When follicles reach the maximum size, the granulosa cells begin to proliferate forming numerous irregular layers around the ovum. The ovum size decreases and eventually granulosa cells fill the entire follicle (Plate II, 8). This undergoes hypertrophy and becomes a connective tissue scar. Immediately after ovulation the follicle shrinks and due to this the walls become thickened and granulosa increases to several cells in thickness. The follicle regresses rapidly and is eventually reabsorbed into the mass of the ovary.

The average body weight {male = (F = 1.027, P value = 0.4887, t= 1.477), Female = (F = 0.6562, P value = 0.1484}, ovary weight (F = 29, P value = 0.0022), oviducal weight (F = 25.02, P value = 0.0001), GSI and hormonal titers {Male (testosterone)= (F = 2.195, P value = 0.2043, t= 11.40), Male (progesterone) = (F = 4.666, P value = 0.0581, t= 2.251), Female (testosterone) = (F = 36.53, P value = 0.5122), female (progesterone) = (F = 115.8, P value = 0.001)} along with the metabolites involved in steriodogenesis in the breeding, non breeding and helpers are represented in Tables 2 and 3 and Figures 3 and 4.

DISCUSSION

In birds, there is a pronounced increase in the size of testes during breeding. Histologically this increase in size occurs primarily as a consequence of the enlargement of the seminiferous tubules in which sperms are produced (Lofts and Murton, 1973). Spermatogenesis is reported to be fairly conserved process throughout the vertebrate series. Spermatogenesis occurs in the seminiferous tubules, that possess a permanent population of sertoli cells and spermatogonia which act as a germ cell reservoir for succeeding bouts of spermatogenic activity (Pudney, 1995). Some testicular histological parameters have been explored in the bird's testis for characterization of testicular cycle such as seminiferous tubule diameter, the thickness of seminiferous epithelium, thickness of tunica albuginea, number of interstitial cells, the number of germinative cells (Fuenzalida et al., 1989; Baraldi-Artoni et al., 1997). The present study establishes the correlation between histological and histomorphometric changes occurring in the testis in breeding and non-breeding males.

The interstitial tissue of breeding testes is known to be tightly packed with the Leydig cells which contain relatively large amounts of lipid droplets related to androgen synthesis; while that in the non-breeding testes, the interstitial tissue contains only occasional Leydig cells with an enlarged intercellular space (Rosenstrauch et al., 1998). In Jungle babbler, the average interstitium diameter in breeders is less and tightly packed with interstitial cells while in non-breeders the diameter of interstitium is more with less number of leydig cells. Breucker (1982) has worked with testis of sexually mature swans Cygnus olor and considered that an adequate parameter to evaluate the variation of spermatogenic activity is the variation of the seminiferous tubular diameter. The seminiferous tubular diameter of the swan shows a progressive increase in winter and maximum values in spring (proliferation period) with a decrease in summer and fall (periods of sexual regression and rest). In Jungle babblers, we found smaller relative values as to the diameter of the seminiferous tubules and thickness of the germinative epithelium in non-breeders while the tubular diameters, thickness of the germinative epithelium was maximum in breeders. The variation in the mean body weight does not show any marked variation in relation to the breeding activities. These observations are also supported by the evaluation of testosterone, cholesterol and ascorbic acid in the testis of both breeders and non-breeders. Jungle babblers are social/co-operative breeders and they rely on daily food supply rather than accumulation of fat prior to breeding as seen in many seasonally breeding birds. Hence, no significant variation was observed in the body weights amongst the breeders and non-breeders males.

The avian ovary consists essentially of an outer cortex containing ova which surrounds a highly vascular medulla composed primarily of connective tissue. The surface of the cortex is covered by the cuboidal germinal epithelium. There is vast number of ova developing in the ovary but only few reach maturity and only a comparative few are ovulated. The surface of the ovary is covered by the germinal epithelium consisting of a single layer of cells. All the ova within the ovary are primary oocytes until before ovulation. Within the cortex there are numerous minute developing ova. Also, in Jungle babbler similar stages were found. A histochemical study has been made of seasonal fluctuations in the follicular atresia and interstitial gland tissue with the ovarian cycles of the house sparrow (Guraya and Chalana, 1976) in crow and myna (Chalana and Guraya, 1979) in grey quail (Saxena and Saxena, 1980) in house swift (Naik and Naik, 1965). The atresia of the primordial oocyte forms the predominant feature of the quiescent-winter ovary. The building up of interstitial gland tissue of the thecal origin, which precedes the breeding activity, is closely related to the atresia of previtellogenic follicles of variable sizes (Chalana and Guraya, 1979). In Jungle babblers the ovaries were covered by single columnar epithelium, containing many follicles in the cortical zone. The type and degree of development varied between breeders, non-breeders and helpers. The inner medullar stroma was composed of well vascularised and innervated connective tissue, with spaces (lacunae). In breeding females maximum number of medium sized follicles with biggest diameter was observed, while in non-breeding females more number of small follicles with least diameter was observed. Helper females showed nearly same number of small and medium sized follicles but maximum number of atretic follicles. In Florida Scrub Jay helpers ovarian follicles were not as developed as breeders, they were not completely regressed (Schoech, 1996). The same thing was observed for helper females in Jungle Babblers too. This data is also supported by the levels of progesterone in these females. In helpers, as intermediate numbers of mature follicles are present, the granulosa cells of which are capable of progesterone synthesis also reflected in the progesterone titers in helpers. Due to the subdue levels of progesterone, functional and active oviduct development in helpers was not seen but the progesterone titers are enough to evoke the allo-parental behavior in them. Same as in case of males, no significant variations was seen in body weights of breeders, non-breeders and helpers. The amount of cholesterol and co-enzyme ascorbic acid also supports the histological and morphometric analysis of the ovaries in these females.

Therefore, the present data concerning seminiferous tubules diameter, thickness of the germinative epithelium and diameter of the interstitium revealed a variation in breeders and non-breeders. The highest seminiferous tubules diameter, thickness of germinative epithelium was detected in breeders demonstrating that the greater spermatozoa production occurs in them. The histological and histomorphometric studies carried out in testis show cyclicity in spermatogenesis wherein it could be roughly divided into 7 stages depending on the type of germ cell present. Histomorphometric variations were also noted for the interstitium diameter which decreases during the breeding state. Observation of breeding ovaries, showed cells at various stages of follicular maturation. It had

maximum number of large follicles with maximum diameter. The non-breeding ovaries had maximum number of small follicles while the helper females showed both small and large follicles. Maximum atretic follicles were observed in helpers.

REFERENCES

Aire TA (1997). The structure of the interstitial tissue of the active and resting avian testes. Onderstepoort, J. Vet. Res., 64(4): 291-299.

Ali S (1993). Book of Indian Birds. Bombay Natural history Society, 53: 421.

Andrews MI (1968). Ecology of the social system of Jungle babbler, Turdoides striatus (Dumont). Thesis submitted to the M.S. University of Baroda.

Baraldi-Artoni SM, Orsi AM, Lamano-Carvalho TL, Lopes RA (1997). The annual testicular cycle of the domestic quail (Coturnix coturnix japonica). Anat. Histol. Embryol., 26: 337-339.

Bharucha B, Padate GS (2002). Ascorbic acid-cholesterol correlation in socially breeding species of bird, Jungle babbler (Turdoides striatus). Pavo. 40-41(1-2): 127-132.

Breucker H (1982). Seasonal spermatogenesis in the mute swan (Cygnus olor). Anat. Embryol. Cell Biol., 72: 1-91.

Chalana RC, Guraya SS (1979). Seasonal fluctuations and histochemical characteristics of the interstitial cells in the ovary of Crow and Myna. Pavo., 17(1-2): 65-70.

Chand D, Arneja DV, Arora KL (1978). Relation of testicular ascorbic acid concentration to testicular development in Desi and White Leghorn poultry. Indian J. Exptl. Biol., 16: 676-678.

Crawford N (1958). An improved method for the determination of free and total cholesterol using ferric chloride reaction. Clin. Chim. Acta., 3: 357-367.

De Kretser DM, Loveland KL, Meinhardt A, Simorangkir D, Wreford N (1998). Spermatogenesis. Hum. Reprod., 13(1): 1-18.

Fuenzalida H, Leyton V, Valencia J, Blanquez MJ, Gonzalez ME (1989). Morfologia del testiculo de Pygoscelis papua (FOSTER) durante el periodo de actividade sexual. Arch. Anat. Embryol., 20: 79-81.

Guraya SS, Chalana RK (1976). Histochemical observations on the seasonal fluctuations in the follicular atresia and interstitial gland tissue in House Sparrow ovary. Poult. Sci., 55(5): 1881-1885.

Hackl R, Bromundt V, Daisley J, Kotrschal K, Hostl E (2003). Distribution and origin of steroid hormones in the yolk of Japanese quail eggs (Coturnix coturnix japonica). J. comp. physiol. B., 173: 327-331.

Ikegwuonu FI, Aire TA (1977). Age influenced variations in levels of cholesterol and ATPase activity in the testes of prepubertal chicks. Poult. Sci., 56: 1158-1160.

Kanwar U, Sheikher C, Mehta HS (1977). Seasonal changes in the lipids of the crow Corvus splendeus: Cytochemical and biochemical studies. Indian J. Exptl. Biol., 15: 1040-1042.

Kitabachi AE (1967). Ascorbic acid in steroidogenesis. Nature, Lond., 215: 1385-1386.

Lofts B, Murton RK (1973). Reproduction in birds. In: Avian Biology. Ed. Farner DS and King JR Academic press Inc, NY., 3: 5-40.

Marsa EM, Aoki A (1976). Differentiation of cholesterol compartments in the immature chick testis. J. Reprod. Fertil., 47: 313-318.

Naik RM, Naik S (1965). Studies on the House Swift (G. E. Gray) 4. Gonadal cycle. Pavo., 3(2): 77-88.

Neuman SL, Orban JI, Lin TL, Latour MA, Hester PY (2002). The effect of dietary ascorbic acid on semen traits and testis histologybof male turkey breeders. Poult. Sci., 81(2): 265-268.

Phillips JG, Butler PJ, Sharp PJ (1985). Environment and Reproduction. In: Physiological strategies in avian biology. Ed. Blackie, Chapman and Hall, New York pp. 113-124.

Porter TE, Hargis BM, Silsby JM, El Halawani ME (1989). Differential steroid production between theca interna and theca externa cells: a three-cell model for follicular steroidogenesis in avian species. Endocrinol., 125: 109-116.

Pudney J (1995). Spermatogenesis in non-mammalian vertebrates.

Microse. Res. Tech., 32(6): 459-497.

Reyer HU, Dittami JP, Hall MR (1986). Avian helpers at nest: are they physiologically castrated? Ethol., 71: 216-228.

Roe JH (1954). Chemical determination of ascorbic acid. In: methods of biochemical analysis. Vol 1. Glick D. (ed.), New York, Interscience publishers.

Rosenstrauch A, Weil S, Degen AA, Friedlander M (1998). Leydig cell function, structure and plasma androgen level during the decline in fertility in aging roosters. Gen. Comp. Endocrinol., 109(2): 251-258.

Saxena A, Saxena AK (1980). Gonadal cycle and secondary sexual characters of the Grey Quail (Coturnix coturnix). Pavo., 18(172): 62-77.

Schoech SJ (1996). Endocrine Mechanisms of Cooperative Breeding in Florida Scrub Jays (Aphelocoma coerulescens). Abstracts from International Symposium on Avian Endocrinology. Chateau Lake Louise, Alberta. p. 4.

Whistler H (1949). Popular hand book of Indian birds. Gurney and Jackson, London.

Wingfield JC, Hegner RE, Lewis DM (1991). Circulating levels of luteinizing hormone and steroid hormones in relation to social status in the cooperatively breeding white-browed sparrow weaver, Plocepasser mahali. J. Zool., 225: 43-58.

18

Neurotoxicity of cassava: Mode of cell death in the visual relay centres of adult Wistar rats

O. M. Ogundele[1]*, E. A. Caxton-Martins[2], O. K. Ghazal[3] and O. R. Jimoh[2]

[1]Science and Technology Complex, Trinitron Biotech LTD, P. M. B. 186, Garki, Abuja, Nigeria.
[2]Department of Anatomy, University of Ilorin, Ilorin, Nigeria.
[3]Unilorin Stem Cell Research Laboratory, Ilorin, Nigeria.

Cassava (*Manihot escculenta*) is an annual tuber root crop cultivated widely in the tropics and subtropics, it serves as a major food crop of low protein but high calorie content. At the cellular level, both free cyanide and hydrogen cyanide has been found to induce degeneration via increased lysosomal activity *in vivo* (Sotoblanco et al., 2002). In this study, we investigated the effects of cassava diet administered to adult Wistar rats at 2, 5, 10, 20 and 30 gms of cassava per animal per day for a period of 60 days using metachromasia of Cresyl fast violet for Nissl substance to demonstrate degenerating neurons. The mode of cell death observed in the V1 was found to be an apoptosis-necrosis continuum for moderate dose treatment and necrosis for extreme dose treatment while the SC and LGB was mainly necrosis irrespective of the dosage.

Key words: Primary cortex, superior colliculus, lateral geniculate body, apoptosis, necrosis, reactive oxygen species, cyanide, nitric oxide.

INTRODUCTION

Cyanide as earlier described is a potent neurotoxic substance that can initiate series of intracellular reactions leading to cell dysfunction and eventually cell death; it enhances NMDA (N-Methyl-D-Aspartate) receptor function (Fiske et al., 2001). In cultured neurons, cyanide neurotoxicity is linked with NMDA receptors, which mediated a rise in calcium that in turn activates series of biochemical reactions leading to generation of reactive oxygen species (ROS) and NO (Nitric Oxide), this anti-oxidants species then mediates peroxidation of lipids (Ha et al., 2010). It is concluded that oxidative stress plays an important role in cyanide induced neurodegeneration; this phenomena can cause cell death in two ways:

1. Apoptosis
2. Necrosis

Depending on the cell type or stimuli, a cell may die in

either of the two distinct ways. Apoptosis considers the physiological form of cell death; it is an acute process with distinct morphological and biochemical features (Bathachanya et al., 2001; Batharcharya and Tulwasami, 2008). Apoptotic cells are characterized by condensed and fragmented nuclei, whereas necrotic cells show less plasma membrane integrity without apparent nuclei damage. This two different forms of cell death can be elicited by the same stimuli depending on the intensity of the effects (Isom et al., 1999; Jerome et al., 2003), suggesting that an initial common event can be shared by both forms of cell death. It has been shown previously that cyanide induces cell death in a varying mode for different brain areas (Lee et al., 2010). Cell death occurs predominantly via apoptosis in the cortical region while necrosis was observed in the substansia nigra after the same dose of cyanide was administered in the mice. Selective vulnerability of different aspect of the brain to cyanide may be explained by the triggering of region specific toxic pathways in which oxidative stress may be a common activator (Isom et al., 1999; Lee et al., 2010; Ming et al., 2003). In the two modes of cell death, oxidative stress was a factor but the level of ROS generated varies with the cell types, other reports that oxidative stress can be involved in apoptosis or necrosis

---

*Corresponding author. E mail: Mikealslaw@hotmail.com.

Abbreviations: **V1,** Primary visual cortex; **SC,** superior colliculus; **LGB,** lateral geniculate body; **CcOX,** cytochrome C oxidase,

(Way et al., 1984), it is possible that when high ROS accumulated in the cell, direct and irreversible damage to cell components can lead to necrosis. Moderate levels of ROS may function as cellular messengers and regulatory molecule which mediates apoptotic cell death (Li et al., 2000), in mesencephalic cells, cyanide induced a progressive but rather small increase in necrosis at the concentrations of 100 – 300 Um and at 400 Um, the data recorded shows a steep increase in necrotic index.

MATERIALS AND METHODS

Fifty adult Wistar rats, cassava tuber, cage, Cresyl fast violet, acetic acid, spring balance, microtome, staining jar and Olympus Research Microscope with JVC image processor were used in this study.

Twenty F1 generation adult Wistar rats were divided at random into five groups. Group 1 - 5 are the major experimental groups. The duration of treatment was 60 days, unprocessed cassava uprooted immediately before administration was peeled to remove the external root cortex; the inner white part was chopped into pieces according to the method of Sotoblanco et al. (2002) (Table 1). The animals were sacrificed by cervical dislocation and the V1, SC and LGB were excised by dissection and fixed in Formolcalcium. The tissues were processed sing the method of Pearse (1960) and stained with CFV (Cresyl fast violet) (Yin et al., 2007).

RESULTS

Group 5

On examination under the microscope, group 5 (Control group) – the six layered cortex appeared intact (Figure 1), (5A). The orientation of the horizontal cells in the molecular layer (Martinoti and Cajal) remains normal and a clear demarcation can be observed between the adjacent layers about 0.1 μm apart. The outer granular layer down to the multiform layer gives a normal orientation and cellular arrangement. The outer pyramidal layer is the most prominent (5A arrow head) illustrating the giant Betz cells. The superior colliculus displayed all the cellular and fibrous layers as they interlace from the stratum Zonale down to the stratum album Profundum, few pyramidal cells are observed in the stratum Griseum Intermedium (Figure 3), (5C).

The parvocellular and magnocellular layers of the lateral geniculate body and the intermediate layer separating them are prominent (Figure 2), (5B). In this low dose group, there are some degree of histological disarray in some regions of the molecular layers as compared with those of the control, the cells appeared slightly reduced in size about 2.5 μm and vacuolar spaces are observed around the cells in the outer granular layer (0.5 μm) (Figure 1 (3A), Figure 2 (3B) and Figure 3 (3C)).

However, the very few pyramidal cells did not show such vacuolar spaces and the outer pyramidal layer showed average sized pyramidal cells (3A). A few of

Table 1. Animal preparation and treatment.

Group	Cassava (gm)/rat	Feed (gm)/rat
1	40	0
2	20	20
3	10	30
4	2.5	37.5
5	0	40

HISTOLOGY I

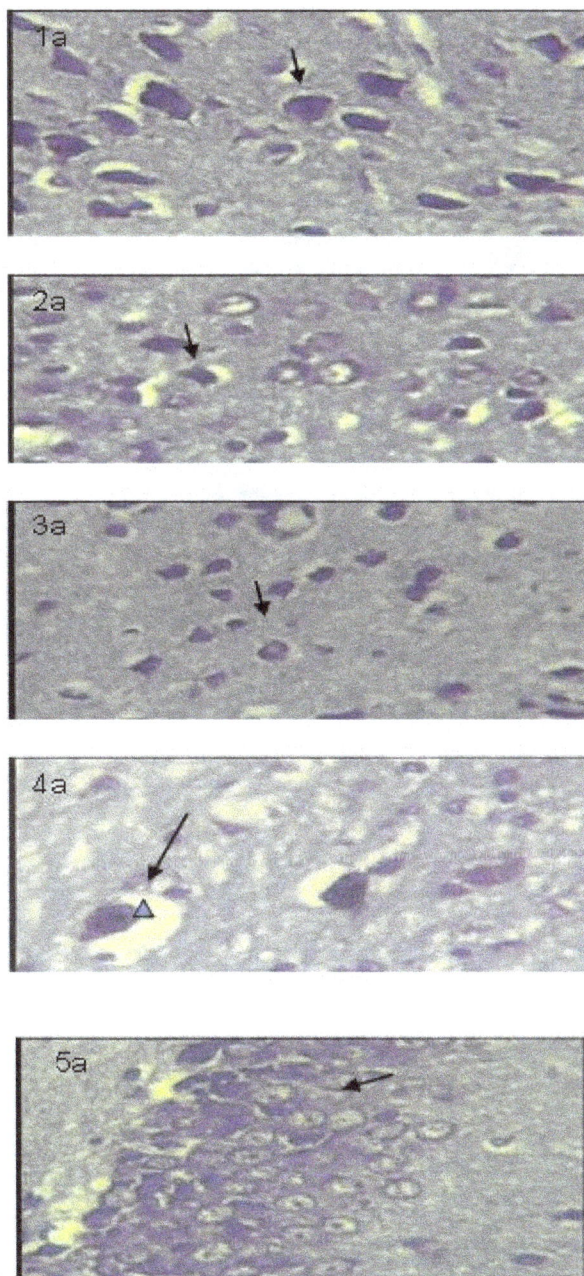

Figure 1. Comparative study of V1 for treatment groups. Magnification: X 1000 (Arrow heads indicates orientation and appearance of pyramidal cells).

HISTOLOGY II

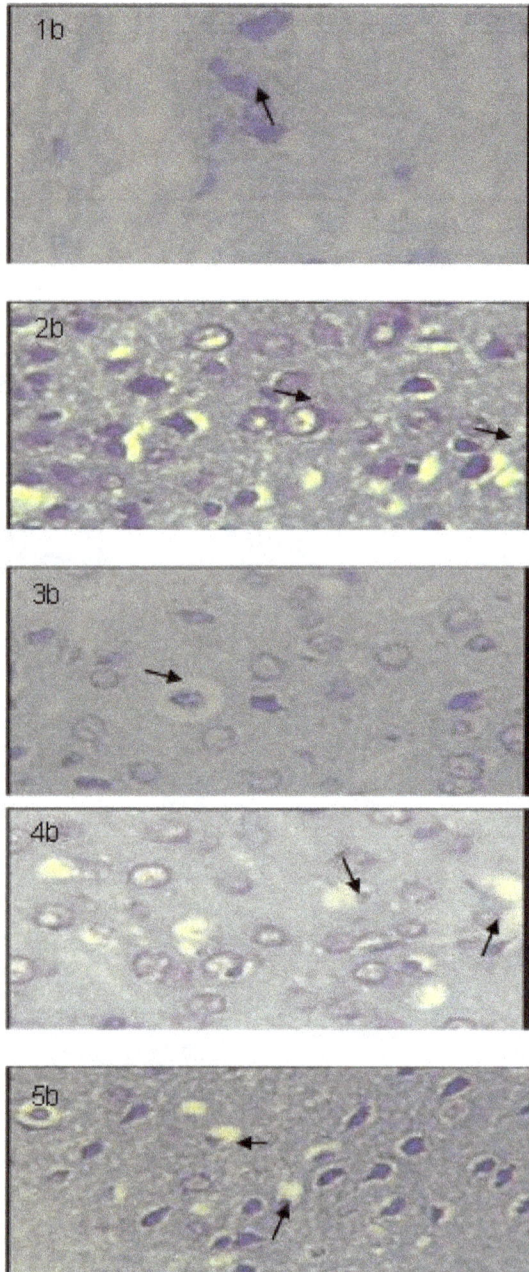

Figure 2. Comparative study of LGB (1b - 5b) for treatment and control groups. Magnification: X1000.

HISTOLOGY III

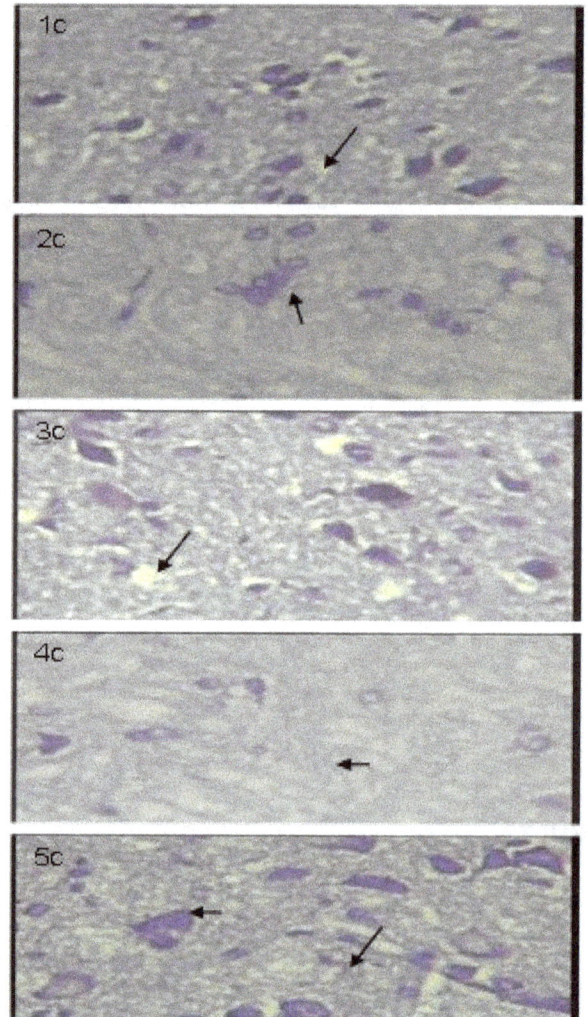

Figure 3. Comparative study of SC (1c-5c) for treatment groups. Magnification: X 1000 (Arrow heads indicates cell architecture and interlacing fibrous layers). Observe the Metachromasia in 5C the prominence of the axons radiating from the basket cells and the connections of the fibrous layers in between the cell rich layers.

them are observed to be smaller than others and are surrounded by vacuolar spaces of 0.5 μm in diameter; 3A. The proportion varied with every position viewed and the degenerating cells appeared prominent in the upper portion of the outer pyramidal layer where cells are smaller and relatively more separated. They were less intensely stained relatively due to cellular degeneration and reduction in Nissl substance. In the superior colliculus, the deep layers showed more degenerative

changes, wherein the quantity of fibrous tissue had surpassed those of other layers (3C). Fibrous layers: The lateral geniculate body showed minimal degenerative changes in the parvocellular part and the magnocellular part remained relatively unchanged as one could distinguish between the spiny and non spiny neurons (3C).

The tufted ones are more distinct than the few bitufted and Chandler cells in the parvocellular layers. The neurons are dilated.

Group 2

There is an overall distortion of the general cellular arrangements as compared with those of the two previous groups. The border between the molecular and the outer granular layer is obscured; arrow head in (Figure 1), 2A. Large vacuolar spaces of 0.8 μm in diameter are seen in the pyramidal and stellate cells of the outer granular and pyramidal layer and cells seemed to be undergoing degenerative changes. The inner granular and pyramidal layers show a decrease in the number of cells with enlarged cells interspersed among abnormally small cells. The fibrous layer is predominant throughout the superior colliculus in the stratum opticam just beneath the stratum Griseum Superficiale.

The cellular component is almost completely altered. Cells are small, with large prominent vacuoles while fibrous strands were observed in both directions rather than being interlaced as seen in the earlier groups. Presumably due to degenerating changes, the stratum Zonale however, remained relatively unchanged (Figures 3 and 2C). The parvocellular and magno-cellular layers (Figure 2), (3B) showed changes similar to those of the superior colliculus, the synapses appeared distorted and are seen as swollen knobs while the few cells observed are small and themselves surrounded by halos such that spaces are seen in between bundles, although certain areas appeared relatively unchanged (2C).

Group 1

The general, histological feature appeared conspicuously distorted and intensely stained, affecting cells in all layers of the visual cortex. This layer displayed large vacuolar spaces of about 1 μm (Figure 1), (1A) and obscured intercellular/interlayer bordering, in the superior colliculus and the lateral geniculate body. Identification is difficult even at x1, 000 (light microscopy). In the superior colliculus, the fibrous strands predominates with small cells scattered in between the fibrous layers. The lateral geniculate body was observed as being more fibrous with fibres directed in one direction. Vacuoles are numerous and neurons are small and spiny (Figure 2 (1B) and Figure 3(1C).

Group 4

This group shows histological features similar to those of the control. The major changes observed were in the outer pyramidal layer where certain pyramidal cells become enlarged compared to the surrounding cells. Vacuoles are more prominent and pale staining (Figure 1), (4A). Vacuoles are greatly decreased and the cells are smaller in size. Physical assessment of the laboratory animals; lethargy, sluggishness, and a progressive general weakness was observed with increasing dosage. Weight loss at the end of the period of administration was almost in high dose treatment group, which was abruptly compared to the low dose treatment and control groups. It is important to note the general decrease in the size of

the vacuolar spaces with reduction in the quantity of cassava in the diet.

DISCUSSION

In correlation with the histological appearance, the fourth group (0.12 mg of cyanide in 10 gms of cassava); showed an increased intercellular distance and implies high membrane transport activity indicated by a rise in alkaline phosphatase (ALP) activity. For the major experimental groups, there is a decreasing trend in (ALP) activity with increasing cyanide concentration. In comparing each major experiment against its corresponding withdrawal group, it was shown that the first two pairs displayed a decreased alkaline phosphatase activity, while the third category showed an increase. Thus, it could be inferred that high doses caused an increased alkaline phosphatase activities. Low doses also increased alkaline phosphatase activities but did not attain levels observed in the high dose treatment groups. In another experiment, withdrawal of group 3 coupled with banana extract administration (Methionine) caused neutralization of the cyanogenic glycoside leading to an increase in the rate of cellular regeneration in the withdrawal groups treated with the aqueous banana extract. As generally observed, increase in vacuolar spaces coupled with an increase in the distribution of such affected areas correlates with an increase in acid phosphatase activity per minute similar to the result of Di-Filipo et al. (2008).

Although, they went further to examine these parameters at higher dosage of cyanide extract but found out that the effects of cyanide at these doses were deleterious and almost irreversible on long term treatment. However, Osuntokun et al. (1981) reported an increase in the level of β–glucoronidase and sulphur excretion. All these structural changes and shift in enzyme activity was observed as a case of induced oxidative stress. However, statistically, the values are deviant from the median such that no regular pattern of change was observed (Oke, 1979; Nishimura et al., 2010), which implies that a multiple mode exist and the smallest value observed was zero. The mean activity at 180 s is 4.500 ± 0.49. The changes in the level of activities of these enzymes were observed to be characteristic of cells undergoing oxidetive stress. The distension observed in the cells of the molecular layer in the (30 gms of cassava) treatment group coupled with the increase in the size of the vacuolar space at first and a fluctuation afterwards will imply a phasic mechanism of neuronal damage, while the initial increase in acid and alkaline phosphatase activity will imply an early response to oxidative stress. Recall that in the (2.5 gms of cassava) treatment group, minimal changes were observed structurally, whereas in the histochemical quantification, they showed the highest fluctuation in enzyme activity. This was similar to the report of Okafor et al. (2002). Lower doses however, are capable of

Table 2. Result summary.

Group	V1	LGB	SC
1	Necrosis	Necrosis	Necrosis
2	Necrosis	Necrosis	Necro/Apoptosis
3	Apoptosis	Necrosis	Necro/Apoptosis
4	Necro/Apoptosis	Apoptosis	Necro/Apoptosis
5	Apoptosis	Necrosis	Necro/Apoptosis

inhibiting mitochondria activity thus causing the cell to heat up and expand as it is in a shift from aerobic to anaerobic metabolism. This is due to the effects of cyanide being capable of binding to the Fe^{2+} and Fe^{3+} present in Cytochrome oxidase thus impairing the electron transport chain. The integrity of the membrane is altered and its ability to mediate proper transport becomes jeopardized thus alkaline phosphatase activity tends to increase at first since it is the enzyme that mainly responsible for membrane transport.

Hubel et al. (1987) reported that a major association occurs in cyanide metabolism and calcium uptake which serves as a form of neutralizing mechanism in neurons. It is noteworthy to emphasize that calcium ions are capable of stimulating vacuole formation as observed in the normal electrophysiology of neurons, but may contain different membrane proteins that are not recognized as the usual cell inclusions, they are thus digested by lysosomal enzymes from the primary stage till the stage of formation of secondary autophagic bodies which were observed as vacuolar spaces in the treatment groups. This can be further elucidated by the results of the histochemical quantification which shows the initial high levels of acid phosphatase (a major lysosomal enzyme). In the moderate dose treatment group, some of the cells (Neurons) of the presumptive visual cortex appeared dead and shrunken and some other ones were found to be undergoing cellular degeneration, This may be attributed to a potential shift to anaerobic metabolism despite the presence of oxygen thus causing a type II Histotoxic anoxia, coupled with an increase and gradual build up of oxygen species (Isom et al., 1999; Nelson, 2006). The mode of cell degeneration of neurons found the high and low dose groups thus have the following features; distortion of membrane and axons, presence of large vacuolar spaces which are characteristic of necrosis while the moderate dose group showed decreased Nissl substance with larger vacuolar spaces and with little distortion in membrane activity compared to the high dose groups; the moderate dose have features of both apoptosis and necrosis which we describe as partial apoptosis and necrosis (Table 2), (Tor-Agbydye et al., 2003; Joshua et al., 2007).

In the superior colliculus and lateral geniculate body, as stated in the result, there was an increased fibrosis because the cells have regressed, and the axons are responding to the anoxic state of its corresponding neuronal cytoplasm characteristic of necrosis, although vacuoles are present but are of negligible size, the prominent observation was obliteration of the cellular layer by fibrous layers. Thus, they showed a marked decrease in length and knobbed ends creating spaces in between cells. Osuntokun et al. (1981) and Oke (1979a) reported similar results but stated that it was non conclusive and that no positive correlation existed between the histological and statistical findings, no mechanism was des-cribed and the only enzyme assayed was β-gluco-ronidase in which an early increase was also reported to demonstrate the initial increase in lysosomal activity (Denison et al., 2009; Sotoblanco et al., 2002; de Haro, 2009), assayed for neurotransmitter hormones and they observed low activity and fluctuations in the level of cerebral calcium.

Osuntokun et al. (1981) treated the Wistar rats with amino acid supplements of tyrosine and methionine (Batharcharya and Tulwasami, 2008). Adopting the methods of De la Cruz et al. (2009), treatment with banana extract which contains sulphur and sulphur (containing amino acids) gave a faster regenerating effects in the withdrawal groups. Statistical analysis showed tendency towards normal in these withdrawal groups compared to the untreated withdrawal group. At this point, it will be emphasized that the cyanide in unprocessed *Manihot escculenta* even at low concentration are capable of inducing oxidative stress and state of histotoxic anoxia which may be restricted to certain regions of the visual relay centers depending on dosage and that this effects depends on the dietary pattern of the individual.

Conclusion

Cassava can be described as being neurotoxic, the gradual deteriorating vision in individuals in cassava endemic region can be attribute to the neurotoxic effects of the cyanogenic glycosides and other phytotoxins found in cassava, the mode of cell death and the morphological changes observed, is dose dependent and most important are protein diet dependent, therefore, if two persons are exposed to the same dose of cassava diet and one has protein supplements in addition, the neurotoxic effects of cassava will be less deleterious in the individual exposed to protein diets (Osuntokun, 1981). Sulphur containing amino acids like cysteine and tyrosine have been found to play an important role in the removal of cyanide as thiocyanate, the presence of this sulphur groups could be said to be the rate limiting factor in the excretion of cyanide as thiocyanate which is the major defense of the body against cyanide intoxication (Ernesto et al., 2002).

The toxicity of the phytotoxins follow a similar mechanism by inhibiting Cytochrome C oxidase, a terminal enzyme in the electron transport chain, thus generation NO(Nitric oxide) tension at complex I and III

(Isom and Way, 1984). NO are endogenous modulators of cell activity, but if present in excess concentration, it could trigger cell death, the concentration also determines the mode of cell death (Isom et al., 1999), other mechanism of toxicity involves cyanide binding to the binuclear centre by displacing molecular oxygen to create oxygen tension, this has been found to induce a state of histotoxic anoxia in cells (Li et al., 2000). The dose of the cassava in diet was found to have played a major role in the mode of cell death. The mode observed in the V1 was found to be an apoptosis-necrosis continuum for moderate dose treatment and necrosis for extreme dose treatment, while the SC and LGB was mainly necrosis irrespective of the dosage.

ACKNOWLEDGEMENTS

The authors wish to thank Trinitron Biotech LTD, Mr. Gerry Nash and Helen Odedairo.

REFERENCES

Bhattacharya R, Tulsawani R (2008). In vitro and in vivo evaluation of various carbonyl compounds against cyanide toxicity with particular reference to alpha-ketoglutaric acid. Drug Chem. Toxicol., 31(1): 149-61.

Bhattacharya R, Lakshmana RPV (2001). Pharmacological interventions of cyanide-induced cytotoxicity and DNA damage in isolated rat thymocytes and their protective efficacy in vivo. Toxicol. Lett., 119: 59-70.

de Haro L (2009). Disulfiram-like syndrome after hydrogen cyanamide professional skin exposure: two case reports in France. J. Agromed., 14(3): 382-384.

de la Cruz CC, Medialdea NP, Romero AM (2009). Amyotrophic lateral sclerosis syndrome-plus and consumption of cassava (Manihot) . Is this a new presentation of the neurotoxic motor-neuron syndrome? Neurologia., 24(5): 342-343.

Denison TA, Koch CF, Shapiro IM, Schwartz Z, Boyan BD (2009). Inorganic phosphate modulates responsiveness to 24,25(OH)2D3 in chondrogenic ATDC5 cells. J. Cell Biochem., 107(1): 155-162.

Di Filippo M, Tambasco N, Muzi G, Balucani C, Saggese E, Parnetti L, Calabresi P, Rossi A (2008). Parkinsonism and cognitive impairment following chronic exposure to potassium cyanide. Mov Disord., 23(3): 468-470.

Ernesto M, Cardosso AP, Nicala D, Mirone E, Massasa F, Cliff J, Haque MR (2002). Persistent Konzo and Cyanogenic toxicity from cassava in Northern Mozambique Acta. Trop., 82(3): 357-362.

Fiske BK, Brunjes PC (2001). NMDA receptor regulation of cell death in the rat olfactory bulb. J. Neurobiol., 47: 223-232.

Ha JS, Lee JE, Lee JR, Lee CS, Maeng JS, Bae YS, Kwon KS, Park SS (2010). Nox4-dependent H(2)O(2) production contributes to chronic glutamate toxicity in primary cortical neurons. Exp. Cell Res.10:316(10):1651-61. Epub 2010 Apr 2.

Hubel DH, Wiesel TN (1972). Laminar and columnar distribution of geniculo-cortical fibers in the macaque monkey. J. Comp. Neurol., 146: 421-450.

Isom GE, Gunasekar PG, Borowitz JL (1999). Cyanide and neurodegenerative disease. In Chemicals and Neurodegenerative Disease (Bondy SC, Ed.), Prominent Press, Scottsdale, AZ, pp. 101-129.

Jerome N, Roger A. Baldwin S, Allen G, Denson GF, Claude G (2003). Wasterlain Hypoxic neuronal necrosis: Protein synthesis-independent activation of a cell death program PNAS, 100(5): 2825-2830.

Joshua JB, John S, Flora MV, Nenad Š, Pasko R (2007). Notch regulates cell fate and dendrite morphology of newborn neurons in the postnatal dentate gyrus PNAS, 104(51): 20558-20563.

Lee B, Park J, Kwon S, Park MW, Oh SM, Yeom MJ, Shim I, Lee HJ, Hahm DH (2010). Effect of wild ginseng on scopolamine-induced acetylcholine depletion in the rat hippocampus. J. Pharm. Pharmacol., 62(2): 263-271.

LI R, Sonik A, Stindl R, Rasnick D, Duesberg P (2000). Aneuploidy vs. gene mutation hypothesis of cancer: Recent study claims mutation but is found to support aneuploidy. USA. Proc. Natl. Acad. Sci., 97: 3236-3241.

Ming Z, Stefan M, Kioumars D, Marie C, Robert MC, Clas BJ, Hjalmar B, Oleg S, Jonas F, Ann MJ (2003). From the Cover: Evidence for neurogenesis in the adult mammalian substantia nigra PNAS 100(13): 7925-7930.

Nelson L (2006). Acute cyanide toxicity: mechanisms and manifestations. J. Emerg. Nurs., 32(4): S8-11.

Nishimura K, Kitamura Y, Taniguchi T, Agata K (2010). Analysis of motor function modulated by cholinergic neurons in planarian Dugesia japonica. Neuroscience., 168(1): 18-30.

Okafor PN, Okoronkwo CO, Maduagwu ON (2002). Occupational and dietary exposure of humans to cyanide from large scale cassava processing and ingestion. Food Chem. Toxicol., 40(7): 1001-1005

Oke OL (1979). Some aspects of the role of cyanogenic glycosides in nutrition. Wld. Rev. Nutr. Diet., 33: 70-103.

Osuntokun BO (1981). Cassava diet, chronic cyanide intoxication and neuropathy in Nigerian Africans. Wld. Rev. Nutr. Diet., 36: 141-73.

Pearse AGE (1960). Histochemistry. Little, Brown and Company. Boston 1960.

Solomonson LP (1981). Cyanide as a metabolic inhibitor. In Cyanide in Biology, VennesLand B, Conn EE, Knowles CJ, Wesley J, Wissing F. Academic Press, London, New York, Toronto, pp. 11-18.

Sotoblanco RA, Aparicio MA (2002). Relationships between dietary cassava cyanide levels and Brain performance. Nutr. Reports Int., 37: 63-75.

Yin J, Pan SY, Zhou L, Lü TM, Luo YF, Lu BX, Nan FYK, Da Xue Xue B (2007). Pathological analysis of heroin spongiform leukoencephalopathy Chin. J. Pathol., 27(6): 881-883.

19

Factors affecting seasonal prevalence of blood parasites in dairy cattle in Omdurman locality, Sudan

Mohammed Safieldin A.[1], Atif Abdel Gadir E.[2*] and Khitma Elmalik H.[2]

[1]Federal Ministry of Animal Resources and Fisheries, P. O. Box 293, Khartoum, Sudan.
[2]Faculty of Veterinary Medicine, University of Khartoum, P. O. Box 32, Khartoum North, Sudan.

This study was conducted in Al-Rodwan project in Omdurman to investigate the prevalence of blood parasites in dairy cattle during different seasons. A total of 290 animals were examined during three seasons: dry cool (100), dry hot (95) and wet hot (95). The results showed that the prevalence of blood parasites during different seasons was 8, 5.25 and 6.32% for dry cool, dry hot and wet hot season, respectively. The prevalence of *Theileria* species infection was found to be 7, 5.25 and 6.32% for dry cool, dry hot and wet hot season, respectively. While the prevalence of *Babesia* species infection was only recorded in the dry cool season as (1%). There was no effect ($\chi^2 = 0.6$, $p > 0.05$) of season on the occurrence of blood parasites. Strong association (t-test= −43.6, $p < 0.05$) was found between presence of blood parasites and milk yield.

Key words: Dairy cattle, blood parasites, season, Sudan.

INTRODUCTION

Intensive and semi-intensive production system of Sudan distributed either within aggregation sites in different locations or in small herds located in different sites around towns. The high needs for animal proteins especially milk and milk products in recent years in Khartoum State oriented the producers to import highly milk producing foreign breeds to face the human consumption. Parasitic diseases affect the milk industry by the direct effect on milk production, difficult control of vectors, high cost of the treatment and financial implications for farms management to prevent the parasitic infestations.

Therefore, this study was planned to investigate the presence of blood parasites in dairy cattle during different seasons in Al-Rodwan project, Omdurman.

The effect of blood parasites and their vectors on cattle productivity differ according to several factors such as the causative agent, breed and the disease status (clinical, sub-clinical or chronic). Many studies were conducted to study the impact of each of these factors on cattle productivity such as milk yield and weight gains (Pholpark et al., 1999; Michael et al., 1989; Gitau et al., 2001;

Muragura et al., 2005).

Many workers conducted research in Sudan on the scope of epidemiological aspects of blood parasites in cattle. Many of these studies discussed the evidence and prevalence of parasitic infections in different parts throughout Sudan (Karib, 1961; El Bihari et al., 1974; Osman, 1992; Abdel et al., 1994; Hassan, 2003).

MATERIALS AND METHODS

Area of study

Al-Rodwan project in Omdurman was chosen to screen dairy cattle for blood parasites. It is located in the North Western site of the locality and considered as the main dairy cattle aggregation site in the area with approximately 5,000 head according to the record of Ministry of Agriculture, Animal Resources and Irrigation, Khartoum State (2003).

Study population

Selected cattle from dairy farms in Al-Rodwan project were investigated during dry cool (February-March), dry hot (May-June) and wet hot (August-September) seasons. A hundred animals from the chosen herds of animals were studied during the above seasons. The majority was of cross breeds (89%) and the rest was a local breeds (11%).

*Corresponding author. E-mail: atifvet@yahoo.com.

Sampling and sample collection

The sampling was done according to cluster sampling method (two stage sampling) as described by Thrusfield (1995). A total of 290 blood samples were collected during the three different seasons from the same animals identified. The blood was collected in the morning from the jugular veins using vacutainers with EDTA. The samples were labeled with animal number, placed in an ice box at 4°C and transported as soon as possible to the laboratory before processing for parasitological examinations.

Parasitological examinations

Wet mount

One drop of fresh blood was placed on a slide, covered with a cover slip and examined microscopically for detection of motile parasites at 10×40 magnification.

Buffy coat examination (Woo, 1970)

A capillary tube was taken; the end of capillary tube was put on a drop of the blood sample, filled to about three-quarters and sealed by plastoseal at one end. It was placed in the haematocrit centrifuge which was run for 5 m. After centrifugation the packed cell volume was read, and then the capillary tube was placed onto a clean slide and covered with one drop of distilled water and examined microscopically at 10×40 magnification to detect trypanosomes and microfilariae.

Thin blood film

A small drop of fresh blood was put in the middle of one end of the slide, and spread right across the slide and then air dried. The slide was labeled using a pencil. Blood films were fixed in absolute methyl alcohol for 2 m, stained in 5% diluted Giemsa's stain for 45 m, and washed in distilled water and then dried. Immersion oil was put on the blood film and examined microscopically for the detection of blood parasites at 10×100 magnification.

Data analysis

Stata 6.0 for Windows 98/95/NT was used for data analysis.

RESULTS

The presence of blood parasites using blood film in Al-Rodwan dairy project was investigated during different seasons. The results showed that a prevalence of blood parasites was 8 (8%), 5 (5.25%) and 6 (6.32%) for dry cold, dry hot and wet hot season, respectively (Table 1). The prevalence of Theileria species infection was 7, 5.26 and 6.32% in dry cold, dry hot and wet hot season, respectively. Prevalence of Babesia species infection was only recorded in dry cool season as 1% (Table 2). There was no effect of season (χ^2 = 3.1, p> 0.05) on the presence of blood parasites. A positive association (t-test= −43.6 - p< 0.01) was found between presence of blood parasites and milk yield of cows resulting in reduction in milk production.

DISCUSSION

The study of blood parasites in dairy cattle during different seasons in Omdurman area revealed a higher prevalence of Theileria species infection compared to Babesia species infection. Similarly, different workers recorded the presence of blood parasites in both intensive and pastoral production systems of Sudan (Abdalla, 1984 and Hassan, 2003). The presence of blood parasites infection in dairy cattle in Al-Rodwan project was attributed to the fact that most of the farms in this area were infested with ticks; particularly, all the farms built of mud and block stones which constitute a suitable environment for that ticks.

There was no effect of season (χ^2=3.145, p>0.05) on the prevalence of blood parasites. This finding disagreed with the results of different researchers. Perez et al. (1994) found that season was a risk factor for presence of Babesia bovis infection. El Mentenawy (2000) found during a study aimed at investigating the parasites infecting cattle blood at Al-Qassim region in Saudi Arabia, that theileriosis prevalence reached a maximum in (84.3%) in both autumn and summer seasons, while it dropped to 59.4% in spring.

The disagreement of this study could be attributed to application of acaricides and administration of anti-piroplasmal drugs by farm owners at intervals, which could have affected the prevalence of blood parasites during different seasons. It could also be due to the mismanagement practiced at Al-Rodwan while allows for continuous tick challenge throughout the year. An association (p < 0.01) was observed between presence of blood parasites and milk yield of producing animals. Similar results were reported by different researches. Michael et al. (1989) studied the effect of theileriosis on milk yield and suggested that it caused decrease in milk yield. Patarroyo et al. (1995) stated that bovine babesiosis caused by Babesia bigemina remains a significant constraint to milk cattle production. Although we could not link PCV with blood parasites, yet this could be one of the major factors that affect milk yield.

Other blood parasites, particularly Trypanosoma or microfilaria were not encountered during this study, although reported in other parts of the capital Khartoum. Possible explanation is that Al-Rodwan project is found in an area where present conditions are not suitable for insect propagation.

This should not be overlooked as micro-climates may be created through negligence and lack of awareness and that used permit the infestation of insect species that are known as mechanical or biological vectors of some parasites. This may come as a result indiscriminate introduction of cattle which may originate from infected herds for example, with Trypanosoma species or microfilaria In conclusion, infection with Theileria species and Babesia species were prevalent in Omdurman. Infection with blood parasites had economic impact due to reduction in milk production.

Table 1. Summary of the blood parasites survey in Al-Rodwan dairy project.

Unit	Season Frequency (%)		
	Dry cool	Dry hot	Wet hot
Total of animal examined	100	95	95
Buffy coat			
Positive	0(0)	0(0)	0(0)
Negative	100(100)	95(100)	95(100)
Wet mount			
Positive	0(0)	0(0)	0(0)
Negative	100(100)	95(100)	95(100)
Thin blood stain			
Positive	8(8)	5(5.26)	6(6.32)
Negative	92(92)	90(94.74)	89(93.68)

Dry cool: February-March   Dry hot: May-June   Wet hot: August-September

Table 2. Prevalence of blood parasites during different seasons in Al-Rodwan dairy project.

Season	No. examined	Prevalence (%)		Over all
		Theileria spp.	*Babesia* spp.	
Dry cool	100	7	1	8
Dry hot	95	5.26	0.00	5.26
Wet hot	95	6.32	0.00	6.32

Dry cool: February-March   Dry hot: May-June   Wet hot: August-September

ACKNOWLEDGEMENTS

The authors thank the staff of Vet Serve Organization in Al-Rodwan dairy project.

REFERENCES

Abdalla HM (1984). Studies on *Babesia bigemina* in cattle in Northern Sudan. MSc thesis, University of Khartoum.

Abdel RMB, Zakia AM, Bakheit HA, Halima MO, Fayza AO, Osman AY (1994). Epidemiology of bovine tropical theileriosis in the Sudan. Proceeding of a workshop held at the Sudan Veterinary Association Residence, Khartoum, Sudan 4-5 May 1994. (Eds. Atelmanan AM and Kheir SM), pp. 6-17.

El Bihari S, Gadir FA, Suleiman H (1974). Incidence and behavior of microfilariae in cattle. Sud. J. Vet. Anim. Husb., 15(2): 82-85.

El Mentenawy TM (2000). Prevalence of blood parasites among cattle at the central area of Saudi Arabia. Vet. Parasitol., 87(2-3): 231-236.

Gitau GK, McDermott JJ, McDermott B, Perry BD (2001). The impact of *Theileria parva* infections and other factors on calf mean daily weight gains in smallholder diary farms in Murag'a District, Kenya. Prev. Vet. Med., 51(3-4): 149-160.

Hassan DAS (2003). Epidemiological Studies on Tropical Theileriosis (*Theileria annulata* infection of cattle) in the Sudan. MSc thesis, University of Khartoum.

Karib AA (1961). Animal trypanosomiasis in the Sudan. J. Vet. Sci. Anim. Husb., 2: 39-46.

Michael SA, el Refaii AH, McHardy N, Rae DG (1989). Effect of treatment of chronic theileriosis with buparvaquone on milk yields. Trop. Anim. Health Prod., 21(4): 218-222.

Muragura GR, McLeod A, McDermott JJ, Taylor N (2005). The incidence of calf morbidity and mortality due to vector-borne infection in smallholder dairy farms in Kwale District, Kenya. Vet. Parasitol., 130: 305-315.

Osman OM (1992). *Theileria annulata* in the Sudan. *In*: Recent development in research and control of *Theileria annulata*. Proceeding of a workshop held at ILRAD, Nairobi, Kenya, 17-19 September 1990 (ed. T. T. Dolan), pp. 125.

Patarroyo JH, Prates AA, Tavares CA, Mafra CL, Varga MI (1995). Exoantigenes of an attenuated strain of *Babesia bovis* used as a vaccine against bovine babesiosis. Vet. Parasitol., 59(3-4): 189-199.

Pholpark S, Pholpark M, Polsar C, Charoenchai A, Paengpassa Y, Kashiwazaki Y (1991). Influence of *Trypanosoma evansi* on milk yield of diary cattle in northern Thailand. Prev. Vet. Med., 42(1): 39-44.

Perez E, Herrero MV, Jimenez C, Herd D, Buening GB (1994). Effect of management and host factor on seroprevalence of bovine anaplasmosis and babesiois in Costa Rica. Preventive Vet. Med., 20(1-2): 33-46.

Thrusfield M (1995). Veterinary Epidemiology. 2nd ed. Blackwell Science Ltd. UK.

Woo PTK (1970). The haematocrit centrifuge technique for the diagnosis of African trypanosomiasis. Acta Tropica, 27: 384-387.

20

In-vitro and *-vivo* anti-*Trypanosoma evansi* activities of extracts from different parts of *Khaya senegalensis*

I. A. Umar*, M. A. Ibrahim, N. A. Fari, S. Isah and D. A. Balogun

Department of Biochemistry, Ahmadu Bello University, Zaria, Kaduna State, Nigeria.

The *in vitro* activities of the aqueous and ethanolic extracts of the leaves, root bark and stem bark of *Khaya senegalensis* on *Trypanosoma evansi* were evaluated. The ethanolic extract of the stem bark was found to possess the highest *in vitro* activity among the six extracts tested; as it eliminated the parasites within 5 min post incubation at concentrations of 0.5 and 1 mg/ml. This extract was therefore used to treat rats experimentally infected with *T. evansi* at concentrations of 20, 40 and 80 mg/kg body weight, beginning 7 days post infection (p.i). At the termination of the experiment on day 13 p.i, the stem bark ethanolic extract significantly ($P < 0.05$) kept the parasitemia lower than was observed in the untreated infected rats, whereas the parasites were eliminated from the bloodstream of Diminal-treated rats at day 9 p.i. All the infected animals developed anaemia whose severity could not be ameliorated by the extract treatment. It was therefore concluded that the stem bark ethanolic extract of *K. senegalensis* possessed both *in-vitro* and *-vivo* anti-*T. evansi* activity but could not prevent the disease –induced anaemia.

Key words: *Khaya senegalensis*, *Trypanosoma evansi*, anti – trypanosomal.

INTRODUCTION

Animal trypanosomiasis is still a major factor retarding the growth of the livestock industry in Africa. The disease has undergone a dramatic and devastating resurgence in recent years especially in sub-Saharan Africa (Welburn et al., 2001); and thus an important priority for biomedical and public agencies, agricultural sector and the scientific community (Aksoy, 2003). One of the important pathogenic trypanosomes in animals is *Trypanosoma evansi*; the causative agent of *Surra* that is highly fatal to a number of domesticated mammals such as camels, horses and water buffaloes among others (Vanhollebeke et al., 2006). Since the adaptation of the parasite to mechanical transmission by blood sucking insects (tabanids), the disease has spread beyond its original distribution in sub-Saharan Africa and is now also present in South America, North Africa and large parts of Asia (Vanhollebeke et al., 2006).

The existing treatments of trypanosomiasis are challenged with problems comprising drug resistance, toxicity and expensive/limited drugs (Gutteridge, 1985, Atawodi et al., 2002). Therefore, there is a need to search for cheaper, more effective, easily available and less toxic chemotherapeutic agents for combating trypanosomiasis. The use of herbal preparations for the treatment of the disease still holds a strong potential in that some ethnomedicinal plants have been demonstrated to contain potent trypanocides (Igweh and Onabanjo, 1989; Owolabi et al., 1990; Nok et al., 1993; Atawodi, 2005).

The indigenous use of *Khaya senegalensis* (Juss), a dry zone mahogany belonging to the family *Meliaceae*, in the treatment of trypanosomiasis has been reported (Atawodi et al., 2002) and *in vitro* anti-trypanosomal activity of the plant against *Trypanosoma brucei* has been demonstrated (Wurochekke and Nok, 2004; Atawodi, 2005). More recently, the *in vivo* action of the stem bark aqueous extract of the plant against *T. brucei* (Ibrahim et al., 2008) has been reported, but information on the *in vitro* and/or *in vivo* action of the plant against *T. evansi*, with broadest host and geographic range among the pathogenic animal trypanosomes is still lacking.

Hence this work was designed to evaluate the *in vitro*

---

*Corresponding author. E-mail: smaumar@yahoo.com.

Abbreviation: PCV, Packed cell volume.

and *in vivo* anti-trypanosomal activities of extracts from various parts of *K. senegalensis* against *T. evansi*.

MATERIALS AND METHODS

Sample collection

The stem bark, root bark and leaves of mature *K. senegalensis* were collected from the botanical garden of Biological Sciences Department, Ahmadu Bello University, Zaria, Nigeria; and were identified at the herbarium of the same Department with a voucher number of 90081 which was deposited. The parts of the plant were thoroughly washed and shade-dried for a week to a constant weight. The dried parts were pounded to fine powder with mortar and pestle and then stored in dry containers until needed.

Experimental animals

A standard protocol was drawn up in accordance with the Good Laboratory Practice (GLP) regulations of the World Health Organization (WHO). The principles of laboratory animal care were also duly followed in this study. Apparently healthy white albino rats of both sexes weighing between 90-172g were used for the work and were obtained from Pharmacology Department, Ahmadu Bello University, Zaria, Nigeria. The animals were kept in well ventilated laboratory cages with 12 hours day/night cycles. The rats were maintained on a commercial poultry feed (ECWA Feeds, Jos-Nigeria) and drinking water *ad libitum*.

Extracts preparation

One hundred grams of the fine powdered plant parts were soaked in 300 ml of either distilled water or ethanol and sequentially extracted by shaking for 6 h on wrist action shaker. The preparations were left to stand for a further 24 h. After filtration through Whatmann's filter paper, samples were concentrated to dryness on a water bath at 40°C, packaged in water-proof polythene bags and stored in the refrigerator at 4°C until required (Atawodi, 2005).

Trypanosome

Trypanosome evansi (Sokoto strain) was obtained from an infected mouse in the Parasitolgy Department, Faculty of Veterinary Medicine, Ahmadu Bello University, Zaria, Nigeria. Parasites harvested from a donor rat at peak parasitaemia were used for both the *in vitro* studies and infection of experimental animals.

In vitro screening

Exactly 1 mg of the different extracts were weighed and dissolved in 1 ml of phosphate buffered saline (PBS). Serial dilution of this stock solution was done using PBS to obtain concentrations ranging from 0.25 to 1 mg/ml. Assessment of the *in vitro* anti-trypanosomal activity was performed in triplicates in 96 well microtitre plates (Flow laboratories Inc., Mclean, Virginia, USA). In wells of the microtitre plates, 20 µl of each extract was incubated at 37°C, with 40 µl of the infected blood (obtained from a donor rat with about 10^9 *T. evansi* per ml of blood), achieving effective extract concentrations of 0.083 to 0.332 mg/ml in the reaction mixtures. For control, the 20 µl of extract was replaced with PBS. Parasite count was then monitored on a glass slide (covered with a covering slip) and observed under a microscope at ×400 magnification. The number of motile parasites was counted at 5 min intervals for 1 h.

In vivo activity of the stem bark ethanolic extract

35 rats were divided into seven groups of five rats each. The rats in five groups were intraperitoneally infected with 10^6 *T. evansi* per 100 g body weight (b.w) and the level of parasitaemia monitored daily by the Herbert and Lumsden (1976) method for the 14 days period of the experiment. Exactly, 20, 40 and 80 mg/kg b.w. of the extract (Inf +EDI, Inf + EDII and Inf + ED III respectively) were administered orally and 0.5 ml/100 g b.w. of diminal® (a standard drug for African Trypanosomiasis that contains 445 mg diminazene aceturate + 555 mg phenazone/g. Eagle Chemical Company LTD, Ikeja, Nigeria) given intraperitoneally to four different groups of the infected rats, starting a day after the parasites were first detected in the bloodstream (Day 7 post infection); the remaining infected group was left untreated (infected control). One group of the uninfected rats was orally administered with 40 mg/kg b. w. of the extract (extract control) whereas the other group was maintained as uninfected untreated (normal) control. The pre-infection and terminal packed cell volumes of all the rats were determined by the microheamatocrit method.

Analysis of data

Paired means were compared using students' t-test.

RESULTS

Both the aqueous and ethanolic extracts of the various parts of *K. senegalensis* showed *in vitro* anti-trypanosomal activity in a dose-dependent fashion (Figures 1 and 2) with the ethanolic extracts seemingly exhibiting a higher activity against the *T. evansi* parasites. However, the stem bark ethanolic extract possessed the highest activity against the parasite and thus was used for the *in vivo* studies.

The parasitaemias of both the infected control and the infected treated groups are presented in Figure 3. While a progressive increase in parasitaemia was observed in the infected controls; treatments with the extract significantly (P < 0.05) lowered the level of parasitaemia when compared to the infected untreated rats, whereas the parasites totally disappeared from the bloodstream of diminal®-treated infected group on day 9, p.i.

The *T. evansi* infections in this work caused significant fall in PCV of infected rats; indicative of anaemia. However, while administration of Diminal® to infected animals significantly reduced the magnitude of decline in PCV, administration of the extract at the various doses had no effect on the disease-induced anaemia. The PCVs of the two groups of uninfected animals remained relatively constant.

DISCUSSION AND CONCLUSIONS

Some plant extracts have been demonstrated to contain potent trypanocidal constituents (Igweh and Onabanjo, 1989; Owolabi et al., 1990; Atawodi, 2005). So far, only aqueous and methanolic extracts of stem bark of *K. senegalensis* have been shown to possess *in vitro*

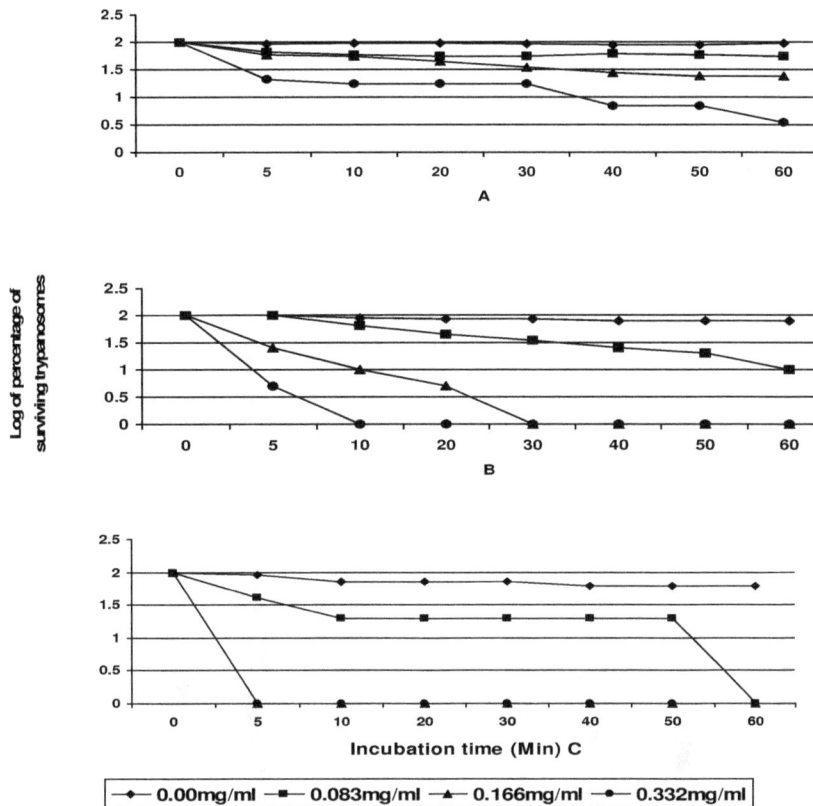

Figure 1. Profiles of log percentage surviving trypanosomes after incubation with various concentrations of ethanolic extracts of the leaves (A) root bark (B) and stem bark (C) of *Khaya senegalensis*.

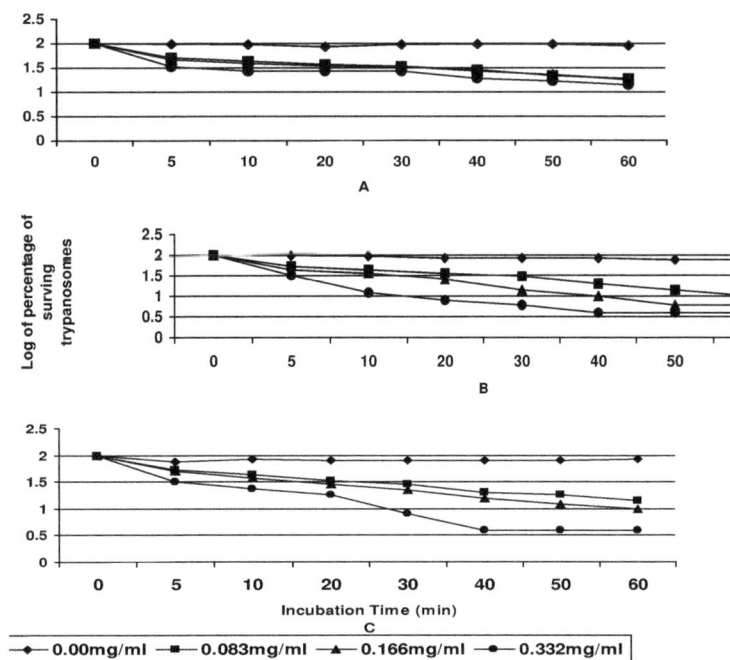

Figure 2. Profiles of log percentage surviving trypanosomes after incubation with various concentrations of aqueous extracts of the leaves (A)root bark (B)stem bark (C) of *K. senegalensis*.

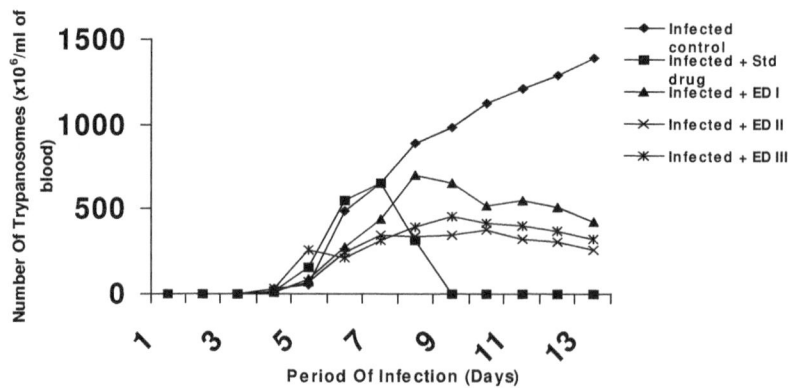

Figure 3. Effect of ethanolic extract of stem bark of *K. senegalensis* on *T. evansi* infected rats.

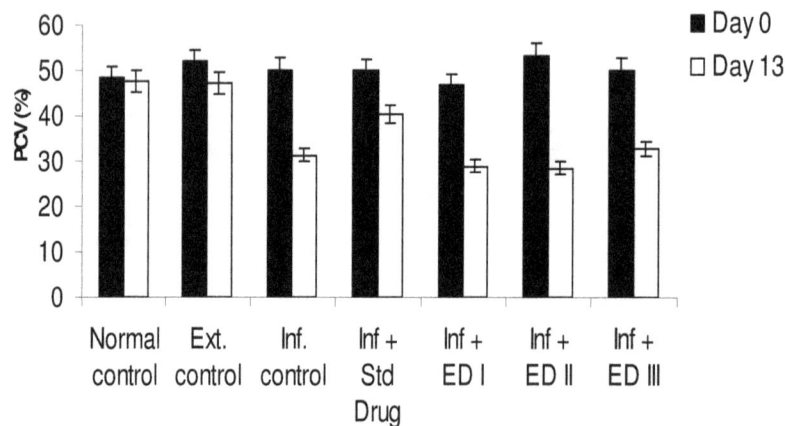

Figure 4. Effect of ethanolic extract of stem bark of *K. senegalensis* on the PCV level of *T. evansi* infected rats.

activity against *T. brucei*. (Wurochekke and Nok, 2004, Atawodi, 2005). This report reveals that the leaves, stem bark and root bark of *K. senegalensis* also contain some water and ethanol-extractable phytochemicals that possess *in vitro* activities against *T. evansi*.

Parasite motility constitutes a relatively reliable indicator of viability of most zooflagellate parasites (Kaminsky et al., 1996). Cessation or drop in motility of trypanosomes may therefore serve as a measure of anti–trypanosomal potential of the crude extract when compared to the control. The quantitative difference in anti–trypanosomal activities among the plant parts could be attributed to the variation(s) in concentration and composition of phytochemicals in the different parts since distinct function(s) is performed by all the parts and hence tend to produce slightly different chemical constituents. Since a plant with high *in vitro* anti–trypanosomal activity may have no *in vivo* activity and vice versa, because of peculiarities in the metabolic disposition of the plant's chemical constituents, we tested

the most active extract (ethanolic extract of the stem bark) for *in vivo* anti–trypanosomal activity so that a definite statement can be made on the anti-*T. evansi* activity of the plant.

The observed *in vivo* anti–*T. evansi* activity of the stem bark ethanolic extract of this plant support earlier reports that some plant extracts possess *in vivo* activities against trypanosomes (Asuzu and Chineme, 1990; Nok et al., 1993; Ibrahim et al., 2008). This could also provide the scientific basis for the traditional use of *K. senegalensis* in the management of trypanosomiasis (Atawodi et al., 2002).

The exert mechanism for the *in vivo* action of this extract is unknown since the active ingredient(s) were not isolated. However, previous reports attributed the trypanocidal activity of a number of tropical plants to the flavonoids (azaanthraquinone), highly aromatic planar quaternary alkaloids, barbarine and harmaine (Hopp et al., 1976, Nok, 2001). Furthermore, Sepulveda-Boza and Cassels (1996) suggested that many natural products exhibit their trypanocidal activity through interference with redox balance of the parasites acting either on the respiratory chain or on the cellular defenses against oxidative stress. This is because natural products possess structures capable of generating radicals that may cause peroxidative damage to trypanothione reductase that is very sensitive to alterations in redox balance. It is also known that some agents act by binding with the kinetoplast DNA of the parasite (Atawodi et al., 2003). Since the stem bark ethanolic extract of *K. senegalensis* has also been shown to contain flavonoids (Makut et al., 2008), it is thus possible that this extract acted through one or more of these mechanisms.

Anaemia is a constant feature of trypanosome infections whose severity is linked to the level of parasitemia (Umar et al., 2000). The extract did not affect the severity of anaemia in infected animals probably because parasitaemia, albeit low grade, was persistent in the animals and the aetiological factors involved in the heamolysis have been established before the extract treatment. The Diminal®-treated infected rats showed significant improvement in PCV from infection levels perhaps because the drug was able to eliminate parasites from the blood to levels undetectable by microscopic examination. We therefore concluded that the stem bark ethanolic extract of *K. senegalensis* possessed both *in-vitro* and *-vivo* anti-*T. evansi* activity. However, the antitrypanosomal activity of this plant reported herein appears to be relatively lower than the previously reported activity against *T. brucei* (Ibrahim et al., 2008) but further work on the toxicology, isolation and identification of the bioactive components would certainly reveal whether this plant could be exploited for the development of new generation of trypanocides. Furthermore, information reported in this study could be useful in assessing the overall antitrypanosomal activity of this plant.

REFERENCES

Aksoy S (2003). Control of tsetse flies and trypanosomes using molccular genetics. Vet. Parasitol., 115(2):125-145.

Asuzu IU, Chineme CN (1990). Effects of *Morinda lucida* leaf extract on *T. brucei brucei* infection in mice. J. Ethnopharmacol., 30: 307-313.

Atawodi SE (2005). Comparative *in vitro* trypanocidal activities of petroleum ether, chloroform, methanol and aqueous extracts of some Nigeria savannah plants. Afri. J. Biotechnol., 4(2): 177-182.

Atawodi SE, Ameh DA, Ibrahim S, Andrew JN, Nzelibe HC, Onyike E, Anigo KM, Abu EA, James BD, Njoku GC, Sallau AB (2002). Indigenous knowledge system for treatment of trypanosomiasis in Kaduna State of Nigeria. J. Ethnopharmacol., 79: 279-282.

Atawodi SE, Bulus T, Ibrahim S, Ameh DA, Nok AJ, Mamman M, Galadima M, (2003). *In vitro* trypanocidal effects of methanolic extracts of some Nigeria savannah plants. Afr. J. Biotechnol., 2(9): 317-321.

Gutteridge WE (1985). Existing Chemotherapy and its limitations. Brit. Med. Bull., 41: 162-168.

Hopp KH, Cunnigham IV, Bromel MC, Schermester LJ, Wahba KSK (1976). In vitro antitrypanosomal activity of certain alkaloids against *Trypanosoma lewisi*. Llyoydia, 39(5): 375-377.

Ibrahim MA, Njoku GC, Sallau AB (2008). *In vivo* activity of stembark aqueous extract of *Khaya senegalensis* against *Trypanosoma brucei*. Afri. J. Biotechnol. 7(5): 661-663.

Igweh AG, Onabanjo AO (1989). Chemotherapeutic effects of *Annona senegalensis* in *T. brucei brucei* infection in mice. J. Ethnopharmacol., 30: 307-313.

Kaminsky F, Nkuna MHN, Brun R (1996). Evaluation of African medicinal plants for their *in vitro* trypanocidal activity. J. Ethnopharmacol., 55: 1-11.

Makut MD, Gyar SD, Pennap GRI, Anthony P (2008). Phytochemical screening and antimicrobial activity of the ethanolic and methanolic extracts of leaf and Bark of *Khaya senegalensis*. Afri. J. Biotechnol. 7(9): 1216-1219.

Nok AJ (2001). Azaanthraquinone inhibits respiration and in vitro growth of long slender blood stream forms of *T. congolense*. Cell Biochem. Func., 20: 205-212.

Nok AJ, Esievo KAN, Hondjet I, Arowosafe S, Onyenekwe PC, Gimba CE, Kagbu JA (1993). Trypanocidal potential of *Azadirachta indica:* In vivo activity of leaf extract against *T. brucei brucei*. J. Clin. Biochem. Nutr., 15: 113-118.

Owolabi OA, Makanga B, Thomas EW, Molyneux DH, Oliver RW (1990). Trypanocidal potentials of Africa woody palnts. In vitro trials of *Khaya grandifolia* seed extracts. J. Ethnopharmacol., 30: 227-231.

Sepulveda-Boza S, Cassels BK (1996). Plants metabolites active against *Trypanosoma cruzi*. Planta Medica, 62: 98-105.

Umar IA, Toh ZA, Igbalajobi FI, Igbokwe IO, Gidado A (2000). The effect of orally administered vitamin C and E on severity of aneamia in *Trypanosoma brucei* – infected rats. Trop. Veterinarian, 18: 71-77.

Vanhollebeke B, Truc P, Poelvoorde P, Pays A, Joshi PP, Katti R, Jannin JG, Pays E (2006). Human *Trypanosoma evansi* infection linked to a lack of Apolipoprotein 1. New Engl. J. Med., 355: 2752-2756.

Welburn SC, Coleman PG, Fevre E, Mandlin I (2001). Sleeping sickness- a tale of two diseases. Trends Parasitol., 17: 19-24.

Wurochekke AU, Nok AJ (2004). *In vitro* antitrypanosomal activity of some medicinal plants used in the treatment of trypanosomiasis in Northern Nigeria. Afri. J. Biotechnol. 3(9): 481-483.

21

Haplotypic diversity of West African populations of groundnut seed-beetle, *Caryedon serratus* ol. (Coloeptera, Chrysomelidae, Bruchinae): Results from geographical and DNA sequences data

Awa Ndiaye[1,2]* and Mbacké Sembène[1]

[1]Faculty of Science and Technology, University Cheikh Anta Diop, P. O. Box 5005 Dakar, Senegal.
[2]Horticultural Development Centre CDH/ ISRA Camberene Dakar Senegal, Box 3120, Senegal.

Caryedon serratus, groundnut seed-beetle, is a major pest of groundnut (*Arachis hypogaea*). Structuring of its biotype is subservient to the host plant that has already been established, while differences in the life history and ecology of the populations have led to the study of their genetic relationship, although a potential structure of these populations is seen in different biotypes. The aim of this work is to identify the different haplotypes circulated in the West African sub-region and to assess the phylogenetic affinities between ecological populations of *C. serratus*. Molecular marker and DNA sequences were used to assess genetic diversity and population structure of *C. serratus* on the West of Africa. Sequences analysis of mitochondrial (Cytb) and ribosomal (28S) DNA and the phylogenetic reconstructions by parsimony and maximum likelihood assigned the presence of haplogroup with maintained gene flow between the sampled populations. The result shows that genetic distance is not significant between localities, even if it appears that the variability which is quite strong and peculiar are retained for populations of Niger and the Senegal-Gambia border. Geographical isolation effect on population genetic structure tested with correlation test confirms this result.

Key words: *Caryedon serratus*, DNA sequences Cyt b, 28S, haplotype, groundnut, West Africa.

INTRODUCTION

Groundnut (*Arachis hypogaea*) was introduced from South America to Africa towards the end of the 16th century. Its cultivation in West Africa remained at a very low level until there was intensification of edible oil refining during the last third of the 19th century. As a result, groundnut farming experienced a dramatic increase, particularly in Senegal. The first infestation of stored groundnuts by the bruchid *Caryedon serratus* were reported in this country at the turn of the 20th century (Davey, 1958; Delobel, 1995). The problem is still spreading in some parts of the continent. About 60 years, after it was first recorded as a pest of groundnut in West Africa (Roubaud, 1916), it has recently become a major primary groundnut pest in Central Africa (Matokot et al., 1987) and Asia (Dick, 1987). The losses occur at all the phonological stages, from harvest to consumption. They reached 70% in 6 months of storage in Burkina Faso (Ouedraogo et al., 2010), 83% for 4 months of exposure in Senegal (Ndiaye, 1991) and over 60% in Congo after 10 months (Delobel, 1988). Weevil populations are structured on host-races depending on the host plant (Sembène and Delobel, 1998). *C. serratus* was adapted by allopatry, secondarily to groundnut; it lives in very different ecosystems and is widely distributed throughout the world. Furthermore, this weevil is one of the pests

*Corresponding Author Email: awa.ndiaye2@ird.fr.

Table 1. Population sampling: Origin, host plant and samples code.

Countries	Localities and geographical coordinates	Host plants	Samples code
Senegal	Samba Dia 14° 07'N / 16° 34' W	*Piliostigma reticulatum*	CsSd or CsSSd
	Karang 13° 35'N /16° 25'W	*Arachis hypogaea*	CsKg or CsSKg
	Kawil 14° 10'N/16° 05'W	*Piliostigma reticulatum*	CsKw
	Keur Ayip 13° 57'N /15° 48'W	*Arachis hypogaea*	CsKa or CsSKa
	Kédougou 12° 56' N/ 12° 21'W	*Piliostigma reticulatum*	CsK or CsSK
Burkina Faso	Tenkodogo 11°30'N / 0°3'W	*Arachis hypogaea*	CsBT
	Ouagadougou 12° 21 'N / 1° 32'W	*Arachis hypogaea*	CsB
Mali	Piama 10°44'N 6°08'W	*Piliostigma reticulatum*	CsMP
	Bawérékoro (Ségou) 14° 36'N / 05° 51'W	*Piliostigma reticulatum*	CsMB
Niger	Youry 13° 17'N / 02°11'E	*Piliostigma reticulatum*	CsN

with a wide range due to unintentional introductions associated with the transport of seeds they infest. Due to this discontinuous distribution of the species, geographically isolated populations exist in Africa and around the world (Delobel, 1992). In cases when the population occupies different geographical areas that are eventually disjoint, the gene flow that normally circulates inside the specie is not more or less interrupted between one or several populations. Thus, it does not induce genetic divergence of these populations which can then be assimilated to sub-species. Parameters such as polymorphism and variability are used to describe population, and to assess genetic differences between populations.

This study attempts to identify the different haplotypes circulating in the West African sub-region by the nucleotide polymorphism. Analysis was carried out on the sequence data of two molecular markers: the gene encoding the protein cytochrome b (Cytb) and that of the D2 domain of the subunit of nuclear ribosomal RNA (28S). A separate analysis of these two genes was performed before they were combined into a single dataset. The hypothesis of a potential role of geographic variation in the genetic structure of *C. serratus* across West Africa was tested by the Mantel test. Different methods of phylogenetic inference were performed and the results were also discussed given the preliminary data on the phylogeny of this weevil in Senegal and the inputs sequences data.

MATERIALS AND METHODS

Study site

Several individuals of *C. serratus* were collected from four countries in West Africa (Table 1). In Senegal, the samples were collected in the villages of Samba Dia (14° 07'N / 16° 34'W), Kawil (14° 10'N

/ 16° 05'W), Karang (13° 35'N / 16° 25'W), Keur Ayip (13° 57'N / 15° 48'W) and kedougou (12° 56'N/ 12° 21'W). Karang and Keur Ayip are situated in Senegal-Gambia border. In Mali, the sampling took place in the locality of Piama (10° 44'N / 6° 08 'W) and in the Bawere (Segou) locality (14° 36'N / 05° 51'W). For Burkina, they were collected in Tenkodogo (11° 47'N / 0° 22'W) and Ouagadougou (12° 21 'N / 1° 32'W) localities, while for Niger, they were collected in the locality of Youry (13° 17'N / 02° 11'E).

Caryedon serratus samples

Beetles used in this study were reared from eggs, larvae or pupae found in *A. hypogaea* or *Piliostigma reticulatum* or in pods collected from *A. hypogaea* or *Piliostigma reticulatum*. Groundnut samples were collected from the field during drying. The term population refers to samples from the same geographic origin. Weevils were observed under the microscope and the identification was based on the elytra and keys used to determine them (Delobel and Tran, 1993). Samples were named accordingly, Cs for *C. serratus*, following their geographic origin (Table 1), whereas adult beetles were genetically analyzed almost immediately after they emerged from pods.

DNA extraction, polymerase chain reaction and sequencing

DNA was extracted, amplified and sequenced with standard protocols described by Sembène et al. (2010b), and each sequence was obtained from the DNA of a single seed-beetle. The abdomen, elytra and antennae were kept apart to avoid contamination by fungi and nematodes and to permit subsequent morphological observations. A partial cytochrome b (Cyt b) end region was amplified using the primers: CB1 (5'–TATGTACTACCATGAGGACAAATATC 3') and CB2 (5'-ATTACACCTCCTAATTTATTAGGAAT-3'). The 28S ribosomal DNA was targeted for PCR, amplified and sequenced with primer D2CF D45F (5'-TAC CGT GAG GGA AAG TTG AAA-3 ') and D2CR D45R (5'-AGA CTC CTT GGT CCG TGT TT-3'). For both markers, PCR amplification was performed in 25 µl reaction volume 2.5 µl enzyme buffer supplied by the manufacturer, 2.5 mM MgCl2, 0.6 unit of Taq polymerase (Promega), 17.5 pM of each primer, 25 nM of each DNTP and 1 µl of DNA extract. After an initial denaturation step at 92°C for 3 min, reaction was subjected to 35 cycles for 1 min at

92°C, 1 min 30 s at 48°C and 1 min at 72°C. PCR products were loaded on a 1.5% agar gel. Purification of PCR products and sequencing of genes Cyt b and 28S were performed by Macrogen Company (South Korea).

Sequence alignment and phylogenetic reconstructions

The obtaining sequences were aligned using ClustalW (Thompson et al., 1993) as implemented in BioEdit. Kimura 2-parameter genetic distance between haplotype was calculated. G-tests (log-likelihood ratio test) were performed to test for genetic distance homogeneity among sites (Sokal and Rohlf, 1981).

Phylogenetic relationships were reconstructed with MEGA4 software v. 4.0.0.162 using the maximum parsimony method (MP) (Fitch, 1971). The MP analysis was carried out with the heuristic search option with 50 random stepwise taxon addition replicates, using the branch swapping tree bisection-reconnection (TBR) option. A bootstrap procedure (500 iterations with the same option of heuristic search) was used to establish the score of each node (Felsenstein, 1985) by retaining groups compatible with the 50% majority rule consensus. A strict consensus tree was computed whenever multiple equally parsimonious trees were obtained.

The method of maximum likelihood (ML) (Felsenstein, 1981) is used to test all the stories that may have led to the current data set analyzed. This analysis of the likelihood was made by submitting data to the interface PHYML. Node stability was evaluated using 100 bootstrapping replications, and the majority-rule consensus trees were conducted.

The reconstructions were rooted with a sequence of *Callosobruchus maculatus* (Bruchidae) in the locality of Fouta, Senegal, and were used as an out-group. Correlation between the genetic and geographic distances was evaluated by a Mantel test with the null hypothesis assuming that the regression slope of genetic distances (Kimura 2-parameter) / log geographical distance (km) was zero.

RESULTS

Cytochrome b sequences

The total number of nucleotides aligned without ambiguity is 427 bp. The alignment was straightforward and it involved no insertions/deletions. 37 haplotypes were identified as a result of 64 polymorphic sites of which 32 are individual haplotypes. Between haplotypes, the percentage of transition is 90.32 against 9.68% of transversions. The cyt b sequences, obtained from 37 sites, contain parsimony information and 27 singleton sites. The amino acid transcript reveals that we still have the same amino acid (alanine, cysteine, glycine and threonine) with different proportions between haplotypes. Haplotype H3 is dominant and has 6 individuals (CsKa1, CsKa3, CsKg5, CsKg13, CsKw2 and CsKw5) from the locality of Keur Ayip, Karang and Kawil, in Senegal. Haplotype H4 is composite of the individuals from the village of Samba Dia (CsSD3, CsSD1), while H7, H12 and H34 are represented by the pairs of individuals (CsMB11, CsSD5), (Cskg15, CsN7) and (Cska5, CsN6), respectively. However, the other haplotypes are individual.

28S sequences

Sequences of 28S involve 7 haplotypes from 60 variable sites on the portion of sequences of 465 bp. Of these 60 polymorphic sites, 42 are carried by individuals belonging to the Niger haplotype H4 (Table 2) lining up to 91.80%. This sequence admits four parsimony informative sites, 53 sites singletons and 408 conserved sites. Deletions are considered as a fifth state character for the 28S counted in 3 sites (9, 375 and 376 pb). It may be noted that 80.95% of the transition mutation-type and 18.41% of transversion are against only 0.64% of the deletion. The nucleotide frequency in 465 pb is as follows: adenine (0.161), guanine G (0.315), cytosine C (0.294) and thymine T (0.23).

Distance matrix and phylogenetic trees

Phylogeny based on Cyt b reconstruction, using the method of maximum parsimony (MP) (Figure 1A) and maximum likelihood (ML) (Figure 1B), are congruent and they confirm two haplo-groups (HG1 and HG2 majority), and a basal clade composed of individual haplotype H37 (CsMP6). HG1 and HG2 are robust (100% bootstrap value) and very heterogeneous, each containing individuals from all the sampled localities. The first group, HG1, of the consensus tree maximum parsimony closes haplotypes H1 to H20, and the second, HG2, closes haplotypes H21 to H37. The same distribution is obtained with the method of maximum likelihood. It reveals two haplotype groups composed of the same haplotype, with CsMP6 (H37) as the only difference swapped with the individual CsMB10 (H19), which is isolated and which formed the basal clade; although H37 is integrated in the haplogroup HG1. The two groups close each haplotype from all points of our sample. The genetic distance within HG2 is equal to 0.017 and is higher than that found inside HG1 (0.015). Between the two haplogroups, the genetic distance is 0.057. The 2 haplo-groups are supported by satisfying the bootstrap values (71% in MP and 75 MV for HG1et, 100% MP and 91% for MV HG2).

The consensus tree of parsimony (Figure 2A) and the maximum likelihood (Figure 2B) for the 28S show a similar topology with a high similarity of haplotypes. However, four haplogroups can be distinguished (Table 2). The haplogroup HG1 is the exact representation of the phylogenetic haplotype H1, which is numerically the majority. It includes individuals from Senegal, the Senegal-Gambia border (Karang, Keur Ayip), Mali, Burkina Faso and Niger. It includes all individuals of Burkina Faso, and almost all individuals from Senegal, Mali and Niger. The haplogroup HG2 includes solely the haplotype H2 and brings together two individuals from two neighbouring localities: (CsSK4) Kedougou and (CsMB6) Bawere, where they are supported by high bootstrap values in both methods of phylogenetic

Table 2. Haplotypes composition and genetic distances between haplotypes for the 28S RNA.

Samples	Haplotypes	Genetic distance between haplotypes						
		H1	H2	H3	H4	H5	H6	H7
CsSKa1, CsSKa2, CsSKa3, CsSKa5, CsSKa6, CsSKg1, CsSKg13, CsSKg10, CsSKg5, CsSKg14, CsSK2, CsSK3, CsSK5, CsSK6, CsSSd1, CsSSd2, CsSSd4, CsSSd5, CsMP2, CsMP3, CsMP5, CsMB8, CsMP8, CsMB9, CsBT1, CsBT2, CsBT3, CsBT4, CsBT5, CsN2, CsN9, CsN10.	H1	–						
CsSK4, CsMB6	H2	0.002	–					
CsN1	H3	0.035	0.038	–				
CsN6	H4	0.006	0.009	0.028	–			
CsN5	H5	0.004	0.006	0.035	0.006	–		
CsN11	H6	0.092	0.094	0.127	0.095	0.097	–	
CsMP6	H7	0.002	0.004	0.033	0.004	0.006	0.090	–

reconstructions (90% for ML, 100% for MP). The individual from Niger stands out in ML reconstitution, associated with an individual from the locality of Piama, and form a clade supported by boostrap values of 70%. However, genetic distances (28S) between haplotypes ranged from 0.002 to 0.121 (Table 2).

The assembly sequences of two genes (Cyt b+ 28S) could be made for some individuals (27), and it gives us a sequence of 915 bp. The parsimony consensus tree (Figure 3A) shows a great strength of the phylogenetic affinities, and the majority of individuals in the sample are grouped in a clade of maximum bootstrap value (100%) and they form the haplo-group HG1. Only individuals of the Senegal-Gambia border, particularly Keur Ayip (CsKa2, CsKa5), associate with an individual from Niger (CSN6) to give the haplogroup HG2 basal clade in the tree. Between these 2 groups obtained for the 2 genes, the genetic distance is 0.026. The genetic distance within the HG1 is 0.018 due to the fact that the inside of it is very low, while that of HG2 is 0.003.

Maximum Likelihood (Figure 3B) shows a different topology with the 2 haplogroups which form clades with the bulk of haplotypes, and which form a clade by one individual, CsMP3, in Piama (Mali).
The MP analysis on Cyt b nucleotide data yielded 37 for most of the parsimonious trees (length = 402), where the consistency index (CI) was 0.559, and the retention index (RI) was 0.882. With the same methods used on the 28S RNA data set, 684 was seen for most parsimonious trees (length = 105), where CI = 0.833 and RI = 0.857. Finally, analysis of the combined (Cytb+28S) data set yielded a tree which was 729 steps long (CI = 0.619; RI = 0.839). All analyses supported the proximity between haplotypes and the remark was still the same; however, all inferences suggest a distribution that did not seem to be dependent on geographical distance. This is assessed by the Mantel test for both genes separately and for the data set of genes when combined (Figure 4). In each case, the correlation gives a p-value (0.743 for Cytb+28S) that is above the level of significance (alpha = 0.05); thus, we can not reject the null hypothesis H0 (H0 = the distance matrices are uncorrelated). Therefore, there is no correlation between genetic and geographic distance between our sampling areas.

DISCUSSION

The objective of this work is to identify the different haplotypes circulated in the West African sub-region and to assess the phylogenetic affinities between allopatric populations in the species of *C. serratus*. Different methods of phylogenetic inferences were conducted not to perform the phylogeny in the proper sense (between species), but to, a priori, deduce genetic affinities between populations of the groundnut bruchid in West Africa. The phylogenetic affinities of haplotypes, inferred with mitochondrial fractions (Cyt b) and nuclear ribosomal gene 28S, assign the presence of haplogroups and/or sub-groups between haplotypes. Bruchids subservient to peanuts and those that infest *Piliostigma reticulatum* have the same nucleotide sequences in their sequence and align perfectly. The genetic identity between bruchids infesting peanuts and those subservient to *P. reticulatum* has already been proven in studies of morphometric, allozyme (Sembène and Delobel, 1996, 1998; Sembène et al., 1998), cytochrome B (Delobel et al., 2003, Sembène et al., 2003) and the ITS1 gene (Sembène, 2004; Sembène et al., 2010a) of the groundnut bruchid.

This justifies that our analysis was done regardless of

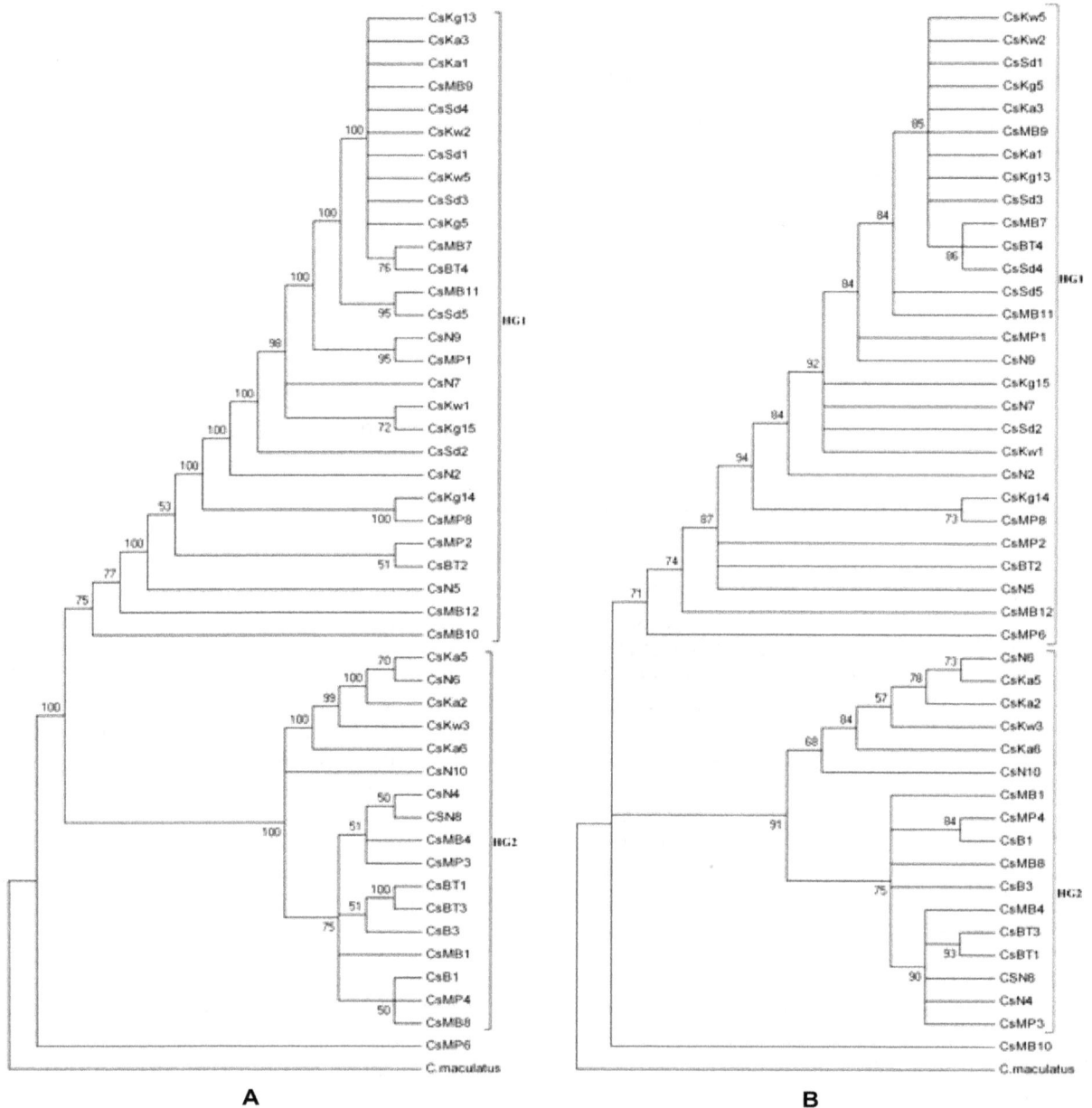

Figure 1. (A) Maximum parsimony consensus tree of partial cytochrome b (427 pb), *Callosobruchus maculatus* are the out group. (B) Maximum likelihood consensus tree of cytochrome b (427 pb), *C. maculatus* are the out group. Numbers above branches show the bootstrap support values. Other abbreviations are seen in Table 1.

the host plant (*P. reticulatum* or peanut).

Haplotype diversity

The number of haplotypes found for Cyt b (37) is much higher than that of 28S (7). This difference is explained by the fact that cytochrome b gene is highly mitochondrial in mutation and rapid in evolution, in that its mutations are preserved unlike the 28S which is not only less mutant, but does not retain its present form and more

mutations and deletions. At this level of intraspecific variability, we can ask if the nuclear 28S gene is a good marker for characterizing strains *C. serratus*. This gene is known as a very slow marker. The fact is that we found 18 variable sites on a portion of 465 bp for 39 sequences. The phylogenetic inferences used gives the same topology for cyt b with 2 groups and each contained different haplotypes, with individuals from all countries sampled, namely: Senegal, Mali, Burkina Faso and Niger. It does not establish grouping by localities. Nonetheless, haplotypes can be found in two remote locations.

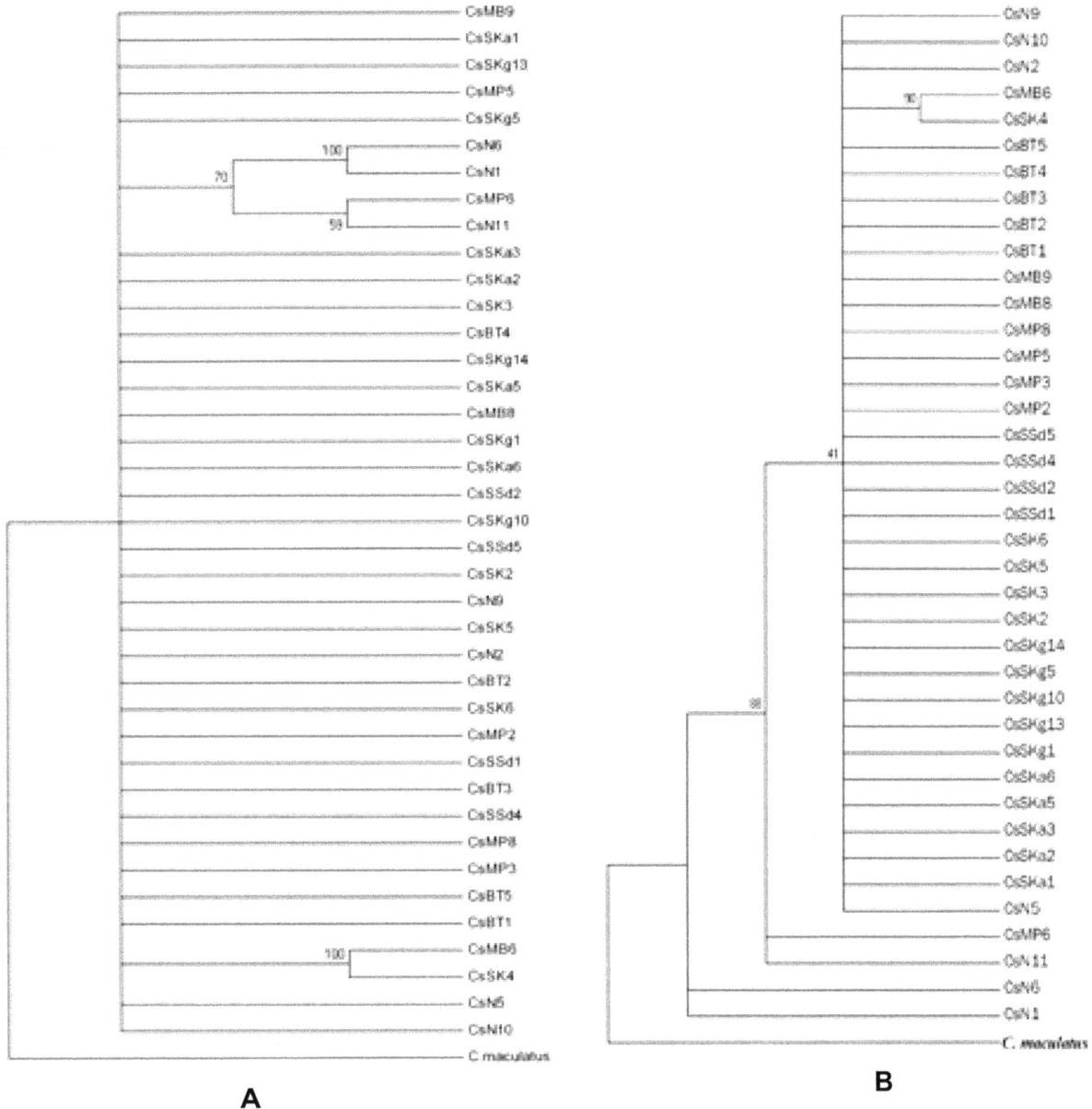

Figure 2. (A) Maximum parsimony consensus tree of partial 28S RNA (467 pb), *C. maculalus* are the out group. (B) Maximum likellhood consensus tree of partial 28S RNA (467 pb), *C. maculatus* are the out group.

Genetic distances and phylogenetic relationship

Genetic distance between groups is equal to 0.057 for cytb, in that the values of intra-group genetic distances range from 0.015 to 0.017. The haplo-group HG1, which is a numerical majority of the 28S gene, also include absolutely identical individuals, from all sampling locations, although the genetic distance within the haplotype is null. Genetic distances for 28S RNA sequences, between haplotypes, varied from 0.002 to 0.038. It also implies that H1, in the history of "Bruchid-peanut association" was dispersed in the area of West

Africa from Senegal, which is the original point of infection in the twentieth century (Davey, 1958; Delobel, 1995; Sembène et al., 2010b) and was given later in Mali and Niger by divergent haplotypes depending on agro-ecological settings. Genetic distances within and between haplotypes for all data sets show the remains in an interval, and no intraspecies exceeded the genetic distance typical of sympatric (Kerdelhule, 2002) and geographic populations (Klein and Seitz, 1994). This result seems to be using the same logic as those obtained by Sembène et al. (2010a) in so far as the genetic distances derived from combining the data

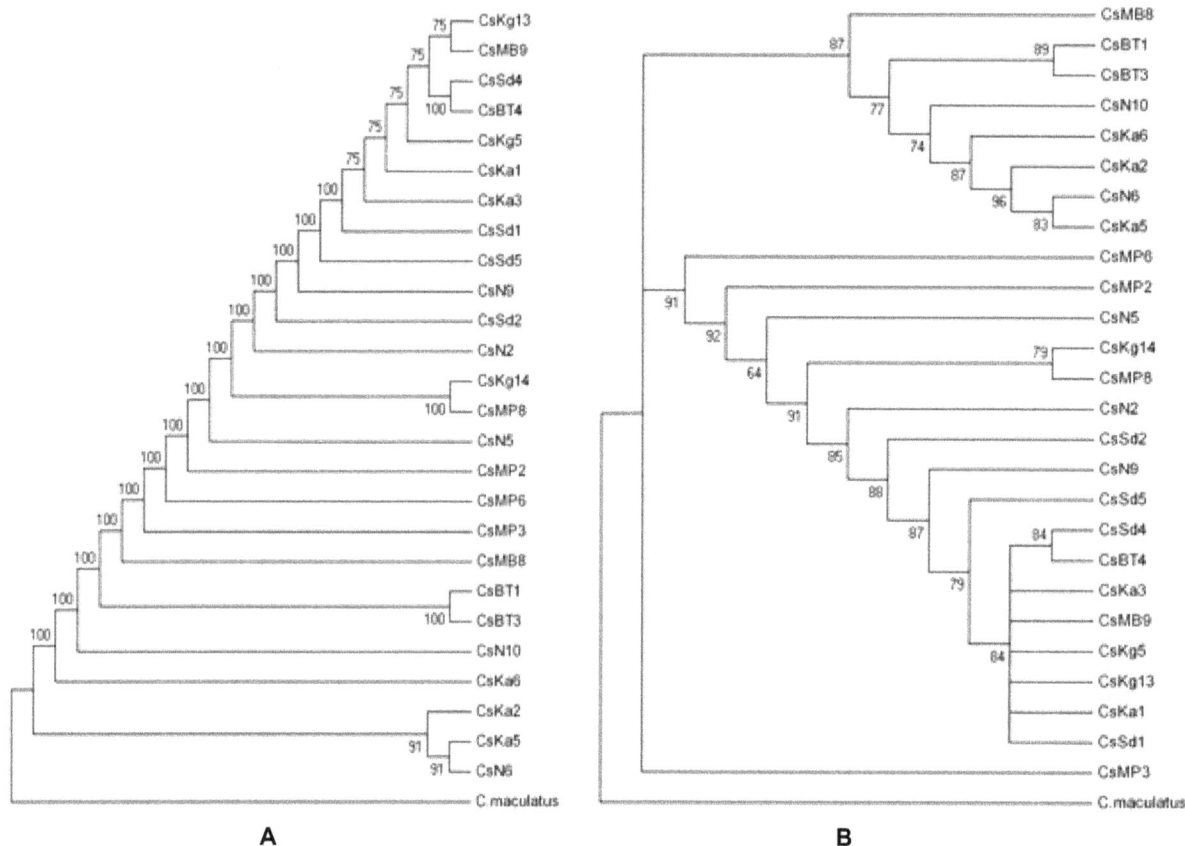

Figure 3. (A) Phylogenetic relationships from maximum parsimony consensus tree of the combined data (915 pb), *C. maculatus* is an out-group. (B) Phylogenetic relationships from maximum likelihood tree of the combined data (915 pb), *C. maculatus* is an out-group.

mitochondrial genes of cytochrome B and ITS1 in *C. serratus* subservient to peanut vary from 0.000 to 0.037. An exception may be retained by the peculiarity of the H6 present in Niger (28S RNA), where genetic distances derived from other haplotypes ranged from 0.090 to 0.127. The distance between biotypes infesting *Cassia sieberiana* from Sembène (2010a), ranges from 0.186 to 0.195. Thus, H6 is almost equidistant between the other haplotypes and individuals subservient to *C. sieberiana* which enhances the peculiarity of the Nigerian population.

The characteristics of the Nigerian haplotypes can be registered in a context of gene flow that would exist with the weevil's tamarind or *Caryedon gonagra* from India, in this Nigerian area, where there is a crossroad of trade between West Africa and North Asia. Gambian-Senegalese border also tends to stand out of the lot by mutations of their specific. For example, the haplotype H3 of cyt b is exclusively composed of individuals (CsKa1, CsKa3, CsKg5, CsKg13, CsKw2 and CsKw5), from Senegal (keur Ayip, Karang and Kawil). Karang and keur Ayip are located on the border of Gambia, which is a country maintaining trade with more distant countries

such as Nigeria and Sierra Leone. Genetic specificity of these individuals from Senegal could then be explained by the introduction of haplotypes from other parts of West Africa not yet sampled. This support is the topology obtained by the assembly of two genes. Moreover, Keur Ayip stands back to the lot, and its association with the individual from Niger could support the hypothesis of an introduction by transportation of the infested seed. In terms of Kawil, a remark drew our attention during sampling (May, 2010) on this site, in that ripening pods were found on *P. reticulatum*. This could confirm the infestation of *P. reticulatum* already explained by Sembène and Delobel (1998), and Sembène et al. (2008, 2010). Adults residual on *P. reticulatum*, would suffer a bottleneck and regenerated Caryedon in the next generation. The result is a selection of genes making the population of that particular area. Despite the large number of mutations for both cyt b and 28S as the particularization of the populations of some areas, gene transfer can be retained between the localities sampled. Indeed in all our analyses, the haplotypes were grouped according to their nucleotide similarity without necessarily taking into account their geographical origin. The

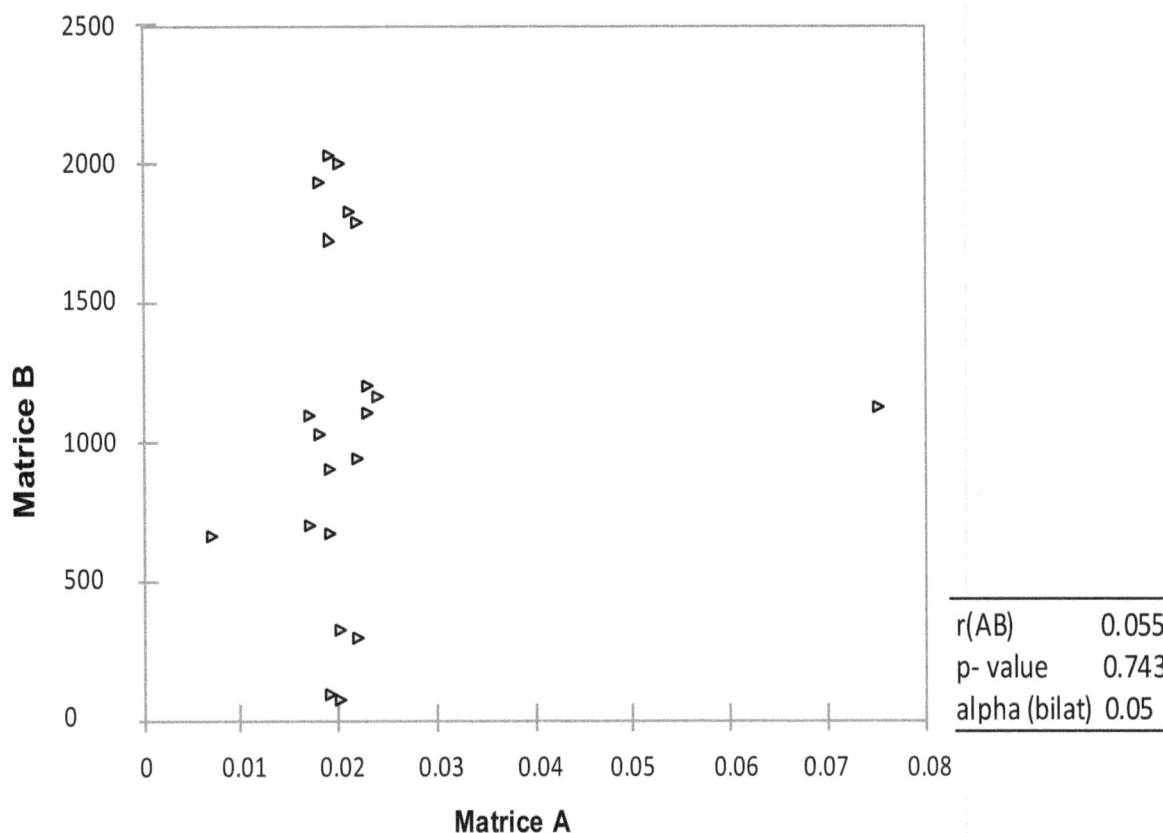

Figure 4. Isolation by distance pattern in West Africa. Regression of genetic distance (Kimura 2 parameters) against the logarithm of geographical distance (in km) for the combined data set of cytb + 28S.

hypothesis of isolation by geographical distance is evaluated by the Mantel correlation test which provides for the genes separately, though the combined p-value was not significant. The null hypothesis assuming no regression between genetic distances and geographical distances could not be rejected in any of our tests. So, there is no correlation between genetic and geographic distance matrices. There is variability between populations of our sampling revealed by the nucleotide polymorphism. However, all the individuals of a species are not similar; though the accumulation of variability should lead to their differentiation, and under certain conditions, they become new species or subspecies. For now, there is a gene flow which is maintained between the populations of *C. serratus* in West Africa. The role of anthropogenic dispersal, through trade, for example, is yet reported to be a genetic identity which is found between individuals from localities whose capacity of natural distribution of the weevil can not rally. In addition, some pests have a wide range due to unintentional introductions associated with the transport of seeds they infest. At this step, we certainly can not speak of isolation in allopatric populations of *C. serratus* infesting peanuts in the sub-region. These results were predictable and are consistent with Sembène et al. (1998), where weevils are

grouped according to the host plant and not to the geographical distance. Influence of gene flow in homogenizing genetic traits or as diversification factor remains controversial (Mayr, 1974; Dobzansky, 1977; Sembène, 1998). The absence of isolation by distance may suggest a population structure by larval food already shown in the works of Sembène, though particularization of certain ecotypes could not come from an ecological specialization.

ACKNOWLEDGEMENTS

This publication was produced with the financial support from the IRD-DSF and the International Foundation for Science.

REFERENCES

Davey PM (1958). The groundnut bruchid, *Caryedon gonagra* (F.). Bulletin. Entomol. Res., 49: 385-404.

Delobel A (1988). Comment résoudre le problème des pertes en cous de stockage : l'exemple de la bruche de l'arachide au Congo. ORSTOM/Brazzaville, p. 3.

Delobel A (1992). La bruche de l'arachide: un exemple contemporain de déplacement du spectre alimentaire. Rapport pour l'Antenne

ORSTOM/Muséum National d'Histoire Naturelle, p. 14.

Delobel A (1995). The shift of *Caryedon serratus* (Ol.) from wild Caesalpiniaceae to groundnuts took place in West Africa (Coleopter : Bruchidae*)*. J. Stored. Prod. Res., 31: 101-102.

Delobel A, Tran M (1993). Les Coléoptères des denrées alimentaires entreposées dans les régions chaudes. Faune tropicale XXXII. Orstom/ CTA, Paris.

Delobel A, Sembène M, Fédière G, Roguet D (2003). Identity of groundnut and tamarind seed-beetles (Coleoptera: Bruchidae Pachymerinae), with the restoration of *Caryedon gonagra* (F.). Annales de la Société Entomologiques de France, 39: 197-206.

Dick KM (1987). Losses caused by insects to groundnuts stored in a warehouse in India. Trop. Sci., 27: 65-75.

Dobzansky TH (1977). Génétique du processus évolutif. Flammarion, Paris, p. 346.

Felsenstein J (1981). Evolutionary trees from DNA sequences: a maximum likelihood approach. J. Mol. Evol., 17: 368-376.

Fitch WM (1971). Toward defining the course of evolution: minimum change for a specific tree topology. Syst. Zool., 20: 406-416.

Kerdelhué C, Roux-Morabito G, Forichon J (2002). Population genetic structure of *Tomicus piniperda* L. (Curculionidae: Scolytinae) on different pine species and validation of *T. destruens* (Woll.). Mole. Ecol., 14: 483-494.

Klein M, Seitz A (1994). Geographic differentiation between populations of *Rhinocyllus conicus* Frölich (Coleoptera: Curculionidae): concordance of allozyme and morphometric analysis. Zool. J. Linnean Soc., 110: 181-191.

Mayr E (1974). Populations, espèces et évolution. Hennann, Paris, p. 496.

Matokot L, Mapangou-Divassa S, Delobel A (1987). Evolution des populations de *Caryedon serratus* (Ol.) dans les stocks d'arachide au Congo. Agron. Trop., 42: 69-74.

Ndiaye S (1991). La bruche de l'arachide dans un agrosystème du centre-ouest du Sénégal: contribution a` l'étude de la contamination en plein champ et dans les stocks de l'arachide (*Arachis hypogaea* L.) par *Caryedon serratus* (Ol.) (Coleoptera-Bruchidae); role des légumineuses hôtes sauvages dans le cycle de cette bruche. Thèse Université de Pau et des Pays de l'Adour.

Ouedraogo L, Traore NS, Guenda W, Dabire LCB (2010). Influence des plantes hôtes sur la fécondité et le développement larvaire de la bruche de l'arachide *Caryedon serratus* Olivier (Coleoptera : Bruchidae) au Burkina Faso. J. Appl. Biosci., 31: 1906-1915.

Roubaud E (1916). Les insectes et la dégénérescence des arachides au Sénégal. Mémoire Comité Etude historique et scientifique Afrique Occidentale Franc- aise 1: 363-438.

Sembène M, Delobel A (1996). Morphometric identification of populations of Sudan-Sahelian groundnut bruchid, *Caryedon serratus* (Olivier) (Coleoptera Bruchidae). J. Afr. Zool., 110: 357-366.

Sembène M, Delobel A (1998). Genetic differentiation of groundnut seed-beetle populations in Senegal. Entomologia Exp. Appl., 87: 171-180.

Sembène M, Brizard JP, Delobel A (1998). Allozyme variation among populations of groundnut seed-beetle *Caryedon serratus* (Ol.) (Coleoptera: Bruchidae) in Senegal. Insect Sci. Appl., 18: 77-86.

Sembène M, Vautrin D, Silvain JF, Rasplus JY, Delobel A (2003). Isolation and characterization of polymorphic microsatellites in the groundnut seed beetle, *Caryedon serratus* (Coleoptera, Bruchidae). Mole. Ecol. Notes, 3: 299-301.

Sembène M (2004). Inter-strain fecundity and larval mortality in the groundnut beetle *Caryedon serratus* (Coleoptera: Bruchidae). Int. J. Trop. Insect Sci., 24: 319-322.

Sembène M, Rasplus JY, Silvain JF, Delobel A (2008). Genetic differentiation in sympatric populations of the groundnut seed beetle, *C. serratus* (Coleoptera: Chrysomelidae): new insights from molecular and ecological data, Int. J. Trop. Insect Sci., 28(3): 168-177.

Sembène M, Kébé K, Delobel A, Rasplus JY (2010). Phylogenetic information reveals the peculiarity of *Caryedon serratus* (Coleoptera, Chrysomelidae, Bruchinae) feeding on *Cassia sieberiana* DC (Caesalpinioideae). Afr. J. Biotechnol., 9(10): 1470-1480.

Sembène M, Ndiaye A, Kébé K, Doumma A, Sanon A, Kétoh KG, Granjon L, Rasplus JY (2010). When DNA sequences and microsatellites loci tell the story of field groundnut infestation by *Caryedon serratus* Ol. (Coleoptera, Chrysomelidae, Bruchinae). J. Cell Anim. Biol., 4(10): 143-150.

Sokal RR, Rohlf FJ (1981). Biometry, second ed.W.H. Freeman & Co, New York, San Francisco.

Thompson LG, Mosley-Thompson E, Davis M, Lin PN, Yao T, Dyurgerov M, Dai J (1994). ``Recent warming'': ice core evidence from tropical ice cores with emphasis on central Asia, Global Planetary Change, 7: 145-156.

22

The role of oxidative stress in the development of congestive heart failure (CHF) in broilers with pulmonary hypertension syndrome (PHS)

Mokhtar, Fathi[1], Kambiz, Nazer adl[1], Yahya, Ebrahim Nezhad[1], Habib, Aghdam Shahryar[1], Mohsen, Daneshyar[2] and Taimor, Tanha[1]

[1]Department of Animal Science, Islamic Azad University, Shabestar Branch, Iran.
[2]Department of Animal Science, Urmia University, Iran.

The present study examined the possible role of reactive oxygen species in the pathogenesis of heart failure in broilers. Our experiment was conducted with 160 one–day-old male broilers (Ross 308) to investigate the mechanisms of cell injury in the pathogenesis of pulmonary hypertension syndrome. The chickens were divided into two groups of four replicates each, with 20 chicks per replicate. One group was raised in normal temperature (NT) while the other group was raised in cold temperature (CT) to induce the pulmonary hypertension syndrome. Mortality was inspected to determine cause of death and diagnose heart failure. Hematological, biochemical and pathological tests were used to determine the incidence of PHS: total red blood cell (RBC), hemoglobin (HGB), hematocrit (HCT), activity of alanine transaminase (ALT), aspartate transaminase (AST) and lactate dehydrogenase (LDH). Malondialdehyde (MDA) was used as an indicator of lipid oxidation, subsequent to generated oxidative stress. Samples of blood and liver tissue were taken at day 21 and 42 of age. At the end of the experiment (week 6), 2 chicks from each replicate were randomly selected and slaughtered. The heart was removed, the right ventricle was dissected away from the left ventricle and septum, and the ratio of right ventricle weight to total ventricle weight (RV/TV) was calculated. The results of our experiment indicated the significant difference between the two groups for RBC and HGB and for RBC, HGB and HCT at days 21 and 42 of age, respectively. However, there was no significant difference for activity of ALT, AST and LDH in plasma between the groups at day 21 of age. but CT birds had higher levels (p < 0.05) AST, ALT and LDH activities in plasma as compared with other birds. MDA content of plasma and liver was significantly higher (p < 0.05) in CT group at both ages. RV/TV ratio and mortality due to ascites were significantly higher in CT birds. In conclusion, heart failure and subsequent PHS can be associated with oxidative stress.

Key words: Oxidative stress, heart failure, ascites, hematological, broiler.

INTRODUCTION

Pulmonary hypertension syndrome (PHS) or ascites is a metabolic disorder that mostly occurs in fast-growing broiler chickens. High altitude, hypoxia, poor ventilation, low temperature and fast growth rate are known to be predisposing factors for the incidence of this syndrome (Huchzermeyer and DeRuyck, 1986; Maxwell et al., 1986; Wideman et al., 1995a, b; Hassanzadeh et al., 1997;

Balog, 2003). Pulmonary hypertension and cardiac dysfunction are the most important features of ascites. Pathological findings indicate that the creation of a cavity on the exterior surface of the right ventricular wall is the first sign of damage in pulmonary hypertension. As the injury progresses, it leads to dilation and hypertrophy of the right ventricle resulting in increased blood viscosity, reduced oxygen supply, congestive heart failure (CHF) and accumulation of fluids in the abdominal cavity (Julian, 1990, 1993; Odum, 1993; Owen et al., 1995).

Oxidative metabolism is a normal process in all tissues.

*Corresponding author. E-mail: fathi_mokhtar@yahoo.com.

Cardiomyocytes require a constant supply of oxygen for normal cardiac functions. However, oxygen associated metabolism in the myocardium sometimes can contribute to cardiac dysfunction, and may ultimately lead to heart failure (Giordano, 2005; Redout et al., 2007). During the normal oxidative metabolic process, various reactive oxygen species (ROS) and reactive nitrogen species (RNS) are produced. During this normal metabolism, 1 to 2% of oxygen is converted to ROS (Sheeran and Pepe, 2006). ROS are implicated in different disorders, including thermal injury, inflammation, sepsis, mutagenesis, carcinoma, autoimmune diseases and ischemia reperfusion injury (Flohe et al., 1985; McCord, 1985; Halliwell, 1989; Farber et al., 1990). The role of ROS in the injury induced by ischemia reperfusion has been convincingly shown in different organs, including the brain, liver, skin, muscle, lung, intestine, kidneys and heart (Halliwell, 1989; Jaeschke, 1991; Diaz-Cruz et al., 1996).

However, under some circumstances, increased ROS/RNS production or decreased antioxidant defenses may lead to oxidative stress, in which case the generated reactive species can alter the properties of lipids, proteins and nucleic acids, leading to cellular dysfunction. Recent research findings from different laboratories suggest that ROS and RNS play a critical role in the development of human heart failure (Andreka et al., 2004; Sam et al., 2005; Nediani et al., 2007). Lipid peroxidation can alter the membrane properties of cellular and sub cellular organelles (mitochondria and sarco-endoplasmic reticulum) crucial for the maintenance of normal cardiomyocyte function. Broilers with congestive heart failure (CHF) show evidence of calcium overload in these sub-cellular components (Maxwell et al., 1993; Li et al., 2006) and evidence of breakdown and release of proteins of the contractile apparatus, such as myosin and troponin T, into the circulation (Maxwell et al., 1994).

The role of oxidative stress has long been debated in the pathogenesis of heart failure in human and animal models of cardiomyopathy. However, limited research has been carried out to investigate the possible involvement of oxidative stress in PHS and CHF in broilers. In order to further understand the physiological and biochemical disturbances leading to PHS and CHF in commercial broilers, we were interested in examining the possible role and molecular mechanisms of oxidative stress in the pathogenesis of these syndromes.

MATERIALS AND METHODS

Birds and diets

One hundred and sixty (160) 1-day-old male broiler chickens (Ross 308) were used in this experiment. Chickens were allocated randomly randomly to 2 treatment groups, with 4 replicates each and 20 chicks per replicate (per cage). The two groups were broilers under normal temperature (NT) and broilers under cold environmental temperature (CT). All chicks were fed a basal corn-soybean meal diet to meet requirement, including 22.04% CP and 3,200 kcal/kg of ME (1 to 21 days), or 20.26% CP and 3,200 kcal/kg ME (22 to 42 days). Feed and water provided as *ad libitum* consumption.

Management and measurements

Broilers of the NT group were reared at 32°C for the first week and then it was reduced 2°C per week up to week 5 which was kept at 22°C until the end of the experiment (Daneshyar et al., 2007; 2009). For inducing ascites, the birds of the CT group were raised under 32°C and 30°C during week 1 and 2, respectively. The room temperature was decreased to 15°C during week 3 and maintained between 10 and 15°C for the rest of the study (Igbal et al., 2001) Daneshyar et al., 2007; 2009). Mortality was recorded daily and all of the dead birds were inspected for diagnosis of ascites. Diagnosis of ascites generally depends on observation of the following symptoms:

1) Right ventricle hypertrophy, cardiac muscle laxation;
2) Swollen and stiff liver;
3) Clear, yellowish, colloidal fluid in the abdominal activity (Geng, 2004).

Sampling

At days 21 and 42 of age, one chick from each replicate was randomly selected. thenblood samples were taken from the wing vein after a 3-h starvation period. At the end of experiment, one bird per replicate was killed and its abdomen opened for signs of heart failure and ascites. About 5 grams of liver tissue were removed, homogenized and used for MDA determination. The heart was dissected and removed from the body to determine the ratio of right ventricular (RV) weight to total ventricular (TV) weight ratio. Birds having RV/TV values more than 0.299% were considered to have ventricular hypertrophy (Jolian, 1987). Blood samples were collected in tubes with EDTA anticoagulant tubes. Portions of each blood sample were immediately used for determining total red blood cell (RBC) count, hematocrit (HCT) and hemoglobin (HGB). The remaining was centrifuged and their plasma was collected and stored at –80°C for further enzymatic and chemical analyses.

Malondialdehyde (MDA): The blood was centrifuged at $1,500 \times g$ for 5 min; plasma was collected in labeled tubes and stored at –80°C until analysis. After thawing, 500 µL of plasma was placed in a labeled glass tube and mixed with the reagents of a commercial kit for the measurement of thiobarbituric acid reactive substances (TBARS). Each tube was covered with a glass marble and incubated at 95°C for 45 min. The tubes were removed from incubation and allowed to cool in an ice bath for 10 min. Once cooled, the tubes were centrifuged at $3000 \times g$ for 10 min and the supernatant carefully removed from the tubes for analysis. The absorbance of the supernatants was measured at 532 nm using a UV/VIS spectrophotometer (Gildford Instrument Laboratories, Inc., Oberlin, OH) and the results were compared against a standard curve made with 100, 50, 25, 12.5, and 0 nmol/mL of malondialdehyde dimethyl acetyl.

Statistical analysis

Data were analyzed based on a completely randomized design using the GLM procedure of SAS (SAS 9.1 institute2002). Duncans multiple range tests were used to separate the means when treatment means were significant ($p \leq 0.05$); thus a probability level of $p \leq 0.05$ was considered statistically significant. Data were presented as means ± SD.

Table 1. RV/TV ratio and mortality percentage of broilers under normal (NT) and cold (CT) environmental temperatures.

Treatment	RV/TV ratio	Total mortality percentage due to ascites (%)
NT	0.02 ± 0.22^{b}	7.5 ± 1^{b}
CT	0.31 ± 0.01^{a}	38 ± 4^{a}

Data presented as the mean ± standard error. Means within columns with different superscript letters are significantly different ($p < 0.05$).

RESULTS

Incidence of heart failure

Total mortality due to PHS of CT birds during the the recent experiment was significantly higher ($p < 0.05$) than in the NT ones (38% versus 7.5%). Moreover, the RV/TV ratio of CT birds was greater ($p < 0.05$) than that of other birds (Table 1).

Hematology

The results of hematological values are summarized in Table 2, Blood RBC and HGB contents of CT birds was greater than that of NT ones at d 21 of age ($P<0.05$). Moreover CT birds had the higher blood RBC, HGB and HCT contents at d 42 of age ($P<0.05$).

Enzymes release

The activities of plasma ALT, AST and LDH are shown in Table 3, there was no significant difference between treatment groups for these enzymes at day 21 of age ($P<0.05$), but at day 42, the CT birds had higher plasma ALT, AST and LDH activity than NT ones ($p < 0.05$).

Plasma and liver MDA contents

The plasma and liver MDA contents of broilers are shown in Table 4. CT birds had the higher MDA content in both the plasma and liver at both ages (day 21 and 42 of age) ($P<0.05$).

DISCUSSION

Cold temperature is one of effective factors for inducing hypoxia and PHS. Hence this method was used for inducing heart failure and PHS in recent experiment. Cold temperature despite increasing demand for oxygen consumption, leads to reduced ventilation and decreased oxygen availability in broiler houses (Buys et al., 1999; Daneshyar et al., 2009). Hypoxia is thought to be the primary cause in the development of ascites; therefore, conditions that impose greater metabolic demand or decreased oxygen consumption increase incidence of PHS (Buys et al., 1999). Cold temperature initiates a cascade of events that results in PHS and death (Julian, 1993; Wideman and Bottje, 1995). Greater RV/TV ratio of CT birds compared to NT ones (0.31 vs 0.22) is a vonsequence of higher oxygen demands. Higher mortality of CT birds compared to NT birds (38 vs 7.5 percent) in present study is a result of cold induced hypoxia. Moreover, this hypoxemia leads to some hematological changes such as hematocrit, hemoglobin, red blood cell, So the increased blood RBC, HCT and HGB of CT birds in our experiment shows these hematological changes due to PHS. Dilation and hypertrophy and increased PCV have been reported by some researchers (Julian, 1990, 1993; Owen et al., 995). Despite the changes in hematological indices, hypoxia and be the major cause of ROS production. The low-flow circulation of blood that occurs in birds predisposed to ascites can induce anoxia in different tissues, including the heart. The reoxygenation induced via compensation efforts may happen continuously in the ischemic tissues, resulting in increased production of ROS (Dawson et al., 1993).

Higher plasma and liver MDA contents of CT birds in this experiment indicates the ROS production and lipid peroxidation. So it is suggested that the pathogenesis of ascites syndrome may be initiated by increased production of ROS. Hypoxia of cold temperature not only induces the ROS production, but even causes the injuries in some internal tissues. ROS may cause lipid peroxidation in the membrane of the cells and hence resulting in tissue injury in organs, including lung, heart and liver (Arab et al., 2006). In plasma and liver tissue started from day 21, rises in enzyme release (ALT, AST and LDH) (Table 3) and HCT (Table 2) were observed at day 42. It is suggested that the pathogenesis of ascites syndrome may be initiated by increased production of ROS. As the injury proceeds, it causes dilation and hypertrophy of the right ventricle, resulting in increased PCV, blood viscosity, and the accumulation of fluids in the abdominal cavity due to heart failure (Julian, 1990, 1993; Owen et al., 1995).

A transient hypoxia and then reoxygenation followed by frequent hypoxia can be the major cause of ROS production. The low-flow circulation of blood that occurs in birds predisposed to ascites can induce anoxia in different tissues, including the heart. The reoxygenation induced via compensation efforts may happen conti-nuously in the ischemic tissues, resulting in increased

Table 2. RBC, HGB and HCT of broilers under normal (NT) and cold (CT) environmental temperature.

Day	treatment	RBC(10^6/µl)	HGB(g/dl)	HCT (%)
21	NT	1.71 ± 0.12^b	6.12 ± 0.31^b	29.02 ± 0.80
	CT	2.42 ± 0.16^a	8.57 ± 0.49^a	34.27 ± 2.04
42	NT	2.0 ± 0.20^b	7.75 ± 0.35^b	32.0 ± 0.65^b
	CT	2.8 ± 0.17^a	11.20 ± 0.35^a	39.3 ± 2.27^a

Data presented as the mean ± standard error. Means within columns with different superscript letters are significantly different ($p < 0.05$).

Table 3. ALT, AST and LDH levels in plasma of broilers under normal (NT) and cold (CT) environmental temperature.

Day	Treatment	ALT(U/L)	AST(U/L)	LDH(U/L)
21	NT	2.65 ± 0.57	211.75 ± 32	2975 ± 202
	CT	4.00 ± 0.41	223.50 ± 1	3100 ± 250
42	NT	3.75 ± 0.41^b	217.5 ± 1^b	3162 ± 320^b
	CT	7.37 ± 0.25^a	240.5 ± 7^a	4920 ± 674^a

Data presented as the mean ± standard error. Means within columns with different superscript letters are significantly different ($p < 0.05$).

production of ROS (Dawson et al., 1993).

As the results indicate, increased production of ROS was shown at day 21 in the CT chickens. These agents may cause lipid peroxidation in the membrane of the cells resulting in tissue injury in organs, including lung, heart and liver (Arab et al., 2006) (Table 4). The increase in the amount of AST, ALT and LDH at day 42 is an indicator of a progressive liver cell injury accompanied by the increased production of ROS, resulting in the induction of a chain of oxidative reactions in the liver and other organs (Arab et al., 2006). As the results indicate, the amounts of ALT, AST and LDH have increased during days 21 to 42. There is evidence that serum values of ALT and AST are elevated before the clinical signs and symptoms of liver disease appear. As the injury proceeds, the gross damage (heart failure, fluid accumulation (ascites)) follows, resulting in death. This process can probably explain the pathophysiology of ascites in broilers.

Nain et al. (2008) reported that the morphological changes observed in myocardial mitochondria are consistent with oxidative damage. Notably, mitochondria are the major source of ROS, but because of their very high component of membranes, they are also a very sensitive target of ROS attack. The membrane lipids are very sensitive to oxidative damage due to the presence of polyunsaturated fatty acids, subsequently leading to lipid peroxidation (Halliwell and Gutteridge, 1985). Currently, one of the most common and well-recognized approaches to measuring the effects of free radicals is the estimation of oxidative damage (lipid peroxidation) to cellular membranes (Lykkesfeldt and Svendsen, 2007). So, the measurements from the lipid peroxidation in the present

study showed that in broilers with PHS, oxidative stress increases correlate with heart failure increases. The biochemical evidence of oxidative damage (elevated MDA) corresponds well with the observed morphological changes in the mitochondria, such as mitochondrial swelling, vacuolization, loss and disintegration of cristae (Nain et al., 2008).

The heart is one of the greatest energy-consuming organs in the body, which requires a constant supply of oxygen to maintain its metabolic functions (Giordano, 2005). In the cardiac tissue, mitochondria comprise 30% of the cardiomyocyte volume (Sheeran and Pepe, 2006). The major steps in ROS formation are complex I and complex III of the electron transport chain in the inner mitochondria membrane (Turrens and Boveris, 1980; Turrens et al., 1985). During normal metabolism, 1 to 2% of oxygen is converted to ROS. Hence, increased ROS or RNS production or decreased antioxidant defenses lead to oxidative stress. α-Ketoglutarate dehydrogenase (α-KGDH), one of the key rate-limiting enzymes of the tricarboxylic acid cycle, is involved in energy synthesis pathways. Studies in rats have demonstrated that α-KGDH is a sensitive target of hydrogen peroxide (H_2O_2).

In an anaerobic situation, LDH contributes to energy synthesis by anaerobic glycolysis. An increased production of ROS/RNS occurs during tissue hypoxia (Chen and Meyrick, 2004), which can negatively affect the activity of energy synthesis and transformation pathways. With hypoxia, activation of LDH enzyme by ROS may work as a force to counter the negative effect of other enzymes on energy synthesis and transformation pathways. Recently, higher LDH activity was observed in broilers developing CHF (Nain et al., 2008). Hence,

Table 4. MDA equivalents levels in plasma and liver tissue of broilers under normal (NT) and cold (CT) environmental temperature.

Day	Treatment	MDA in plasma (nm/m lit)	MDA in liver(nm/m lit)
21	NT	1.3 ± 0.31^b	0.85 ± 003^b
	CT	2.5 ± 0.33^a	1.32 ± 0.23^a
42	NT	1.6 ± 0.2^b	1.1 ± 0.04^b
	CT	6.27 ± 0.43^a	2.6 ± 0.25^a

Data presented as the mean ± standard error. Means within columns with different superscript letters are significantly different ($p < 0.05$).

increased activity of LDH in broilers developing CHF is most probably due to generated oxidative stress in the broilers. Insufficiency of creatine phosphate and ATP leads to deterioration in heart pump function in broilers (Nain et al., 2008; Olkowski et al., 2007). This suggests that the observed decline in energy phosphates with deterioration in heart functions might be associated with the decreased activity of these enzymes during oxidative stress.

In conclusion, the results of this study suggested that heart failure in broilers with hypoxia and subsequent PHS can be associated with ROS production during oxidative stress. So, oxidative stress due to hypoxia is the most initial problem with PHS and CHF. ROS can cause cell injury and increase the release of enzymes in plasma, including ALT, AST and LDH. ROS can also contribute to the deterioration of ATP synthesis in myocardium, subsequently leading to lowered energy reserve in themyocardium.

REFERENCES

Andreka P, Tran T, Webster KA, Bishopric NH (2004). Nitric oxide and promotion of cardiac myocyte apoptosis. Mole. Cell. Biochem., 263: 35-53.

Arab HA, Jamshidi R, Rassouli A, Shams G, Hassanzadeh MH (2006). Generation of hydroxyl radicals during ascites experimentally.

Buys N, Scheele CW, Kwakernaak C, Van Der Klis JD, Decuypere E (1999). Performance and physiological variables in broiler chicken lines differing in susceptibility to the ascites syndrome: Changes in blood gases as a function of ambient temperature. Br. Poult. Sci., 40: 135-139.

Balog JM (2003). Ascites syndrome (pulmonary hypertension syndrome) in broiler chickens: Are we seeing the light at the end of the tunnel? Avian Poult. Biol. Rev., 14(3): 99-126.

Chen JX, Meyrick B (2004). Hypoxia increases Hsp90 binding to eNOS via PI3K-Akt in porcine coronary artery endothelium. Lab. Investig., 84: 182-190.

Daneshyar M, Kermanshahi H, Golian AG (2009). Changes of biochemical parameters and enzyme activities in broiler chickens with cold-induced ascites. Poult. Sci., 88: 106-110.

Daneshyar M, Kermanshahi H, Golian AG (2007). Changes of blood gases, internal organ weights and performance of broilerchickens with cold induced ascites. Res. J. Biol. Sci., 2: 729-735.

Dawson TL, Gores GJ, Nieminen AL, Herman B, Lemasters JJ (1993). Mitochondria as a source of reactive oxygen species during reductive stress in rat hepatocytes. Am. J. Physiol., 264: C961.

Diaz-Cruz A, Nava C, Villlanueva R, Serret M, Guinzberg R, Pina AE (1996). Hepatic and cardiac oxidative stress and other metabolic changes in broilers with the ascites syndrome. Poult. Sci., 75: 900-903.

Flohe L, Beckmann R, Giertz H, Loschem G (1985). Oxygen-centered free radicals as mediators of inflammation, in: SIES, H. (Ed.) Oxidative Stress, (New York, Academic Press), pp. 403-415.

Geng ALYM, Guo I, Yang Y (2004). Reduction of Ascites Mortality in Broilers by Coenzyme Q10. Poult. Sci., 83: 1587-1593.

Giordano FJ (2005). Oxygen, oxidative stress, hypoxia and heart failure. J. Clin. Invest., 115: 500-508.

Halliwell B, Gutteridge JM (1985). The importance of free radicals and catalytic metal ions in human diseases. Mole. Aspects Med., 8: 89-193.

Halliwell B (1989). Current status review, free radicals, reactive oxygen species and human disease; A critical evaluation with special reference to arteriosclerosis. Br. J. Exp. Pathol., 70: 737-742.

Huchzerumeyer FW, Deruyck AM (1986). pulmonary hypertension syndrome associated with ascites in broilers. Vet. Rec., 119: 94.

Hassanzadeh M, Buys N, Vander Pooten A (1997). Myocardial _-adrenergic receptor characteristics in T3 induced ascites and in broiler lines differing in ascites susceptibility. Avian Pathol., 26: 293-303.

Iqbal M, Cawthon D, Wideman RF, Bottje WG (2001a). Lung mitochondrial dysfunction in pulmonary hypertension syndrome I. Site specific defects in electron transport chain. Poult. Sci., 80: 485-495.

Julian RJ (1993). Ascites in poultry. Avian Pathol., 22: 419-454.

Julian RJ (1990). Pulmonary hypertension: A cause of right heart failure ascites in meat type chickens. Feeds Stuffs, 29: 19-21.

Jaescheke H (1991). Reactive oxygen and ischaemia/ reperfusion injury of the liver. Chem. Biol. Interact., 79: 115-136.

Lykkesfeldt J, Svendsen O (2007). Oxidants and antioxidants in disease: Oxidative stress in farm animals. Vet. J., 173: 502-511.

Li K, Qiao J, Zhao L, Dong S, Ou D, Wang J, Wang H, Xu T (2006). Increased calcium deposits and decreased Ca2_-ATPase in right ventricular myocardium of ascitic broiler chickens. J. Vet. Med. A, 53: 458-463.

Mccord JM (1985). Oxygen-derived free radical in postischemic tissue injury. N. Engl. J. Med., 312: 159-162.

Maxwell MH, Robertson GW, Moseley D (1994). Potential role of serum troponin I in cardiomyocyte injury in the broiler ascites syndrome. Br. Poult. Sci., 35: 663-667.

Maxwell MH, Robertson GW, Spence S (1986). Studies on ascites syndrome in young broilers: 1. Haematology and pathology. Avian Pathol., 15: 511-524.

Maxwell MH, Robertson GW, Mitchell MA (1993). Ultrastructural demonstration of mitochondrial calcium overload in myocardial cells from broiler chickens with ascites and induced hypoxia. Res. Vet. Sci., 54: 267-277.

Nain S, Ling BB, Wojnarowicz C, Laarveld B, Alcorn J, Olkowski AA (2008). Biochemical factors limiting myocardial energy in a chicken genotype selected for rapid growth. Comparative Biochem. Physiol., Part A, 149(1): 36-43.

Nediani C, Borchi E, Giordano C, Baruzzo S, Ponziani V, Sebastiani M, Nassi P, Mugelli A, d'Amati G, Cerbai E (2007). NADPH oxidase-dependent redox signaling in human heart failure: relationship between the left and right ventricle. J. Mole. Cell. Cardiol., 42(4): 826-834.

Olkowski AA, Nain S, Wojnarowicz C, Laarveld B, Alcorn J, Ling BB (2007). Comparative study of myocardial high energy phosphate substrate content in slow and fast growing chicken and in chickens with heart failure and ascites. Comparative Biochem. Physiol., Part A, 148: 230-238.

Odum TW (1993). Ascites syndrome: Overview and update. Poult. Digest, 52(1): 14-22.

Owen RL, Wideman RF, Cowen BS (1995). Changes in pulmonary arterial and femoral arterial blood pressure upon acute exposure to hypobaric hypoxia in broiler chickens. Poult. Sci., 74: 708-715.

Redout EM, Wagner MJ, Zuidwijk MJ, Boer C, Musters RJ, van Hardeveld C, Paulus WJ, Simonides WS, (2007). Rightventricular failure is associated with increased mitochondrial complex.

Sam F, Kerstetter DL, Pimental DR, Mulukutla S, Tabaee A, Bristow MR, Colucci WS, Sawyer DB (2005). Increased reactive oxygen species production and functional alterations in antioxidant enzymes in human failing myocardium. J. Cardiac Failure, 11(6): 473-480.

Sheeran FL, Pepe S (2006). Energy deficiency in the failing heart: linking increased reactive oxygen species and disruption of oxidative phosphorylation rate. Biochem. Biophys. Acta., 1757: 543-552.

SAS Institute (2002). SAS Users Guide: Statistics. SAS Institute Inc.,

Turrens JF, Boveris A (1980). Generation of superoxide anion by the NADH dehydrogenase of bovine heart mitochondria. Biochem. J., 191: 421-427.

Turrens JF, Alexandre A, Lehninger AL (1985). Ubisemiquinone is the electron donor for superoxide formation by complex III of heart mitochondria. Arch. Biochem. Biophys., 237: 408-414.

Wideman RF, Ismail JRM, Kirby YK, Bottje WG, Varderman RC (1995a). Furosemide reduces the incidence of pulmonary hypertension syndrome (ascites) in broilers exposed to cool environmental temperatures. Poult. Sci., 74: 314-322.

23

Purification and characterizations of NAD dependent isocritrate dehydrogenase from human kidney mitochondria

Mukaram Shikara

Biotechnology Division, Department of Applied Sciences, University of Technology, Baghdad, Iraq.
E-mail: mukaramshikara2010@yahoo.com.

NAD-IDH (Nicotinamide adenine dinucleotide socitric dehydrogenase) has been purified in 1216-fold with a total recovery of 7.5% from the mitochondria of human kidney by using a combination of affinity chromatography with the anion-exchange matrix that allowed obtaining a preparation of high purity. The shape of the peak from Sephacryl S-100 and the results from SDS-PAGE confirm that the enzyme is a tetramer with subunits of 80,000 each, and a native molecular mass of about 320,000. The enzyme shows activity in the absence of any divalent metal ions, but Mn^{+2} is a better activator than Mg^{+2} at lower concentrations (0.5 mM), but it will inhibit the enzyme at higher concentrations (2 mM). ATP (Adenosine triphosphate) and NADH inhibit the enzyme competitively according to Lineweaver-Burk plot. The NAD-IDH does not indicate a homotropic cooperative effect of isocitrate in either the absence or presence of ADP. Increasing concentrations of NAD decrease the Km of isocitrate while increasing levels of isocitrate lower the Km of NAD. Km of either isocitrate or NAD is lowered further in the presence of ADP. The inhibition by ATP (or NADH) cannot be counteracted by ADP in the presence of isocitrate, so ADP cannot enhance NAD-IDH activity nor reverse inhibition by ATP (or NADH), while isocitrate will bind to the enzyme and prevent it from interacting with ADP. The activity of NAD-IDH in mitochondria is probably controlled in a complex way by NADH, ATP and divalent ions.

Key words: Nicotinamide adenine dinucleotide socitric dehydrogenase, isocitrate, Adenosine triphosphate, Nicotinamide adenine dinucleotide hydrogenase, human kidney, purification.

INTRODUCTION

Isocitrate dehydrogenases (IDH) are enzymes in tricarboxylic acid cycle (TCA) that catalyze oxidative decarboxylation of isocitrate to a-ketoglutarate using NAD or NADP as cofactor and widely distributed in the three domains of life: archaea, bacteria and eukaryotes (Karlstrom et al., 2005). NAD-dependent IDH is localized in mitochondria matrix, while NADP–IDH is localized in both mitochondria and cytoplasm. The biochemical knowledge of NAD-IDH is much less advanced comparing with NADP-IDH since it has been difficult to work with NAD-IDH than NADP-IDH enzyme. The distinct function of NAD-IDH is its ability to recognize its respective substrates and coenzyme (Chen and Jeong, 2000). To our knowledge, no NAD-IDH from eukaryotic organisms has yet been purified to homogeneity but several researchers purified partially NAD-IDH from the

mitochondria of several eukaryotes such as potato mitochondria (Tezuka and Laties, 1983), mitochondria from blowfly muscles (Wadano et al., 1989), mitochondria of higher plants (Chen and Gadal, 1990), and pea mitochondria (Oliver and McIntosh, 1995).

The regulatory importance of NAD-IDH is still under debate. The present paper is dealing with the properties of NAD-IDH enzyme purified from the mitochondria that was isolated from human's kidney.

MATERIALS AND METHODS

50 g of human kidney from 20 years old assassinated female had been obtained from Department of Forensic Medicine and Pathology, Faculty of Medicine, University of Jordan, Amman, Jordan. Standard proteins, DEAE-Sepharose, DEAE-Sephacel,

Sephacryl S-100 were purchased from Sigma. Molecular weight markers were obtained from Boehringer, Mannheim, Germany. Motor driven tightly fitting glass/Teflon Potter Elvehjem homogenizer (30 mL). Buffer A consisted of 0.05 M Tris/MOPS and 12.1 g of 1 M sucrose. The mixture was brought to 1 L of distilled water and adjusted to pH 7.5. It was stored at 4°C, while buffer B consisted of 0.05 M Tris/MOPS and brought to 1 L of distilled water and adjusted to pH 7.5 and stored at 4°C (if ADP is used, the pH will be adjusted to 8.2). All glassware were washed three times with distilled water to avoid Ca^{2+} contamination. Ca^{2+} overloaded was the most common cause for the dysfunction of isolated mitochondria.

All buffers were prepared in the same day of the experiment to avoid bacterial/yeast growth in stored buffers, and since pH depended on temperature, the pH must be measured in all solutions at 25°C.

Purification of mitochondria

Mitochondria were purified according to the method of Frezza et al. (2007) as modified by Gregg et al. (2009) from human kidney. The kidney was rinsed free of blood by using ice-cold buffer A and minced into small pieces using scissors. The buffer used during the mincing was discarded and replaced with 5 mL of ice-cold fresh buffer A. The suspension was transferred into a Teflon pestle and homogenized at 1,600 g at 4°C to minimize activation of damaging phospholipases and proteases for few minutes. All other harsher techniques, including glass pestle in a glass potter, could easily damag the mitochondria. The optimal ratio between tissue and isolation buffer ranges from 1:5 to 1:10 (w: v). The homogenate was transferred into a 50 mL polypropylene Falcon tube and centrifuged at 600 g for 10 min at 4°C. The pellet (contained the cell debris and nuclei) was discarded. The supernatant (contained all the lighter cellular fractions) was centrifuged at 5,000 g for 20 min at 4°C in chilled centrifuge tubes. The supernatant (contained microsomes, membrane fragments, ribosomes and cytoplasmic enzymes) was decanted and the pellet (contained crude mitochondria) was resuspended in 12 mL of ice-cold buffer A.

The re-suspension was enriched in mitochondria which mixed with other organelles such as "golgi" apparatus, endoplasmic reticulum and vacuoles. In order to obtained pure mitochondria, this re-suspension was loaded onto a 4 mL each of 15, 23, 32 and 60 wt: v step sucrose density gradient prepared in the aforementioned buffer and centrifuged at 100,000 g in a SW60Ti rotor (Beckman Ultra centrifuge tube) for 5 h at 4°C. Fractions were then collected from the bottom of the gradient. The intact mitochondria form a brown band at the 60 to 32% sucrose interface and removed gently by using a pipette with a cut tip and placed into a separate Beckman centrifuge tube which was filled with buffer A. The semi-pure mitochondria were centrifuged for 30 min at 10,000 g at 4°C, and the supernatant was decanted, while the precipitate consisted of pure mitochondria was used for the next step. The mitochondria should not be diluted with buffer in order to retain their functionality for a longer time.

The concentration of mitochondria in this preparation is about 80 mg.mL^{-1} with the total volume is about 1 mL.

Assays for mitochondria

To identify the mitochondria, slides were prepared from all samples and a drop of methyl green pyronin was added and all slides observed at 400X with bright field: nuclei should stain green, cytoplasm red or pink, and mitochondria can be seen as small dots and protein concentration of each fraction was measured (Kurnick and Mirsky, 1950). Another method for identification of mitochondria

is to use the blue dye 2,6 dichlorophenol indolphenol (DCPIP). A mixture of 1 mL of 0.1 M succinate, 1 mL of 5 mM KCN, 1 mL of mitochondrial preparation and 4 mL 0.05 M phosphate buffer (pH 7.5) were mixed and left for 5 min at 25°C. 1 mL of 70 uM DCPIP, 0.3% (w: v) was added to the mixture and read after 5 min at 600 nm. DCPIP will accept an electron from NADH (inside mitochondria) and becomes a reduced colourless form.

Isolation and purification of NAD-IDH

Mitochondrial preparation was thawed at room temperature and was layered onto a 1.5 x 40 cm DEAE-Sepharose column that was previously equilibrated with 4 L of buffer A. About 70% of the enzyme activity was eluted in the second peak at 0.55 M NaCl. The active fractions from the second peak were pooled and dialyzed for 5 h against 4 L of buffer B and then was loaded onto a Sephadex G-200 column (1.5 x 30 cm) that was previously equilibrated with buffer B. The enzyme was eluted with 120 mL linear gradient of 0 to 1 N NaCl in buffer B and one major peak was observed. Solid ammonium sulphate was added to the "peaked pooled fractions" to form 0 to 35%, 35 to 70% and 70 to 90% saturation fractions respectively. Each fraction was dialyzed separately against buffer B with two changes for 24 h and measured for NAD-IDH activity. The 35 to 70% saturation fraction was found to have a high enzyme activity and was layered onto a DEAE-Sephacel (0.7 x 20 cm) which was previously equilibrated with buffer B at a flow rate of 15 mL.h^{-1}.

Bound enzyme was eluted with 60 mL linear gradient (0 to 1 N NaCl in buffer B). Fractions (0.5 mL) containing NAD-IDH activity were pooled, dialyzed as before and applied onto a Sephacryl S-100 column (0.5 x 30 cm) equilibrated with buffer B at a flow rate of 15 mL.h^{-1} and the enzyme was eluted with 100 mL linear gradient (0 to 1 N NaCl in the same buffer) and 1 mL fractions were collected. The pooled peak-activity fractions were dialyzed as before and then concentrated by ultrafiltration and kept at 4°C for immediate use or stored at -20°C.

Enzyme assay and protein determination

The NAD-IDH activity was determined in a 1 mL reaction containing buffer B, 0.5 mM $MnCl_2$ (or 1.0 mM $MgSO_4$), 1.5 mM NAD, 4 mM DL isocitrate, 0.5 mM ADP and the appropriate amount of the enzyme. All the assays were performed at 37°C in a final volume of 2.5 mL. The reaction was started by the addition of the isocitrate and followed by the increase in A_{340}. One unit of activity was defined as the amount of enzyme that catalyzed the production of 1 umole of NADH per mL of reaction mixture per min under the standard conditions. Specific activity is defined as units per mg of protein. All assays were performed in triplicate. Protein was estimated by the method of Bradford (1976) with BSA as a standard.

Enzyme kinetics

A kinetic analysis was done on the purified enzyme following Sephacryl S-100 chromatography. Determination of K_m values for the enzyme was carried out in 0.1 mM tris-HCl buffer, pH 7.5, and 1 mM $MgSo_4$ at different concentrations.

Gel electrophoresis

Sodium dodecyl sulfate-polyacrylamide gel electrophoresis (SDS-

PAGE) was performed as described by Laemmli et al. (1970) as modified by Maizel (1970) using 12% polyacrylamide gel. Proteins were located in the gel by staining with 0.1% (w/v) Coomassie brilliant blue R-250 in 25% (v/v) ethanol and 10% (v/v) acetic acid. NAD-IDH activity was located by running a separate lane with pooled active fractions from Sephacryl S-100 which was cut out from the gel before staining. The lane is cut horizontally into 0.25 mm pieces and submerged in buffer B and assayed for activity as in methods.

Estimation of native molecular mass

The molecular mass of NAD-IDH was estimated by SDS-PAGE. Protein markers of known sizes: ferritin from equine spleen (364kD), catalase from *Aspergillus niger* (240 kD), myosin from rabbit muscles (205 kD), B-galactosidase from *Escherichia coli* (116 kD), bovine serum albumin (67 kD), fumarase from porcine heart (48 kD), ovalbumin (45 kD) and cytochrome C (12.5 kD) were used as standards.

RESULTS

Purification

The purification of the enzyme is shown in Table 1. The NAD$^+$-dependent IDH was purified to near homogeneity (about 1216-fold) with specific activity of 2250 units.mg^{-1} and a recovery of 7.5%. DEAE-Sepharose, Sephadex G-200 and ammonium sulphate fractionation did remove most of the contaminating proteins (Table 1). The shape of the peak of Sephacryl S-100 suggested heterogeneity of the enzyme, or that the enzyme consists of several units (Figure 1). The active fractions of the column preserved in -20°, so that little loss of enzyme activity occurred.

Optimum pH

The pH optimum of NAD-IDH was 7.5, but with the addition of ADP, the pH optimum shifted to the alkaline side and becomes 8.2.

Optimum temperature

The optimum temperature was at 37°C, but dropped quickly and disappeared completely at 50°C.

Enzyme stability

The enzyme in its purified form was stable up to 30 min at room temperature and for 12 h at 4°C. The addition of 20% glycerol to the enzyme stabilized the enzyme substantially (up to 21 h at 4°C), and up to three weeks at -20°C, but any higher concentration will interfere with the chromatography steps. BSA (up to 10%) did not stabilize or activate the enzyme, and shows no effect at all.

Divalent metals effects

The enzyme showed activity in the absence of any divalent metal ions, but Mn^{+2} was a better activator than Mg^{+2} at lower concentrations (0.5 mM), but it would inhibit the enzyme at higher concentrations (2 mM). Mg^{+2} did not inhibit the enzyme over a wide range of concentrations, so it was preferably used instead of Mn^{+2}. Cobalt would activat the enzyme slightly but other divalent ions would have inhibitory or no effect on it (Table 2).

Effects of substrates

ADP enhanced the activity of the enzyme by 60 to 80%, while AMP had no effect at all, but it inhibited some activity if mixed with ADP. GMP, GDP and GTP have no effect on the activity of NAD-IDH. A number of substrates and some intermediates of the TCA cycle (such as glutamate, fumarate, pyruvate, succinate and citrate) showed no stimulation effect on the enzyme. EGTA and EDTA addition inhibited the enzyme by about 20%. For this reason, 1 mM Mg was used in all experiments with 0.5 mM ADP.

Estimation of molecular mass

The pooled active fractions from 35 to 70% desalted ammonium sulphate fraction, DEAE-Sephacel and Sephacryl S-100 were electrophoresed on SDS-PAGE at pH 7.5. Several bands were extending from top to bottom. Four bands were appeared in Sephacryl S-100 lanes (lanes 3, 4, Figure 2) which had strong NAD-IDH activities. These activities were identified by running a separate lane with pooled active fractions from Sephacryl S-100 which was cut out from the gel before staining. The lane is cut horizontally into 0.25 mm pieces and submerged in 0.1 M tris-HCl buffer, pH 7.5 and assayed for activity as in methods. By comparing the positions of the four bands with standards used as described in methods, NAD-IDH from Sephacryl S-100 was clearly a tetramer with subunits of 80,000 and the native molecular mass is about 320,000.

Inhibition by ATP and NADH

ATP and NADH inhibited the enzyme competitively according to Lineweaver-Burk plot (1934). The value of Ki was measured to be 1.8 and 0.6 mM for ATP and NADH respectively under standard conditions (Figures 3 and 4).

Table 1. Purification of NAD$^+$-IDH from human kidney.

Purification step	Volume (mL)	Total activity (units)	Total protein (mg)	Specific activity (unit.mg^{-1})	Purification (-fold)	Yield (%)
Mitochondrial preparation	20	120	65.0	1.85	1	100
DEAE-Sepharose	30	96	5.2	18.5	10	80.0
Sephadex G-200	45	65	0.8	81.2	44	54.1
35 to 70% (NH$_4$)$_2$SO$_4$ fraction	10	43	0.1	430	232	35.8
DEAE-Sephacel	25	15	0.01	1500	811	12.5
Sephacryl S-100	10	9	0.004	2250	1216	7.5

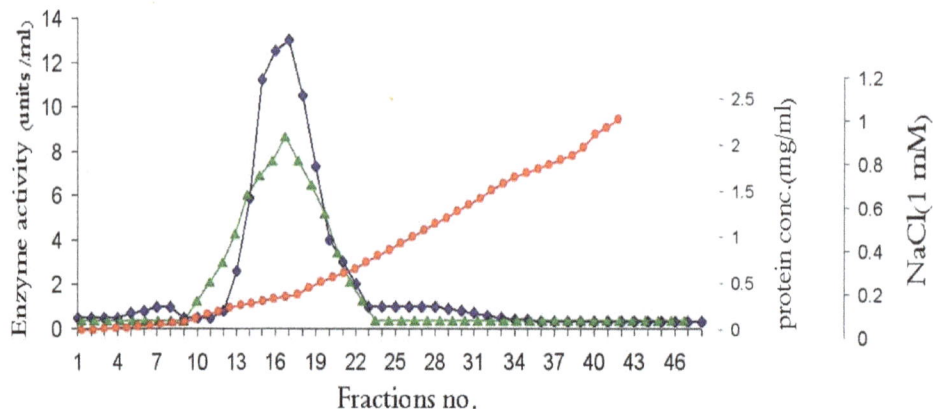

Figure 1. Elution profile of NAD-IDH activity from Sephacryl S-100 column (●) with its protein profile (▲) and the linear gradient of sodium chloride (0-1mM) (■).

Table 2. Effect of divalent metal ions on NAD$^+$-IDH activity from the pooled fractions from Sephacryl S-100.

Ion (mM)	Effect (100%)
Control	100
Manganese (0.5)	140
Manganese (1.0)	320
Manganese (1.5)	110
Manganese (2.0)	70
Magnesium (0.5)	120
Magnesium (1.0)	250
Magnesium (1.5)	180
Magnesium (2.0)	130
Cobalt (0.5)	130
Cobalt (1.0)	170
Cobalt (1.5)	130
Cobalt (2.0)	110
Copper (0.5)	60
Copper (1.0)	50
Calcium (0.5)	100
Calcium (1.0)	100
Zinc (0.5)	40
Zinc(1.0)	17

Effect of ADP on NAD-IDH

It seemed possible that a homotropic cooperative effect of isocitrate in either the absence or presence of ADP might occur if concentrations of NAD-IDH were limiting. A hill plot was plotted with different amounts of isocitrate and fixed amounts of NAD-IDH. Straight parallel lines with slopes approaching a value of less than 1 were obtained with or without ADP. All the values of the slopes were less than 1 indicated that a cooperative effect of isocitrate had not been caused by changes of NAD-IDH concentration (Monod et al., 1963; Jacob et al., 2005). A lineweaver-burk plot was plotted either with different amounts of isocitrate and a fixed amount of NAD-IDH (Figure 5), or different concentrations of NAD-IDH and fixed concentration of isocitrate (Figure 6) in the absence (top) or presence (bottom) of ADP. The plots gave patterns of intersecting straight lines, indicated no deviation from Michaelis-type kinetics under these conditions for either isocitrate or NAD-IDH with the kidney enzyme. In the absence of ADP, the increased concentrations of NAD-IDH decreased the Km of isocitrate while the increased levels of isocitrate lowered the Km of NAD-IDH.

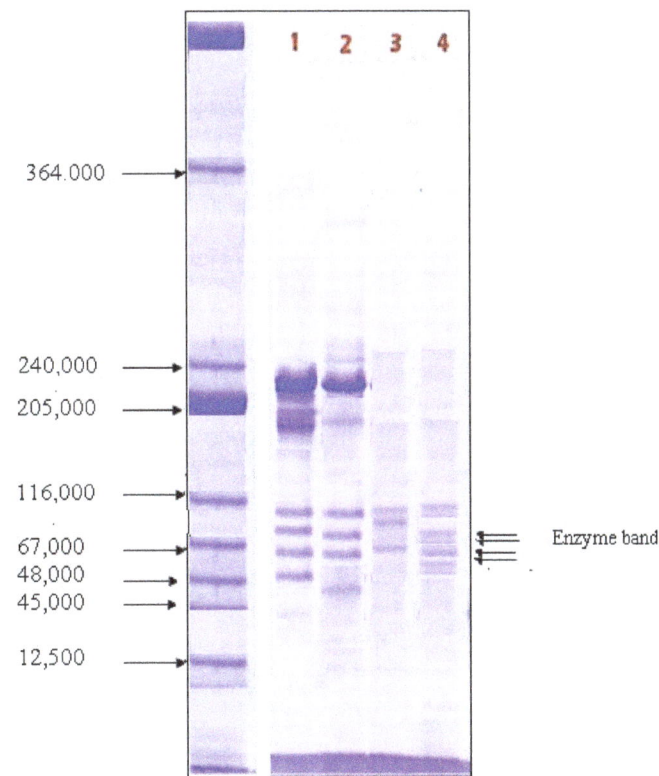

Figure 2. SDS-polyacrylamide gel with four lanes; lane 1, A 35 to 70% desalted ammonium sulphate fraction, lanes 2, the pooled peaked fractions from DEAE-Sephacel column, lanes 3, 4, the pooled peaked fractions from Sephacryl S-100. The arrows indicate the bands that show NAD-IDH activity.

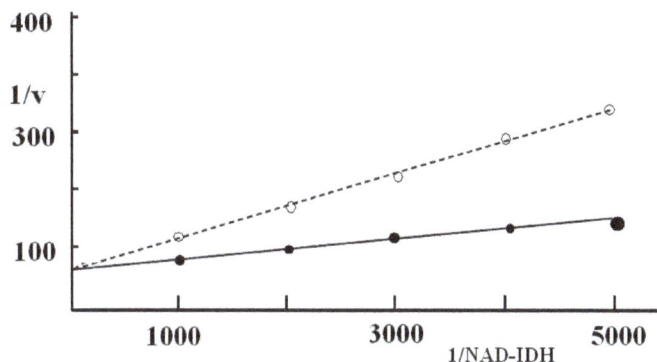

Figure 3. The effect on NAD-IDH in the presence (○) or absence (●) of ATP. The incubation mixture is as described under methods. NAD concentrations are varied while ATP concentration is fixed.

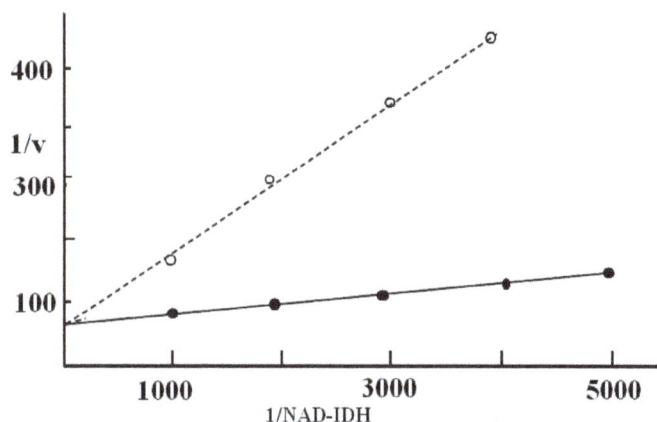

Figure 4. The effect on NAD-DH in the presence (○) or absence (●) of NADH. The incubation mixture is as described under methods. NAD concentrations are varied while NADH concentration is fixed.

The values of Km of either isocitrate or NAD-IDH were lowered further in the presence of ADP. The values of limiting Michaelis constants of isocitrate and NAD-IDH were calculated by the graphic method of Dalziel (1957) for two-substrates system where $\varnothing_1$, $\varnothing_2$, and $\varnothing_3$, are substituted by $\varnothing_{NAD\text{-}IDH}$, $\varnothing_{IC}$ and $\varnothing_{NAD.IC}$ (IC represents isocitrate):

$$e/V_0 = \varnothing_0 + \varnothing_1/[S_1] + \varnothing_2/[S_2] + \varnothing_3/[S_1][S_2]$$
$$e/V_0 = \varnothing_0 + \varnothing_{NAD\text{-}IDH}/[S_1] + \varnothing_{IC}/[S_2] + \varnothing_{NAD.IC}/[S_1][S_2]$$

A secondary plot was made of the intercepts at the ordinate (Figure 7, top) and the slopes (Figure 7, bottom) of the primary plots (Figures 5 and 6) against the reciprocals of the concentrations of isocitrate (or NAD-IDH). The value, furthermore, of the constants $\varnothing_0$, $\varnothing_{IC}$, $\varnothing_{NAD\text{-}IDH}$ were affected very little by the presence of ADP, even when ADP caused a marked decrease in the values of the complex constant $\varnothing_{NAD.IC}$ (Figure 7, bottom). Therefore, the values of V_{max}, K_{IC} and $K_{NAD\text{-}IDH}$ had been calculated and the maximal velocities K_{IC} and $K_{NAD\text{-}IDH}$ were essentially unaffected by ADP (K_{IC} = 0.32 mM and 0.26 nM, $K_{NAD\text{-}IDH}$ = 0.06 mM in the absence and presence of ADP respectively). However, the values of the ratios of the constants $\varnothing_{NAD.IC}/\varnothing_{IC}$ and $\varnothing_{NAD.IC}/\varnothing_{NAD\text{-}IDH}$ were 4 to 7 folds larger in the absence of ADP.

DISCUSSION

NAD-IDH had been purified from the mitochondria of human kidney 1216-fold with a total recovery of 7.5%. The combination of the affinity chromatography with the anion-exchange matrix allowed obtaining a preparation of high purity, and most contaminated proteins were removed. The shape of the peak from Sephacryl S-100 suggested that the enzyme consisted of four subunits. This observation was confirmed by SDS-PAGE which show that only four bands of protein from Sephacryl S-100 column had strong NAD-IDH activities. The enzyme

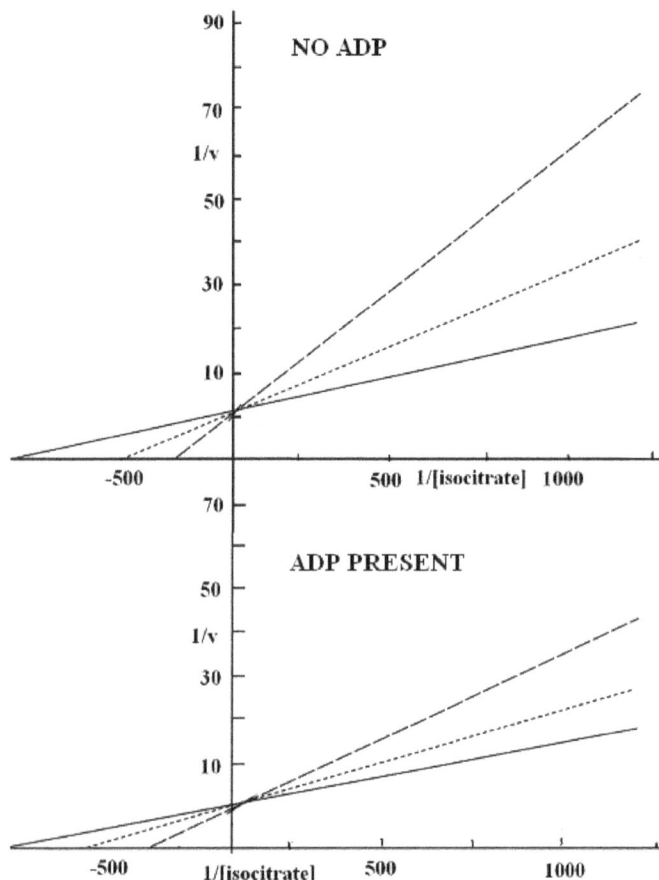

Figure 5. Primary plots of velocity which are using isocitrate concentrations at varying levels of NAD (0.5, 1.0 and 1.5 Mm) in the absence (top) and presence (Bottom) of ADP. The reactions mixture is described under methods.

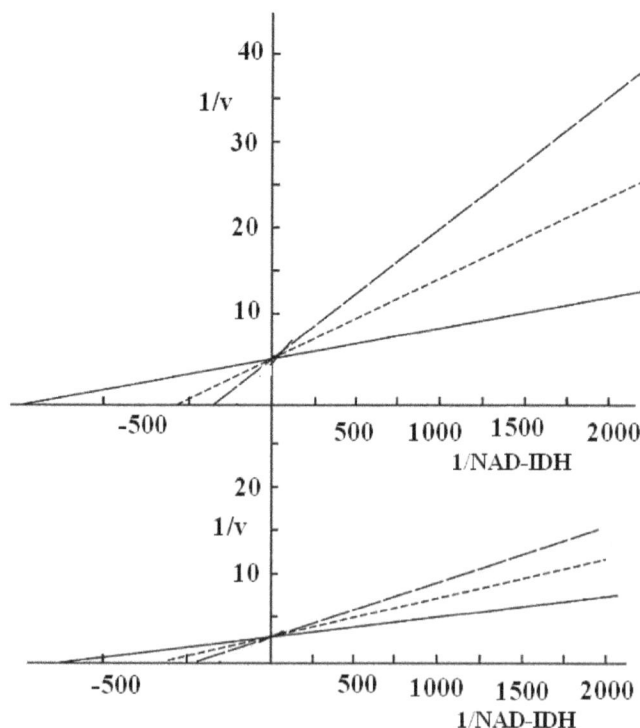

Figure 6. Primary plots of velocity which are using NAD concentrations at varying levels of isocitrate (0.5, 1.0 and 1.5 Mm) in the absence (top) and presence (bottom) of ADP. The reactions mixture is described under methods.

seemed to be a tetramer with four subunits of about 80,000 kD and a native molecular mass of about 320,000. These results agreed to some extents with results obtained from previous NAD-IDHs purifications such as Ehrlich et al. (1981), Shorrosh and Dixon (1992), Alvarez-Villafane et al. (1996), Cornu et al. (1996), Bzymek and Colman (2007) and Dangee et al. (2008).

AMP could not activate NAD-IDH enzyme, even when some researchers claimed the opposite (Hathaway and Atkinson, 1963; Plaut and Aogaichi, 1968). The author believed that such activation was of doubtful physiological significance since the concentration required differed by several orders of magnitude from that present in tissues. The response of NAD-IDH in human kidney to varying concentrations of isocitrate and NAD-IDH corresponded essentially to the kinetics of two substrate reactions observed with a number of other pyridine nucleotide dehydrogenases. ADP did not modify the limiting Michaelis constants of NAD-IDH and isocitrate, but had a marked effect on the slopes of the

lines of Lineweaver Burk plots (Figures 5 and 6) and the secondary plots derived there from (Figure 7). Thus ADP lowered and decreased the ratios of the complex constants $\emptyset_{NAD \cdot IC}/ \emptyset_{IC}$ and $\emptyset_{NAD \cdot IC}/ \emptyset_{NAD-IDH}$ and the symmetry of the effect of ADP on values of $\emptyset_{NAD \cdot IC}$ (Figure 6) seemed to imply that the nucleoside diphosphate did not preferentially modify one of the interacting sites for substrate and coenzyme over the other.

The significance of the complex constants represented by the ratios $\emptyset_{NAD \cdot IC}/ \emptyset_{IC}$ and $\emptyset_{NAD \cdot IC}/ \emptyset_{NAD-IDH}$ were somewhat difficult to evaluate. Hartong et al. (2008) had shown that a number of pyridine nucleotides linked dehydrogenasess followed an ordered sequential reaction pathway in which the addition of coenzyme to the enzyme preceded interaction with substrate (Lee, 2002). The lack of reversibility of the NAD-IDH reaction precluded a definitive kinetic analysis the intersection points of the primary plots were calculated (Figures 4 and 5) according to the method of Frieden (2007). The values were quite comparable to those calculated from those of the rations of the $\emptyset\Phi$ constants (Abiko et al., 2005). Inhibition of NAD-IDH by NADH competitively was consistent with an ordered sequential mechanism of addition of NAD-IDH followed by oxidizing substrate in

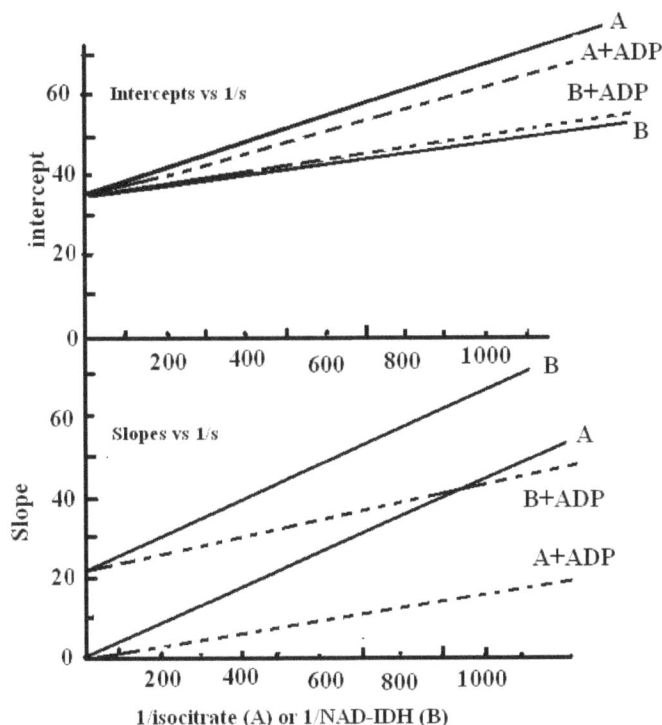

Figure 7. Secondary plots of intercepts (top) and slopes (bottom) of values with respect to the reciprocal of NAD or isocitrate concentration.

which NADH is the last product detached from the enzyme. The oxidizing and reducing forms of the pyridine nucleotide competed, presumably, for the same site on the enzyme and on this case a similar effect of ADP on the dissociation constants of NAD-IDH and NADH may be expected (Bzymek and Colman, 2007; Dangee et al., 2008). Other possibility was that the interaction of the enzyme with NADH may cause the protein to assume a conformation in which a catalytic site was no longer affected by ADP while such interaction still occurs in the enzyme-NAD complex (Lancien et al., 1999). In terms of physiological role of the human kidney, it was significant that isocitrate influenced the value of km of NAD-IDH at limiting substrate and coenzyme concentrations and vice versa. In addition, ADP lowered both constants (Weiss et al., 2000).

The activity of NAD-IDH was influenced by a number of factors such as interacting substrates such as isocitrate and NADH, and the divalent activators (Mn^{+2} or Mg^{+2}). ADP influenced the enzyme also, while ATP and NADH inhibited it competitively. The intermediates of TCA had no effect which suggests that lack of reversibility of the reaction may be due partly to poor binding of TCA intermediates to the enzyme in addition to the product inhibition by ATP (or NADH). Future studies are aiming to the cloning of the gene and required to understand the regulatory mechanism of NAD-IDH.

ACKNOWLEDGMENTS

The author is grateful for the valuable assistance of Dr. Hasan A. Abder-Rahman, PhD, JMC from Forensic Medicine and Pathology Department, Faculty of Medicine, University of Jordan, Amman, Jordan. My gratitude to professor Musa H. Abu Zarga, Department of Chemistry, Faculty of Science, The University of Jordan for his valuable advices. I am debited to Professor Julia Bejar Alvarado, Department of Cellular Biology, Genetics and Physiology, University of Malaga and Professor Manchikatla Venkat Rajam, Department of Genetics, University of New Delhi India for their valuable suggestions and criticism in kinetics studies. My deep gratitude is to the staff in laboratory 1, College of Pharmacy, Al-Zaytoonah University for their cooperation and help.

REFERENCES

Abiko T, Obara M, Ushioda A, Hayakawa T, Hodges M, Yamaya T (2005). Localization of NAD-Isocitrate Dehydrogenase and Glutamate Dehydrogenase in Rice Roots: Candidates for Providing Carbon Skeletons to NADH-Glutamate Synthase. Plant Cell Physiol., 46(10): 1724-1734.

Alvarez-Villafañe E, Soler J, del Valle P, Busto F, de Arriaga D (1996). Two NAD+-isocitrate dehydrogenase forms in *Phycomyces blakesleeanus*: induction in response to acetate growth and characterization, kinetics, and regulation of both enzyme forms. Biochemistry, 35: 4741-4752.

Bradford MM (1976). A rapid and sensitive method for the quantization of microgram quantities of protein utilizing the principle of protein-dye binding. Anal. Biochem., 72: 248-254.

Bzymek KP, Colman RF (2007). Role of α-Asp181, β-Asp192 and γ-Asp190 in the Distinctive Subunits of Human NAD-Specific Isocitrate Dehydrogenase. Biochemistry, 46: 5391-5397.

Chen RD, Gadal P (1990). Structure, function and regulation of NAD and NADP dependent isocitrate dehydrogenases in higher plants and in other organisms. Plant Physiol. Biochem., 28: 411-427.

Chen R, Jeong SS (2000). Functional prediction, identification of protein orthologs and paralogs. Protein Sci., 9: 2344-2353.

Cornu S, Pireaux JC, Gerard J, Dizengremel P (1996). NAD(P) dependent isocitrate dehydrogenases in mitochondria purified from Picea abies seedlings. Physiol. Plant., 96: 312-318.

Dalziel K (1957). Initial steady state velocities in the evaluation of enzyme-coenzyme substrate reaction mechanism. Acta Chem. Scand., 11(1): 706-723.

Dange M, Hartong DT, McGee TL, Berson EL, Dryja TP, Colman RF (2008). Mutations of the beta-subunit of human NAD-specific isocitrate dehydrogenase in patients with Retinitis pigmentosa. FASEB J., 22: 1032.1

Ehrlich RS, Hayman S, Ramachandran N, Colman RF (1981). Re-evaluation of molecular weight of pig heart NAD-specific isocitrate dehydrogenase. J. Biol. Chem., 256: 10560-10564.

Frezza C, Cipolat S, Scorrano L (2007). Organelle isolation: functional mitochondria from mouse liver, muscle and cultured fibroblasts. Nat. Prot., 2: 287-295

Frieden C (2007). Protein aggregation processes: In search of the mechanism. Protein Sci., 16: 2334-2344.

Gregg C, Kyryakov P, Titorenko VI (2009). Purification of Mitochondria from Yeast Cells. http//www.jove.com/index/details.stp?id=1417. JOVE: J. visual. Exp., p. 30.

Hartong DT, Dange M, McGee TL, Berson EL, Thaddeus P Dryja, Colman RF (2008). Insights from retinitis pigmentosa into the roles of isocitrate dehydrogenases in the Krebs cycle. Nat. Genet., 40: 1230-1234.

Hathaway JA, Atkinson DE (1963). The effect of adenylic acid on yeast nicotinamide adenine dinucleotide isocitrate dehydrogenase, a possible metabolic control mechanism. J. Biol. Chem., 238: 2875-2881.

Jacob F, Perrin D, Sánchez C, Monod J, Edelstein S (2005). The operon: a group of genes with expression coordinated by an operator. C.R. Acad. Sci. Paris, 250(1960): 1727-1729". Comp. Rendus Biol., 328(6): 514-520.

Karlstrom M, Stokke R, Steen IH, Birkeland NK, Ladenstein R (2005). Isocitrate dehydrogenase from the hyperthermophile Aeropyrum pernix: X-ray structure analysis of a ternary enzyme-substrate complex and thermal stability. J. Mol. Biol., 345: 559-577.

Kurnick NB, Mirsky AE (1950). Methyl green pyronin II. Stoichiometry of reaction with nucleic acids. J. Cell Biol., 33(3): 265-270.

Laemmli UK (1970). Cleavage of structural proteins during the assembly of the head of bacteriophage T4. Nature, 227: 680-685

Lancien M, Ferrario-Méry S, Roux Y, Bismuth E, Masclaux C, Hirel B, Gadal P, Hodges M (1999). Simultaneous Expression of NAD-Dependent Isocitrate Dehydrogenase and Other Krebs Cycle Genes after Nitrate Resupply to Short-Term Nitrogen-Starved Tobacco. Plant Physiol., 120: 717-726.

Lee P, Colman RF (2002). Implication by site-directed mutagenesis of Arg314 and Tyr316 in the coenzyme site of pig mitochondrial NADP-dependent isocitrate dehydrogenase. Arch. Biochem. Biophys., 401(1): 81-90.

Lineweaver H, Burk D (1934). The determination of the enzyme dissociation constants. J. Am. Chem. Soc., 56: 658-666.

Maizel JV (1971). Polyacrylamide gel electrophoresis of viral protein. In: Maramorosch K, Koprowshi H (Eds.). Methods in Virology 5, Academic Press, New York, pp. 179-246.

Monod J, Changeux J (1963). Allosteric proteins and cellular control systems. J. Mole. Boil., 6: 306-329.

Oliver DJ, McIntosh CA (1995). The biochemistry of the mitochondrial Matrex. In CS Levings III, IK Vasil, eds, The Molecular Biology of Plant Mitochondria: Advances in Cellular and Molecular Biology of Plants. Kluwer Academic Publishers, Dordrecht, Netherlands, 3: 237-280.

Plaut GWE, Gabriel JL (1983). Role of isocitrate dehydrogenase in animal tissue metabolism. In: Lennon DLF, Stratman FW, Zahlten RN eds. Biochemistry of metabolic processes. Amsterdam Elsevier Sci., pp. 129-142.

Shorrosh BS, Dixon RA (1992). Molecular characterization and expression of an isocitrate dehydrogenase from alfalfa (Medicago sativa L.). Plant Mol. Biol., 20: 801-807.

Tezuka T, Laties GG (1983). Isolation and characterization of inner-membrane associated and matrix NAD-specific isocitrate dehydrogenase in potato mitochondria. Plant Physiol., 72: 959-963.

Wadano A, Miura K, Ihara H, Kondo N, Taniguchi M (1989). Purification and some properties of Isocitrate dehydrogenase of a blowfly Aldrichina graham. Comp. Biochem. Physiol., 94B: 189-194.

Weiss C, Zeng Y, Huang J, Sobocka MB, Rushbrook JI (2000). Bovine NAD+-dependent isocitrate dehydrogenase: alternative splicing and tissue-dependent expression of subunit 1. Biochemistry, 39(7): 1807-1816.

24

Testicular biometry and its relationship with body weight and semen output of black Bengal bucks in Bangladesh

Sanjoy Kumar Kabiraj[1], S. A. Masudul Hoque[2], M. A. M. Yahia Khandoker[3]* and Syed Sakhawat Husain[3,4]

[1]Reproductive Biotechnology Laboratory, Department of Animal Breeding and Genetics, Bangladesh Agricultural University, Mymensingh-2202, Bangladesh.
[2]Department of Animal Breeding and Genetics, Bangabandhu Sheikh Mujibur Rahman Agricultural University, Gazipur-1706, Bangladesh.
[3]Department of Animal Breeding and Genetics, Bangladesh Agricultural University, Mymensingh-2202, Bangladesh.
[4]Patuakhali Science and Technology University, Dumki, Patuakhali-8602, Bangladesh.

Biometrical study of testes were done in twelve black Bengal bucks (*Capra hircus*) of three different age groups, 0.5 to 1.0 years (group A), 1.5 to 2.0 years (group B) and 2.5 to 3.0 years (group C) to find out the age depended changes in the biometry of testes and their relationship with semen quality. Before slaughtering the bucks, the semen quality of the bucks of three different age groups was evaluated in terms of volume (ml), live sperm (%) and sperm concentration (billion/ml) for a period of 45 days. The semen volume and sperm concentration of age group C (0.68 ± 0.04 ml and 3.04 ± 0.10 billion/ml respectively) were significantly ($p<0.05$) higher than those of age group A (0.32 ± 0.04 ml and 76.46 ± 2.65 billion/ml respectively) but no significant difference was observed with that of age group B. Whereas, live sperm percentage of age group B (85.64 ± 0.87) was higher than those of other age groups but the difference was not significant ($p>0.05$). Testicular measurements were increased with the advancement of age and body weight. The size and weight of left testis were higher than those of right testis at same age. Semen volume and sperm concentration were highly significant ($p<0.01$) and positively correlated with almost all testicular measurements.

Key words: Biometry, black Bengal bucks, body weight and semen output.

INTRODUCTION

The Black Bengal goat is the second most economically important ruminant in Bangladesh after cattle. Unfortunately, there are severe shortfalls of the stud bucks all over the country (Husain, 2004). It is established that due to haphazard breeding system the genetic merit of Black Bengal goat is under threat. For the better propagation of the species of goat, superior buck selection seems to be very important and alternative approach to boast up the production potential. This has led to the development of methods for predicting potential sperm production and particularly for identifying bucks with high sperm output potential at an early age. To our knowledge, scant information exists on these aspects in the black Bengal buck (Rahman, 2009; Islam, 2001). It is established that due to haphazard breeding system the genetic merit of Black Bengal goat is under threat. In this regard, superior buck selection seems to be very important and alternative approach to boast up the production potential. Therefore, during selection of breeding buck special attention should be given on age, body weight, soundness of the sexual organ and quality of ejaculated semen.

Considering the aforementioned facts in mind the present study was undertaken to ascertain the relationship of age and body weight to testicular growth in black Bengal bucks and to examine the relationship between certain testicular measurements with semen

*Corresponding author. E-mail: khandokerabg@yahoo.com.

production for the establishment of some norms for breeding soundness evaluation of black Bengal bucks.

MATERIALS AND METHODS

The experimental bucks were maintained in the nucleus breeding flock (NBF) of the research project entitled "Conservation of Black Bengal goat as the Potential Genetic Resource in Bangladesh" at AI centre, Bangladesh Agricultural University, Mymensingh-2202.

Animals used

Twelve black Bengal bucks aged between 6 to 36 months were selected on the basis of morphometric characterization and the potentiality to produce quality semen. The selected bucks were again divided into three different age groups as A (0.5 to 1.0 years), B (1.5 to 2.0 years) and C (2.5 to 3.0 years).

Feeding and other managemental procedures

The bucks were fed with Napier and/or German grass twice daily as per requirement. The feed was supplemented with commercial concentrate (crude protein content: I20 g/kg DM and energy content: 10.4 MJ ME/kg DM) in the morning and again in the afternoon at the rate of 100 gm/ buck of 20 kg live weight. The breeding bucks were also supplied with germinated gram (20 gm/buck/day). Clean and safe water was made available at all times. Throughout this study the nutrition of bucks remained uniform and constant. All bucks were vaccinated against Peste des Petits Ruminants (PPR) and dewormed routinely with Ivermectin[®] thrice a year. Individual pen was provided for each buck (10 sq ft) in the shed with the provision of sufficient access to fresh air and their movement freely and they were allowed to graze outside for one hour daily.

Semen collection

The bucks were trained to ejaculate in artificial vagina (AV). Semen was collected within 8.30 AM twice a week from each buck after cleaning the prepuce with antiseptic (savlon) solution. Collection of semen was done with artificial vagina maintaining optimum pressure and temperature about 41 to 43°C. The bucks received homosexual stimulation by being exposed to the teaser male. The semen was collected twice a week up to 45 days thus 12 records were made available for this study.

Evaluation of semen

Immediately after collection the volume of each ejaculates was measured directly from the reading of the graduated collection vial in millimeter. Then semen samples were prepared according to the method described by Herman and Madden (1963) and the concentration of spermatozoa per ml of semen was counted by haemocytometer method and expressed as billion per ml. The following formula was used for calculating total number of spermatozoa per ml of fresh semen.

Total number of spermatozoa per ml of semen = $[\{(C \times 400)/S\} \times D] \times 10^{-3}$

Where,
C = Number of sperm counted in given number of small squares.

S = Number of small squares counted.
D = Dilution ratio.

Eosin-nigrosin staining method has been used as a routine staining in order to evaluate sperm viability (World Health Organization, 1992), briefly, after washing sperm samples in saline solution at 37°C, one drop of the suspension containing 35×10^6 sperm/ml was placed on a tempered glass slide, which was mixed with one drop of Eosin-nigrosin solution. The mixture was smeared on the glass slide and let air dried. The samples were observed under a light microscope. Eosin penetrated in non viable cells which appear red. Nigrosine offers a dark background facilitating the detection of viable, non stained cells. Four smears were performed from each ejaculate, in fresh samples. A total of 333 sperms were counted randomly from different field of the slide. Number of dead spermatozoa deducted from total number of spermatozoa gave the number of live spermatozoa and expressed in percentage.

Measurement of body weight

The body weight of each buck was recorded in kg in the morning before the animals were slaughtered. The weights were taken with a top loading balance.

Measurement of scrotal circumference

The scrotal circumference was measured as per method recommended by the Society of Theriogenology (Ball et al., 1983). The testes were first retracted into the lower part of the scrotum for measurement of scrotal circumference. To prevent separation of the two testes, the thumb and the fingers were placed on the sides rather than on the front or back of the scrotum. Then a flexible metal tape (Scrotal tape; Lane Manufacturing, Co., Denver, USA) was looped and placed around the greatest diameter of the scrotum and pulled snugly so that the tape was firmly in contact with the entire circumference. Repeated measurement was done and the mean of the measures was recorded to ensure the accuracy.

Slaughtering and biometrical study

Immediately after slaughter, both the left and right testes were collected from the bucks of three different age groups. Immediately after collection, the length and width of testes were measured by measuring tape and weights were measured in digital balance.

Statistical analysis

Simple ANOVA was performed considering the age of buck and to observe the significant differences among the mean values, Duncan's multiple range test (DMRT) was done. Lastly correlations among the traits were also performed. Analysis was performed with the help of statistical analysis system (SAS, 1998).

RESULTS AND DISCUSSION

Semen quality evaluation

Semen quality of the bucks of three different age groups were evaluated in terms of semen volume (ml), live sperm percentage and sperm concentration (billion/ml). The evaluation results are summarized in Table 1.

Table 1. Evaluation of semen of black Bengal bucks (Mean ± SE).

Age group (Year)	Volume (ml) (n = 48)	Sperm concentration (billion/ml; n = 48)	Live sperm (%) (n = 48)
Group-A (0.5 to 1.0)	$0.32^b \pm 0.04$	$2.07^b \pm 0.12$	76.46 ± 2.65
Group-B (1.5 to 2.0)	$0.55^{ab} \pm 0.05$	$3.12^a \pm 0.14$	85.64 ± 0.87
Group-C (2.5 to 3.0)	$0.68^a \pm 0.04$	$3.04^a \pm 0.10$	83.88 ± 1.96

Means with different superscripts within the column differ significantly ($p<0.05$). SE = Standard error.

Semen volume

The semen volume of age group C was significantly ($p<0.05$) higher than that of age group A but statistically similar with that of age group B (Table 1). Higher semen volume in age group C might be due to older age with higher scrotal circumferences of the animals. The semen volume obtained in the present study ranged from 0.32 ± 0.04 to 0.68 ± 0.04 ml (Table 1) which strongly supports the findings of other researchers (Karim, 2008; Vilar et al., 1993; Singh et al., 1985). Semen volume of buck may vary according to breed. Furstoss et al. (2009) reported 0.48 ± 0.10 ml of semen produced from Alpine bucks at 7 months of age which was slightly higher than the result of the present study (0.32 ± 0.04 ml at 0.5 to 1.0 year of age). These might be due to breed and nutritional differences. Semen production largely depends on the factors like age, sexual maturity, nutritional status, general health condition, endocrine balance and soundness of the sex organs and season (Peters, 2002; Karagiannidis et al., 2000).

Das et al. (2006) reported that the volume ranges from 0.16 to 0.51 ml in black Bengal buck, which is lower than that of the value of this study. These differences might be due to the aforementioned reasons.

Sperm concentration

The concentration of spermatozoa of age group C was almost similar with that of age group B but significantly ($p<0.05$) higher than that of age group A. These could be due to higher testicular size with higher spermatogeic activity in the animals of age group C. The finding of the present study strongly agrees with the finding of Karim (2008) who reported the average sperm concentration ranged from 2.75 ± 0.28 to 3.24 ± 0.37 billion/ml in black Bengal buck. According to Apu (2007) and Afroz (2005), the averages of sperm concentrations of buck semen were 2678.33 ± 30.59 to 2913.33 ± 46.23 and 2434.00 ± 52.81 to 2853.00 ± 90.12 million/ml respectively which are almost similar to the present study. The present study also collaborates with the studies of other researchers (Das et al., 2006; Mittal, 1982). On the other hand, Khan (1999) reported the average sperm concentration of 3777.93 ± 142.76 million/ml which is higher than the result of the present study.

Leon et al. (1991) and Sharma et al. (1991) reported that sperm concentration might vary according to variation in age, breed, collection frequency, feeding regime and climatic condition. This difference in sperm concentration might be due to the aforementioned reasons.

Sperm viability

In the present study, the variation in live sperm percentage did not differ significantly ($p>0.05$) among the three different age groups (Table 1). The highest percentage of live sperm was found in age group B than that of other two age groups although there was no significant difference ($p>0.05$) among the age groups. Highest percentage of live spermatozoa of the animals of age group B might be attributed to their higher physiological fitness. The present observation of the percentage of live spermatozoa strongly agrees with the findings of Kamal et al. (2005) who reported 84.50% live spermatozoa in buck semen. On the other hand, Karagiannidis et al. (2000) who reported 89.50 to 95.27% live spermatozoa in buck semen which was higher than that of the present study. A variety of factors affect the viability of sperm such as variations in age, breed, feeding regime (Leon et al., 1991; Graves, 1978), pH and osmolarity of semen (Makler et al., 1981), season and ambient temperature (Lincoln and Short, 1980).

The difference in live sperm percentage might be due to these reasons. However, the values obtained in this study were to some extent similar with those of many other investigators (Karim, 2008; Ahmed et al., 1997).

Biometrical study

The results of the biometrical studies were summarized in Table 2.

Body weight

Among the three different age groups, it was observed that body weight of age group C was significantly ($p<0.05$) higher than that of age group A, but no significant

Table 2. Comparative measurements of different parameters in left and right testes of buck with regard to age, body weight and scrotal circumference (n = 4), Mean ± SE.

Age group (Year)	Body weight (kg)	Scrotal circumference (cm)	Testicular parameter					
			Weight (gm)		Length (cm)		Breadth (cm)	
			Left	Right	Left	Right	Left	Right
Group-A (0.5 to 1.0)	$12.41^b \pm 1.80$	$17.50^b \pm 0.65$	$39.06^b \pm 3.39$	$38.11^b \pm 3.20$	$6.10^b \pm 0.13$	$5.85^b \pm 0.13$	3.88 ± 0.24	3.75 ± 0.10
Group-B (1.5 to 2.0)	$23.95^a \pm 1.66$	$21.38^a \pm 0.43$	$63.85^{ab} \pm 5.19$	$62.73^{ab} \pm 5.22$	$7.20^{ab} \pm 0.31$	$6.90^{ab} \pm 0.33$	4.63 ± 0.31	4.38 ± 0.24
Group-C (2.5 to 3.0)	$27.81^a \pm 0.46$	$22.88^a \pm 0.66$	$66.37^a \pm 3.78$	$65.16^a \pm 3.90$	$7.85^a \pm 0.22$	$7.35^a \pm 0.18$	4.85 ± 0.12	4.68 ± 0.12

Means with different superscripts within the column differ significantly ($p<0.05$). SE = standard error.

difference ($p>0.05$) was observed with that of age group B (Table 2). These might be due to higher age with higher body lengths and heights in the animals of age group C. Rahman (2007) obtained the body weight of black Bengal buck at 11.5 to 12.0, 14.0 to 16.0 and 17.5 to 19.0 months of age as 16.62±0.12, 17.62±0.22 and 20.86±0.25 kg respectively which support the results of the present study. The results of the present study also collaborate with the results of Alam (2006) and Herbert et al. (2003). On the other hand, the body weight of black Bengal buck at 2.5 to 3.0 years of age is more than the findings of Fajemilehin and Salako (2008) who reported the average body weight of West African dwarf (WAD) goats, was 14.59 kg at the similar age.

In another study, Mittal and Ghosh (1985) reported the average body weight of Parbatsar breeds of goats at adult age was 48.80±2.45 kg which is much higher than the result of the present study. This might be due to breed and/or sex difference, physical condition of the selected animals, agro-climatic condition, nutritional level, housing, disease prevalence and other managemental procedure.

Scrotal circumference

From Table 2, it was observed that the scrotal circumference of age group C was significantly ($p<0.05$) higher than that of age group A but there had no significant difference ($p>0.05$) with that of age group B. This might be due to difference in age and body weight of bucks, dam's age at first breeding, pregnancy rate and days to rebreeding after kidding. Shamsuddin et al. (2000) reported that the mean scrotal circumference of black Bengal buck at puberty ranged from 14.0 to 16.0 cm which supports the results of the present study. The present study also collaborates with the studies of other researchers (Rahman, 2007; Igboeli, 1974). On the other hand, the finding of the present study was somewhat lower than that of the findings of Keith et al. (2009) and Mekasha et al. (2008). This may be due to breed difference, post-weaning feed level, contemporary group/feed level, age of dam, and covariates age, weight and height of bucks (Bourdon and Brinks, 1986).

Shape and size of testes

In the present study the testes of buck were found ovoid in shape. The testes represented two surfaces, medial and lateral; two borders free (ventral) and attached (dorsal) and two extremities, head extremity and tail extremity. This is very much usual. The size of the testes varied in different age groups and even in between left and right testes of same age group in the present study. In the present study, the average length of the testes of age group C was significantly ($p<0.05$) higher than that of age group A, but the average length in age group B, did not differ significantly ($p>0.05$) with that of other two age groups (Table 2). On the other hand, the average breadth of the testes of age group C was higher than that of other two age groups but there exists no significant difference ($p>0.05$) among the age groups (Table 2). This study indicated that length and breadth of testes were increased with the advancement of age of animals which made a similar agreement with the findings of Gofur et al. (2007) and Islam (2001).

Testes weight

The weight of the testes were varied in different age groups and even in between left and right testes of same age group in the present study. In the present study, the average weight of the testes of age group C was significantly ($p<0.05$) higher than that of age group A but the average weight in age group B did not differ significantly ($p>0.05$) than that of other two age groups. Higher testes weight in age group C might be due to their higher body weight and size (Table 2). This finding made a strong agreement with the results of Islam (2001). On the other hand, Raji et al. (2008) reported the average testicular weight at 1.0, 2.0, and 3.0 years of age were 55.00±2.87, 77.28±1.88 and 103.01±2.23 gm respectively in red Sokoto goats, which was much higher than the present result. This might be due to breed and/or sex difference, physical condition of the selected animals, agro-climatic condition, nutritional level, housing, disease control and other managemental procedure.

The results of the present study also collaborate with the results of Gofur et al. (2007).

Table 3. Correlation coefficients (r) between age, body weight, testes weight, testes length, semen volume and sperm concentration in black Bengal buck.

Parameter	Age	Body weight	Testicular weight	Testicular length	Semen volume
Body weight	0.898 (**)				
Testicular weight	0.778 (**)	0.936 (**)			
Testicular length	0.848 (**)	0.787 (**)	0.668(*)		
Semen volume	0.886 (**)	0.924 (**)	0.867 (**)	0.793 (**)	
Sperm concentration	0.754 (**)	0.855 (**)	0.840 (**)	0.600 (*)	0.769 (**)

**Correlation is significant at the 0.01 level. *Correlation is significant at the 0.05 level.

Correlation coefficients (r) between age, body weight, testicular weight, testicular length, semen volume and sperm concentration in black Bengal buck

The correlation coefficients (r) between age, body weight, testicular weight, testicular length, semen volume and sperm concentration in black Bengal buck are presented in Table 3. It was observed that age had strong correlation coefficient ($p<0.01$) with body weight, testicular weight, testicular length, semen volume and sperm concentration (Table 3). All the parameters studied in this experiment found to be highly ($p<0.01$) correlated with each other (Table 3). The result of the present study agrees with the findings of Raji et al. (2008) who reported a significant correlation coefficient ($p<0.01$; r = 0.78) between testicular weight and body weight in indigenous goats of Nigeria. He also reported a linear relationship between age, body weight and the entire testicular dimension in this goat of Nigeria. In the present study, testicular measurements were significantly correlated with semen volume and sperm concentration which were to some extent similar with other published works (Pant et al., 2003; Vasquez et al., 2003).

Conclusion

The final observation of this study was that testicular biometry and semen parameters were increased with the advancement of age and body weight. The size and weight of left testis were higher than right testis of the same individual of different groups. Semen quality, scrotal circumference and testicular biometry were highly ($p< 0.01$) correlated with each other. However, further study is recommended to further clarification. This study might be helpful in selection of breeding buck.

ACKNOWLEDGEMENT

The authors are very much grateful to USDA for providing the financial support to conduct the experiment.

REFERENCES

Afroz S (2005). Cryopreservation of buck semen. MS thesis, Department of Animal Breeding and Genetics, Bangladesh Agricultural University, Mymensingh, pp. 33-47.

Ahmed MMM, Makawi S, Gadir AA (1997). Reproductive performance of Saanen bucks under tropical climate. Small Rum. Res., 26(1-2): 151-155.

Alam MK (2006). Characterization and performance evaluation of white goat in some selected areas of Bangladesh. MS thesis. Department of Animal Breeding and Genetics. Bangladesh Agricultural University, Mymensingh, pp. 43-52.

Apu AS (2007). Frozen and liquid semen production and assessment of conception rate in Black Bengal goat. MS thesis. Department of Animal Breeding and Genetics. Bangadesh Agricultural University, Mymensingh, pp. 57-68.

Ball L, Ott RS, Mortimer RG, Simons JC (1983). Manual for breeding Soundness examination of bulls. J. Soc. Theriogen., 12: 1-65.

Bourdon RM, Brinks JS (1986). Scrotal circumference in yearling Hereford bulls: adjustment factors, heritabilities and genetic, gnvironmental and phenotypic relationships with growth traits. J. Anim. Sci., 62: 958-967.

Das SK, Husain SS, Amin MR, Munim T, Hoque MA, Khandoker MAMY (2006). Growth performance of progeny using selected Black Bengal bucks. Bangladesh J. Anim. Sci., 35: 27-35.

Fajemilehin, Salako AE (2008). Body measurement characteristics of the West African Dwarf (WAD) Goat in deciduous forest zone of Southwestern Nigeria. Afr. J. Biotechnol., 7(14): 2521-2526.

Furstoss V, David I, Leboeuf B, Guillouet P, Boue P, Bodin L (2009). Genetic and non-genetic parameters of several characteristics of production and semen quality in young bucks. Anim. Reprod. Sci., 110: 25-36.

Gofur MR, Khan MZI, Karim MR, Islam MN (2007). Biometry of testis of indigenous bull (Bos indicus) of Bangladesh in relation to body weight and scrotal circumference. J. Bangladesh Soc. Agric. Sci. Tech., 4(1&2): 205-208.

Graves CN (1978). Semen and its component In: Salisbury, G.W.; VanDemark N.L. and Lodge, J.R. (Eds.), Physiology of Reproduction and Artificial Insemination of Cattle. W.H. Freeman and Company, San Francisco, pp. 247-285.

Herbert S, Sourdaine P, Moslemi S, Plainfosse B, Gilles-Eric S (2003). Immunolocalization of Aromatase in Stallion Leydig Cells and Seminiferous Tubules. J. Histochem. Cytochem., 51: 311-318.

Herman HA, Madden FW (1963). The Artificial Insemination of dairy and beef cattle. A hand book and laboratory manual. Locas Brothers, Columbia, Missouri, USA, pp. 95-105.

Husain SS (2004). Preservation of buck semen and their use in Artificial Insemination for rapid genetic improvement of rural goat population, pp. 11-17.

Igboeli G (1974). A comparative study of the semen and seminal characteristics of breeds of goats. Agric. Fores. J. Zambia, 40: 132-137.

Islam N (2001). Anatomical studies of the male genital system of Black Bengal goat. MS. thesis, Department of Anatomy and Histology, Bangladesh Agricultural University, Mymensingh, pp. 41-49.

Kamal A, Gubartallah A, Ahmed A, Bakhiet O, Babiker A (2005). Comparative studies on reproductive perfrormance of Nubian and Sannen bucks under the climatic conditions of Khartoum. J. Anim.

Vet. Adv., 4(11): 942-944.

Karagiannidis A, Varsakeli S, Karatzas G (2000). Characteristics and seasonal variations in the semen of Alpine, Saanen and Damascus goat bucks born and raised in Greece. Theriogen, 53: 1285-1293.

Karim MF (2008). Comparison of different diluters for frozen semen production in Black Bengal bucks. MS thesis. Department of Animal Breeding and Genetics, Faculty of Animal Husbandry, Bangladesh Agricultural University, Mymensingh, pp. 49-56.

Keith L, Okere C, Solaiman S, Tiller O (2009). Accuracy of predicting body weights from body conformation and testicular morphometry in Boer goats. Res. J. Anim. Sci., 3(2): 26-31.

Khan RA (1999). A quantitative study on semen characteristics of Black Bengal buck. MS thesis submitted to the Department of Animal Breeding and Genetics, Bangladesh Agricultural University, Mymensingh-2202, Bangladesh, pp. 78-92.

Leon H, Porras AA, Galina CS (1991). Effect of the collection method on semen characteristics of Zebu and European type cattle in the tropics. Theriogen, 36: 349-355.

Lincoln GA, Short RV (1980). Seasonal breeding, nature's contraceptive. Recent Prog. Hormone Res., 36: 1.

Makler A, David R, Blumenfeld Z, Better OS (1981). Factors affecting sperm motility & sperm viability as affected by change of pH and osmolarity of semen and urine specimens. J. Fertil. Steril., 36(4): 507-11.

Mekasha Y, Tegegne A, Abera A, Rodriguez-Martinez H (2008). Body size and testicular traits of tropically adapted bucks raised under extensive husbandry in Ethiopia. Reprod. Dom. Anim. 43: 196-206.

Mittal JP (1982). Seasonal variation in semen quality of Barbari bucks. Indian Vet. J., 59: 957-959.

Mittal JP, Ghosh PK (1985). Characteristics of Parbatsar breed of goat from Rajasthan desert. Indian J. Anim. Sci., 55: 673-678.

Pant HC, Sharma RK, Patel SH, Shukla HR, Mittal AK, Kasiraj RK, Misra AK, Prabhakar JH (2003). Testicular development and its relationship to semen production in Murrah buffalo bulls. Theriogen, 60(1): 27-34.

Peters KJ (2002). Evaluation of goat populations in tropical and subtropical environments. http://kinne.net/fertbuck.html.

Rahman S (2007). Morphometric characterization of Black Bengal buck. MS thesis. Department of Animal Breeding and Genetics, Faculty of Animal Husbandry, Bangladesh Agricultural University, Mymensingh, pp. 71-82.

Raji AO, Igwebuike JU, Aliyu J (2008). Testicular biometry and its relationship with body weight of indigenous goats in a semi arid region of Nigeria. ARPN J. Agric. Bio. Sci., 3(4): 35-38.

SAS (1998). Statistical Analysis System, Version 6.03. SAS Institute Inc. Cary NC, 25-109 USA.

Shamsuddin M, Amiri Y, Bhuiyan MMU (2000). Characteristics of buck semen with regard to ejaculate numbers, collection intervals, dilution and preservation periods. Reprod. Dom. Anim., 35: 53-57.

Sharma ML, Mohan G, Sahni KL (1991). Characteristics and cryopreservation of semen of Holstein friesian bulls under tropics. Indian J. Anim. Sci., 61: 977-979.

Singh DH, Sinha MP, Singh CSP, Singh RA, Singh KK (1985). Comparative study on seminal quality of pure and cross-bred bucks. Indian Vet. Med. J., 9: 50-58.

Vasquez L, Vera O, Arango J (2003). Testicular growth and semen quality in peripuberal Brahman bulls. Livestock Res. Rural Dev., 15(10).

Vilar AC, Barnabe VH, Birgel FH, Barnabe RC, Visintin JA (1993). Testis and semen characters in goats reared in a semi-arid area in Pariaba State. Revista Bras. Reprod. Anim., 17: 23-32.

World Health Organization (WHO) (1992). WHO laboratory manual for the examination of human semen and semen-cervical mucus interaction (3rd edn), Cambridge, The Press Syndicate of the University of Cambridge.

25

Somatic cell count, total bacterial count and acidity properties of milk in Khartoum State, Sudan

AdilM. A. Salman[1] and Hind A. Elnasri[2*]

[1]Department of Veterinary Preventive Medicine and Veterinary Public Health, College of Veterinary Science, University of Bahr El Ghazal, Sudan.
[2]Department of Biochemistry, College of Veterinary Science, University of Bahr El Ghazal, Sudan.

This work was conducted in Khartoum State to study the limits of the somatic cell count (SCC), the total bacteria count (TBC) and the acidity of the raw cow milk produced in the three geographical areas of the Khartoum State.A total of 644 stratified random raw milk samples were collectedduring summer and winter. The different counts and acidity were evaluated in the farm milk and compared to that sold in the market. Total bacterial count was carried out using the pour plate count.The bacterial count of equal or less than 9×10^4cfu/ml in the state was 23.9% with a higher percentage in winter (35.4%) compared to 19.4% in summer. The majority of the samples (55.3%) had a count of less than or equal 9×10^5cfu/ml. The percentage in winter was (71.9%) while in the summer it was (48.1%).Regarding the SCC, it was done using new man stain. The percentage of samples of less than 5×10^5 were (27%) in the state. The percentage was higher in winter (43.3%) than summer (20.8%).The majority of the samples (83.4%) were equal to or less than 7.5×10^5 (93.8%) were in winter and (81.5%) in summer. Regarding acidity which was carried out using the titration method, the percentage of samples of 0.2 titratable acidity were 64.3% in the state. The percentage in winter was 73.6% while in summer it was 60.7%. Statistically significant correlations at 0.01 levels between SCC, acidity and TBC were determined. The differences between the counts of SSC and TBC in winter and summer were statistically significant while the difference between the regions was insignificant.

Keywords:Sudan, milk hygiene, somatic cell count (SCC), total bacteria count (TBC), acidity.

INTRODUCTION

In Sudan milk is produced mostly in non-organized way and usually it is being supplied to the consumers from the urban and rural areas by milk vendors or from the groceries. The distribution of milk to the consumers is completely in poor hygienic conditions. On the other hand, milk is an excellent media for growth of a wide variety of bacteria.One of the requirements of production of the high quality milk is maintaining the bacteria count level of microorganisms in a product and to study the

hygienic and sanitary conditions, under which milk was produced, handled, transported and processed (Murphy, 1997; FAO/WHO, 1992). Both temperature and storage time influence the multiplication of the micro-organisms, (Jayarao et al., 2004). Acatincai et al. (2008) stated that the TBC was higher during the summer months it reaches > 7.2×10^4 cfu/ml while it was 6.3×10^4 during winter time. In New York State 50% of the can samples had a count of > 1×10^4 during winter (Boor et al., 1998).Somatic cell count (SCC) is used by milk quality laboratories to determine quality and acceptability of milk (Schallibaum, 2001). At the cow or quarter level the normal SCC is generally below 2×10^5 but, may be below 1×10^5 cells/ml (Muhammad et al., 2009) SCCs are the

*Corresponding author. E-mail: hindelnasrihiab2000@yahoo.com.

Table 1. Number and percentage of raw milk samples.

Parameter	Summer				Winter			
	Kh (%)	Kh. N (%)	Omd (%)	Total (%)	Kh (%)	Kh. N (%)	Omd (%)	Total (%)
Individual	5(07.9)	12(22.2)	17(27.9)	34(19.1)	33(24.8)	19(09.3)	13(10.1)	65(13.9)
Bulk	13(20.6)	15(23.8)	19(31.1)	47(26.4)	35(26.3)	109(53.4)	53(41.1)	197(42.3)
Market	45(71.4)	27(50)	25(41)	97(54.5)	65(48.9)	76(37.2)	63(48.8)	204(43.8)
Total	63	54	61	178	133	204	129	466

Key to areas: Kh: Khartoum, Kh. N: Khartoum North, Omd. Omdurman.

lowest in a clean dry comfortable environment (Acatincai, 2008) that usually includes adequate shelter against sun and rain, absence and type of bedding, free or closed stalls and dry lots which minimize possible contamination of the teats ends from environmental organisms(Khan et al., 2008; Duane and Gerald, 2003). The seasonal variations of the SCC were studied by Sawa and Pwczynski (2002). They revealed a significant influence of season on the SCC which is higher in summer and lower in winter.

Titratable acidity plays a fundamental role and represents a very important parameter for the technical evaluation of the technological quality of milk (Harris and Bachman, 1988). The milk components that are acidic and contribute to normal acidity value are carbon dioxide, protein, phosphate and citrates (Harris and Bachman, 1988). High bacterial count which can convert lactose to lactic acid leads to the elevation of the titratable acidity. Simona et al. (2010) showed that the titratable acidity of milk typically varies from 0.15 to 0.19% lactic acid depending on the composition especially on protein content. In Sudan, the acidity values were studied by many workers (Ibrahim, 1973; Idris et al., 1975; El Zubeir and Ahmed, 2007). They reported that the mean titratable acidity was in the range of 0.18 to 020 but acidity of more than 0.22 was found in different milk samples.The main objective of this study is to evaluate the status of milk hygiene in the state of Khartoum from different sources (farms and market). The main measures that are studied are somatic cell count, total bacterial count and titratable acidity. The samples were collected during two seasons, winter and summer.

MATERIALS AND METHODS

In this study,644 raw cow milk samples were collected from the three regions of Khartoum state (Khartoum region, Khartoum North and Omdurman) in winter and summer. Samples were collected from the farm (Individual + bulk tank) and market (vendors or shops) during the period between April 2008 to February 2009 (Table 1). 50 ml of raw milk was collected using clean sterile glass bottles. The samples were put in an ice box and delivered to the laboratory of Veterinary Preventive Medicine at the Faculty of

Veterinary Science, Khartoum University for analysis.

Somatic cell count (SCC)

According to IDF, (1984) the milk was mixed thoroughly before a final amount of 0.01 ml of milk was pipetted and spread evenly on the entire area of the special slide, (Special circular slide with an area of 1 cm² circle from Bellco Glass inc. Edrudo Road, Vine Land, U.S.A. (5638 to 01930) stock number) were prepared. Every slide is suitable for 4 samples. After drying the slide was stained with the prepared stain (New Man stain) for two minutes and then the cellswere counted under oil Immersion.

Total bacterial count test (TBC)

Total bacterial count was determined as described by ISO (1991), serial dilution (10^{-1} to 10^{-8}) of the milk samples was made and aliquots of 1ml were added to each duplicate Petri dish. Plate count agar was added to each Petri dish and incubated at 35°C for 48 h ±2, after incubation colonies were counted by colony counter and result was expressed as cfu/ml.

Acidity test

Bacteria that normally develop in raw milk produce lactic acid. In the acidity test the acid is neutralized with 0.1 Nsodium hydroxide and the amount of alkaline is measured. From this the percentage of lactic acid can be calculated (Foley et al., 1974). The number of milliliters of sodium hydroxide solution divided by 10 expresses the percentage of lactic acid.

Statistical analysis

All the data obtained during the study were analyzed statistically to find out the level of significance. The analysis of variance was determined by F-test. The mean differences were evaluated at 1% level of significance and the Pearson Correlation coefficient was calculated using the SPSS and Microsoft Excel programmes.

RESULTS

Total bacterial count (TBC)

The percentage of the total samples with counts of less than 1 × 10^5 cfu/ml was found to be 23.9% in state in the

Table 2. Seasonal TBC in the three regions of Khartoum State.

Range	Khartoum			Khartoum North			Omdurman		
	Winter {No (%)}	Summer {No (%)}	Total {No (%)}	Winter {No (%)}	Summer {No (%)}	Total {No (%)}	Winter {No (%)}	Summer {No (%)}	Total {No (%)}
$1\times10^1 - 9\times10^4$	23(36.5)	28(21.1)	51(26.0)	14(25.9)	41(20.1)	55(21.3)	26(42.6)	22(17.1)	48(25.3)
$1\times10^5 - 9\times10^5$	21(33.3)	38(28.6)	59(30.1)	22(40.7)	53(26.0)	75(29.1)	22(36.1)	46(35.7)	68(35.8)
$1\times10^6 - 91\times0^6$	18(28.6)	54(40.6)	72(36.7)	09(16.7)	55(26.8)	64(24.8)	12(19.7)	46(35.7)	58(30.5)
$1\times10^7 - 9\times10^7$	01(01.6)	10(07.5)	11(05.6)	05(09.3)	31(15.2)	36(13.9)	01(01.6)	11(08.5)	12(06.3
$> 10^8$	0(0.00)	04(03.0)	04(02.0)	04(07.4)	22(10.8)	26(10.1)	00(0.00)	05(03.9)	05(02.6)
Total	63	133	196	54	204	258	61	129	190

Table 3. Seasonal total TBC count in Khartoum State at different levels of collection.

Range	Winter				Summer			
	Individual {No (%)}	Bulk {No (%)}	Market {No (%)}	Total {No (%)}	Individual {No (%)}	Bulk {No (%)}	Market {No (%)}	Total {No (%)}
$1\times10^1 - 9\times10^4$	21(61.8)	22(46.8)	20 (20.8)	63(35.4)	25(38.5)	32(16.2)	34(16.7)	91(19.5)
$1\times10^5 - 9\times10^5$	09(26.5)	18(38.3)	38 (39.6)	65(36.5)	15(23.0)	60(30.5)	62(30.4)	137(29.4)
$1\times10^6 - 9\times10^6$	04(11.7)	07(14.9)	28 (29.2)	39(21.9)	25(38.5)	60(30.5)	70(34.3)	155(33.3)
$1\times10^7 - 9\times10^7$	00(0.00)	00(0.00)	7 (7.2)	07(3.9)	00(0.00)	26(13.2)	26(12.7)	52(11.1)
$>10^8$	00(0.00)	00(0.00)	4 (4.2)	04(2.3)	00(0.00)	19(09.6)	12(5.9)	31(06.7)
Total	034	047	96	178	065	197	204	466

two seasons (Table 2). In winter it was found to be 35.4% and in the summer it was 19.5%. The total count of less than or equal to 9×10^5cfu/ml was found to be 55.3% in the state during the two seasons. During winter 71.9% of the samples were within the limits of this count and 44.6% during summer (Table 2). In Khartoum state during the two seasons the percentage of samples with count of more than 10^5 was 44.7%, in the state, with 28.1% in winter and 45.4% during summer (Table 2).The difference in the TBC between winter and summer was found to be statistically significant at 0.05, but the differences in the TBC between the three regions of the state were insignificant at 0.05 level of significance (Table 2).The percentages of counts of less than 10^5 in individual cows' milk, farm bulk tank milk, and market milk during winter season were 61.8, 46.8 and 20.8%, respectively; during summer season the percentages were 38.5, 16.2, and 16.7%, respectively (Table 3). The percentage of counts of less than 9×10^5 in individual cow's milk, bulk tank milk and market milk during winter season were 88.3, 85.1, and 60.4% respectively, during summer season the percentages were 61.5, 46.7 and 47.1%, respectively (Table 3). The percentage of counts of more than 10^6 in individual cow's milk, bulk tank milk, and market milk during winter season was 11.7, 14.9 and

40.6% respectively (Table 3), during summer season the percentages were 38.5, 53.3 and 52.9%, respectively (Table 3). The correlations between individual cow's milk and farm bulk tank milk and between bulk tank milk and vendor milk were significant at 0.05 level (Table 4). At 0.05 level of significance there was no difference between the three regions. At 0.05 level of significance there is a significant difference between the two seasons.

The somatic cell count

As shown in Table 5, the percentages of samples with count less than 5×10^5 Khartoum State were 27.0%. The percentages in winter and in summer were 43.3 and 20.8%, respectively. The percentage of samples of cells less than 7.5×10^5 were 55.9% in Khartoum State. The percentages in winter and in summer were 69.7 and 47.7% during the two seasons, respectively. The percentage of sample, which has a count of over 1×10^6 were 15.1% in Khartoum State. The percentages in winter and in summer were 6.2 and 18.5% in two seasons respectively. Statistically at 0.05 level the differences in count were significant between the three regions of the state and also between the two seasons.

Table 4. Correlations of TBC between individual, bulk, vendor and market milk.

Parameter	Correlation	Individual	Bulk	Vendor	Market
Individual	Pearson correlation	1	0.931(*)	0.470	0.395
	Correlations	0.0	0.021	0.424	0.511
	N	5	5	5	5
Bulk	Pearson correlation	0.931(*)	1	0.956*	0.677
	Correlations	0.021	0.0	0.160	0.209
	N	5	5	5	5
Vendor	Pearson correlation	0.470	0.956(*)	1	0.987(**)
	Correlations	0.424	0.160	0.0	0.002
	N	5	5	5	5
Market	Pearson correlation	0.395	0.677	0.987(**)	1
	Correlations	0.511	0.209	0.002	0.0
	N	5	5	5	5

* Correlation is significant at the 0.05 level (2 - tailed), ** Correlation is significant at the 0.01 level (2 - tailed).

Table 5. Seasonal SCC count in the different three geographical areas of Khartoum State.

$\times 10^3$	Khartoum			Khartoum N.			Omdurman		
	Winter {No (%)}	Summer {No (%)}	Total {No (%)}	Winter {No (%)}	Summer {No (%)}	Total {No (%)}	Winter {No (%)}	Summer {No (%)}	Total {No (%)}
100 to < 200	12(19)	06(04.5)	18(9.2)	04(7.4)	07(3.4)	11(4.2)	01(01.6)	01(0.7)	02(01.1)
200 to < 500	18(28.6)	29(21.8)	47(24)	16(29.6)	40(19.6)	56(21.7)	26(42.6)	14(10.9)	40(21.1)
500 to < 750	14(22.2)	36(27.1)	50(25.5)	17(31.5)	70(34.3)	87(33.7)	16(26.3)	33(25.6)	49(25.8)
750 to <1000	17(26.9)	41(30.8)	58(29.6)	12(22.2)	47(23.0)	59(22.9)	14(22.9)	56(43.4)	70(36.8)
> 1000	02(03.3)	21(15.8)	23(11.7)	5(09.3)	40(19.6)	45(17.4)	04(06.0)	25(19.3)	29(15.3)
Total	63	133	196	54	204	258	61	129	190

Table 6. The averages of seasonal SCC in the three regions of Khartoum State.

Region	Winter $\times 10^5$			Summer $\times 10^5$		
	Market	Bulk	Individual	Market	Bulk	Individual
Khartoum	6.2	5.3	5.2	7.4	6.0	5.6
Omdurman	6.8	5.6	5.6	7.8	6.8	6.0
Khartoum N.	7.8	7.6	6.4	7.8	8.0	6.4

The average counts in Khartoum region for the individual, bulk and market milk were 5.2×10^5, 5.3×10^5, 6.2×10^5 during winter and 5.6×10^5, 6.0×10^5, 7.4×10^5 during summer (Table 6).The average counts in Omdurmanregion for the individual, bulk and market milk were 5.6×10^5, 5.6×10^5, 6.8×10^5 during winter and 6.0×10^5, 6.8×10^5, 7.8×10^5 during summer, respectively (Table 6). The average counts in Khartoum North region for the individual, bulk and market milk were 6.4×10^5, 7.6×10^5, 7.8×10^5 during winter and 6.4×10^5, 8×10^5,

Table 7. Seasonal acidity values in the three regions of Khartoum State.

| Range | Khartoum | | | Khartoum N. | | | Omdurman | | |
	Winter {No (%)}	Summer {No (%)}	Total {No (%)}	Winter {No (%)}	Summer {No (%)}	Total {No (%)}	Winter {No (%)}	Summer {No (%)}	Total {No (%)}
≤ 0.20	47 (74.6)	81 (60.9)	128	38 (70.4)	122 (59.9)	160 (62)	46 (75.4)	80 (62)	126 (66.3)
0.21- 0.22	06 (9.5)	28 (21)	34 (17.3)	08 (14.8)	37 (18.1)	45 (17.4)	06 (9.8)	21 (17.3)	27 (14.2)
> 0.22	10 (15.9)	24 (18.1)	34 (17.3)	08 (14.8)	45 (22)	53 (20.5)	09 (14.8)	28 (21.7)	37 (19.5)
Total	63	133	196	54	204	258	61	129	190

Table 8. Seasonal acidity values in Khartoum State at different levels of collection.

| Range | Winter | | | | Summer | | | |
	Individual {No (%)}	Bulk {No (%)}	Market {No (%)}	Total {No (%)}	Individual {No (%)}	Bulk {No (%)}	Market {No (%)}	Total {No (%)}
≤ 0.20	29 (85.3)	35 (74.6)	67 (69.1)	131 (73.6)	50 (76.9)	127 (64.5)	126 (61.8)	303 (65)
0.21 - 0.22	02 (5.9)	05 (10.6)	11 (11.3)	18 (10.1)	08 (12.3)	31 (15.7)	32 (15.7)	71 (15.2)
> 0.22	03 (8.8)	07 (14.9)	19 (19.6)	29 (16.3)	07 (10.8)	39 (19.8)	46 (22.5)	92 (19.7)
Total	34	47	97	178	65	197	204	466

7.8×10^5 during summer, respectively (Table 6). At 0.05 level of significance there were significant differences between the three regions. At 0.05 level of significance there is a significant difference between the two seasons.

Acidity of milk

The percentages of samples with titratable acidity of less than or equal to 0.20 in Khartoum State were 64.3% .In winter it was 73.6%and in summer it was 60.7%. The percentages of samples with titratable acidity of more than 0.22 were 15.2% in winter, 20.8% in summer and 19.3% during the two seasons in Khartoum State (Table 7).The percentages of samples with titratable acidity of less than or equal to0.20 in Khartoum region were 74.6% in winter, 60.9% in summer and 65.3% during the two seasons. The percentages of samples with titratable acidity of less than or equal 0.20 in Khartoum North were 70.4% in winter, 59.9% in summer and 62% during the two seasons (Table 7). The percentages of samples with titratable acidity of less than or equal 0.20 in Omdurman were 75.4% in winter, 62% in summer and 66% during the two seasons (Table 7).

Statistically (at 0.05 level) the differences in acidity were of no significance between the three regions, but the difference was significant between the two seasons.The percentages of samples with acidity of less than or equal to 0.20 in individual, bulk, and market milk in Khartoum State were 79.8, 66.4and 63.7% in the two

seasons. In winter these percentages were 85.3, 74.6 and 69.1% and in summer, 76.9,64.5 and 61.8% in individual, bulk and market milk,respectively (Table 8). At 0.05 level there was a correlation between individual and farm bulk tank milk and between bulk and market milk, there was a correlation in the acidity values between individual cow milk and bulk tank milk (Table 9).Statistically at 0.05 level there were significant correlations between acidity, SCC, and TBC and at 0.01 there was correlation between TBC and SCC (Table10).Statistically (at 0.05 level) the differences in titratable acidity were of no significance between the three regions, but the difference was significant between the two seasons.

DISCUSSION

The major observations is the wet poor hygienic practices in the farm and during marketing which contributes a lot to the quality of raw milk before it reaches the consumers. The milk was collected from the milking bucket into a plastic or aluminum containers which were not well washed, no cooling system was applied at any level of the milk chain which may last for five hours till milk reaches the consumer. Accordingly it was expected that milk would have a moderate to poor hygienic quality. Smiddy et al. (2007) stated that TBC count greater than 1 recommended that Grade A milk should not exceed 1 × 10^5 and Grade two milk should be less than 3 × 10^5.

Table 9. Correlations of titratable acidity values between different sample sources.

Parameter	Correlation	Individual	Bulk	Market
Individual	Pearson correlation	1	0.889**	0. 791*
	Sig. (2 - tailed)	0.0	0.018	0.500
	N	6	6	6
Bulk	Pearson correlation	0.889**	1	0.821*
	Sig. (2 - tailed)	0.018	0.0	0.019
	N	6	6	6
Market	Pearson correlation	0. 791*	0.821*	1
	Sig. (2 - tailed)	0.500	0.019	0.0
	N	6	6	6

*Correlation is significant at the 0.05 level (2-tailed), ** correlation is significant at the 0.01 level (2-tailed).

Table 10. Correlations between acidity, SCC and TBC in Khartoum State.

Parameter	Correlation	Acidity	SCC	TBC
Acidity	Pearson correlation	1	0.121*	0.009*
	Sig. (1 - tailed)	0.0	0.064	0.000
	N	643	643	643
SCC	Pearson correlation	0.121*	1	0.142**
	Sig. (1 - tailed)	0.064	0.0	0.000
	N	641	642	642
TBC	Pearson correlation	0.009*	0.142**	1
	Sig. (1 - tailed)	0.000	0.000	0.0
	N	643	644	644

* Correlation is significant at the 0.05 level (1-tailed), ** correlation is significant at the 0.01 level (1-tailed).

Raw milk ready for pasteurization must be within the count rate of 1×10^5 to 3×10^5 (Coast et al., 2004 and Jayarao et al., 2001). During this study only 23.9% were within this limit, (equal to or less than 9×10^4), 35.4% in winter compared to 19.5% in summer, it was the highest in Khartoum region (26.0%), followed by Omdurman (25.3%) then Khartoum North (21.3%) during the two seasons. Elevated bacterial counts in summer months are generally due to warm moist environment that increases pathogen exposure and number (Duane and Gerald, 2003).

During this study the majority of samples 55.3% were with a count of less than or equal 9×10^5, higher in winter 71.9% compared to 48.1% in summer this finding is almost in line with Duane and Gerald (2003).In Poland, Marian (2001) reported that raw milk is acceptable if it contains 4×10^5 cell/ml. The acceptable bacterial counts in Pakistan as reported by Muhammadet al.(2009) was 1×10^6,he also stated that most of the milk had a count of more than 1×10^7.Higher counts of more than 1×10^6 reported by many researchers in many countries such as Mali (Bonfoh et al., 2003), Sudan (Elmagli et al., 2006), Malaysia (Chye et al., 2004), and India (Chatterjee et al., 2006). The percentage of samples of count of less than or equal 9×10^5 were highest in Omdurman (61.1%) followed by Khartoum region (56.1%) then Khartoum North (50.4%) during the two seasons, at winter time the percentage were higher than summer. In summer the count of less than 9×10^5 in Omdurman was the best (52.8%) followed by Khartoum (49.7%) then Khartoum North (46.1%), but when comparing the higher counts of more than or equal to 1×10^8 Khartoum North was the worst (10.1%); followed by Omdurman (2.6%) then Khartoum (2%). These differences between the regions were statistically insignificant but the differences were significant between the seasons, this was in agreement with Ahmed and Elzubier (2007) who compared the count between winter and summer in Khartoum State and found higher counts during summer compared to winter season (5.3×10^{10} and 7.5×10^7 cfuml, respectively).In this study the average counts of individual, bulk, vendor and market milk in summer were also higher than those in winter, the highest count was found among vendor and market milk at the range of 1.5×10^7 in Khartoum region and 9.4×10^8 in Khartoum North. This was almost in agreement with Beniwal et al. (1998) who found the change in TBC at the beginning to the end of the channel to be 1×10^5 to 8.1×10^7.

The correlation between the individual TBC and farm bulk tank TBC milk were significant at 0.05, for the bacteria found in milk will be significantly affected by the holding time of the milk and the storage temperature of the milk (Khan,2008). Ombui et al. (1995) compared TBC from farmer cans and distributors cans, he found that 44% of samples with count of more than 10^5 were from farmers cans compared to 86% from distributors cans,this was in direct agreement with Shojaei and Yadollahi (2008) who reported a count of more than 13×10^7 in market milk. Mariana (2001) showed that in Poland milk should not exceed the count of 4×10^5. In this study the percentage of samples with count less than or equal to 5×10^5 were 31.7% higher in winter (43.6%) compared to 20.9% in summer in Khartoum state. But in the USA a count of 7.5×10^5 is acceptable the majority of samples (60.0%) during this study were within this limit. The percentages of samples with count less than 7.5×10^5 was also higher in winter (69.7%) compared to 47.7% in summer, this was in line with Duane and Gerald (2003). Since extreme heat and humidity were among the most important factors which affect the count of somatic cell, the count in summer were expected to be higher, the elevated SCC of raw milk raise the suspicion that the raw milk is produced under poorer standard of hygienic condition and from unhealthy cows. The percentage of samples with a count over 1×10^6 was lower in winter (6.2%) compared to 18.4% in summer, this was lower than what found by Nada (2000) who reported 26% of the samples were of SCC of about 1×10^6 while 32% were between 2×10^5 to 1×10^6. Assgad (2002) reported that 51% of samples in Kordofan State (western Sudan) were with SCC of more than 5×10^7 which is lower than the percentage calculated in this study (68.3%). The seasonal variation of SCC was studied by Sawa and Pwczynski (2002) and revealed a significant influence of season on SCC; higher in summer and lower in winter.Khartoum region was the best in the cell count during the two seasons and also has a lower unacceptable count during the two seasons followed by Omdurman then Khartoum North. These differences were

statistically significant at both the season and area levels at 0.05. This was in agreement with Marian (2001) who found a significant difference between seasons of the year on the SCC. The maximum count of more than1×10^6 was found among summer milk in Khartoum North but generally market milk has almost higher counts compared to individual and farm bulk milk. This may be due to mixing of milk from different farms for probability of getting herd with sub clinical mastitis which was known to elevate SCC. The lowest minimum counts were within Khartoum region milk in winter time this was similar to the TBC result in Khartoum region which suggest the better hygienic condition and relatively better mastitis control programs followed in Khartoum region compared to Khartoum North and Omdurman.

The titratable acidity of milk is very important parameter for the evaluation of milk quality Harris and Beach man (1988) they accepted milk with 0.19% acidity. In the present study, the percentage of samples with titratable acidity of less than or equal 0.20 in Khartoum State were 64.3% during the year, higher in winter (73.6%) compared to 60.7% in summer (Acatincai et al., 2008) reported average acidity ranging between 0.18 and 0.185 during both seasonsbut if the range reported by Idris et al. (1975) of not more than 0.22 was adopted we find that in this study 80.7% of the samples satisfy this rangein the State higher in winter compared to summer. Idris et al. (1975) found the upper limit of titratable acidity to be 0.22% (Mohamed and El Zubeir, 2007) reported acidity ranging between 0.17 to 0.26. In this study 15.9% of the samples in the state were in agreement with this upper limit. Normal range of acidity is affected by total solid content of milk, so Harris and Bachman (1988) suggested that each supplier or area should establish its own limits which when exceeded might indicate high bacterial count. Asaminew and Eyassu (2010) in Ethiopia reported an acidity values ranging between 0.22 to 0.23 in bulk while Ismail et al. (2010) reported acidity ranging between 0.14 to 0.16 in raw cow milk in Egypt.In this study the percentage of samples with acidity more than 0.20 were higher in market milk (29.1%) compared to farm bulk milk (27.0%) during the two seasons, this difference might indicate high bacterial count in marketmilk as shown in this study compared to farm bulk milk; this was in agreement with Ibrahim (1973) who reported an average of 0.18 titratable acidity among farm bulk milk and 0.20 in marketraw milk and also in agreement with Ammar et al. (2008) whofound an average acidity of raw milk in KhartoumState to be about 0.19 in the farm while it was0.22 at the sale points. The maximum limit of titratable acidity was found among samples from Omdurman market (0.26) during summer,this is again in agreement with Ahmed and El Zubeir (2007). This may be due to high bacterial count, for most of the milk sold in Omdurman markets was brought from Khartoum North (the time between milking and distribution ranges from 3 to 5 h) without proper cooling to the milk and the effect of direct sunlight.The correlation between individual and farm bulk milk was significant at 0.05 but that between farm bulk milk and market milk was significant at 0.01 levels. This is in agreement with Ammaretal.(2008) who found this difference to be of significance. This may be due to long time spent by the vendor distributing milk house to house.Statistically there is correlation between the acidity, SCC and total bacterial count (TBC). This correlation shows the direct effect of TBC on SCC and the effect of both TBC and SCC on the acidity.

Conclusion

Most of the raw milk sold in Khartoum State is of poor hygienic quality. So, the Ministry of Agriculture and Animal Resources of Khartoum State should enforce all the regulations needed for producing and purchasing raw milk with acceptable hygienic, chemical land physical quality.

REFERENCES

Acatincai S, Adela M, Cziszter L, Stanciu G,Gavojdiand D, Simona B (2008). Study regarding the correlation between Total Gernms Count and Chemical composition in raw milk. Lucrăristiin Ñifice Zootehniesi Biotehnologii, 41(2): 345-349.

Ahmed M, El Zubeir I (2007). The Compositional Quality of Raw Milk Produced by Some Dairy Cow's Farms in Khartoum State, Sudan.Res. J. Agric. Biol. Sci., 3(6): 902-906.

Ammar A, Ibtisam E, Osman A, Mohamed A (2008).Assessment of Microbial Loads and Antibiotic Residues in Milk Supply in Khartoum State, Sudan.Res. J. Dairy Sci., 2(3): 57-62.

Asaminew T, Eyassu S (2010). Microbial quality of raw cow's milk collected from farmers and dairy cooperatives in Bahir Dar Zuria and Mecha district, Ethiopia . Agric. Biol. North Am., 2(1): 29-33

Assgad HH (2002). Milk quality and aerobic bacteria in raw fluid milk, in Elobeid. M.V.Sc. thesis, University of Khartoum, Faculty of Vet. Sci., 40-45

Beniwal BS, Srivastva DN, Bhardwaj PK (1998). Change in bacterial quality in raw milk during distribution. Ind. J. Anim. Prod. Manage.,(3): 14-21.

Bonfah BA, Wasem AN, Traore AF,Spillman H (2003).Quality of cows milk taken at different intervals from the udder to the selling point at Bamko- Mali. Food Control, 14: 495-500.

Boor KJ, Brown DP, Murphy SC, Bandler DK (1998). Microbial and chemical quality of raw milk in New York State, J. Dairy Sci., 81: 1743-1748.

Chatterjee SN, Bhattacharjee SK, Chandra G (2006). Microbiological examination of Milk in Tarakeswar . India with special reference to coliform. Afr. J. Biotech., 5: 1383-1385.

Chye F, Abdulahb A, Khan M(2004). Bacteriological quality and safety of raw milk in Malaysia. Food Microbiol., 21: 535-541.

Coast D, Reinmann O, Cook N, Ruegg P (2004). The changing Face ofMilk Production, Milk Quality and milking Technology in Brazil(2004) Board of Reagents of the University of Wisconson System. 1st ed. Babcock Institute DiscussionPaper 2004-2 ISBN 1-59215-088-8

Duane N, Gerald RB (2003). The Somatic Cell Count and milk quality –

Neb Guid. University of Nebraska –cooperative extension institute of Agric, and natural resources, G93-1151-A

El Zubeir I, Ahmed M (2007). The hygienic quality of raw milk produced by some dairy farms in Khartoum State, Sudan. Res. J. Microbiol., 2: 988-991.

Elmagli A, Ibtisam E, Zubeir E (2006). Study on the Hygienic quality of Pasteurized milk in Khartoum State (Sudan). Res. J. Anim. Vet. Sci., (1): 12-17.

FAO/WHO (1992). Food Standard Programs, (Codex) Alimentarious Commission. Rome. FAO.

Foley J, Buckley J, Murphy MF(1974). Commercial testing and product control in the dairy industry. Dept. of Dairy and Food Technology. UniversityCollege Cork. p. 312.

Harris B, Bachman KC (1988). Nutritional and management factors affecting solids-not fat, acidity and freezing point of milk, Florida cooperative Extension service, Dairy Sci. Dept., Ds 25.

Ibrahim EA (1973). A note on some characteristics of the raw fluid milk available in the three towns. Sud. J. Vet. Sci. Anim. Husb.,14(1):36-41.

IDF(1984). Recommended methods for somatic cell counts in milk, Doc. No. 168, International Dairy Federation, Belgium, pp.15-30.

Idris OF, Mustafa AA, Wahbi AA (1975). Physiochemical and bacterial composition of raw milk supply to the three towns. Sud. J. Vet. Sci. and Ani. Husb., 16: 87-93.

Ismail M, Ammar E, El-Shazly E, Eid M (2010). Impact of cold storage and blending different lactationsof cow's milk on the quality of Domiati cheese. Afr. J. Food Sci., 4(8): 503-513

ISO(1991). Pour plate method. International Organization for Standardization, pp.4833-1991.Geneva.

Jayarao BM, Pillai SR, Sawant AA, Wolfgang DR, Hegde NV (2004). Guidelines for monitoring bulk tank milk somatic cell and bacterial counts. J Dairy Sci., (87): 3561-3573.

Khan M, Zinnah M, Siddique M,Rashid M, Islam M,ChoudhuryBK(2008).Physical and microbial qualities of raw milk collected from Bangladesh agricultural university dairy farm and the surrounding villages.Bangl. J. Vet. Med.,6(2): 217-221.

Mariana K (2001). Inter relation between Year, Season, and raw milk hygiene quality indices. Electronic J. Polish Agric.,4(1): 17-25.

Mohamed NNI, El Zubeirl EM(2007). Evaluation of the hygienic quality of market milk of Khartoum state (Sudan). Int. J. Dairy. Sci., 2: 33-41.

Muhammad K, Altaf I,Hanif A, Anjum A, Tipu M(2009). Montoring of Hygenic Status of Raw Milk marketed in Lahore City, Pakistan.J. Animal Plant Sci., 19(2): 74-77.

Murphy SC(1997). Raw milk bacteria test. Standard plate count. Proc. National Mastitis Council Regional Meeting. Syracuse. N. Y.,pp. 34-42.

Nada A (2000). Studies on the Sanitary Quality of Raw Fluid Marketed Milk in Khartoum State. M.V.Sc. thesis. University of Khartoum.

Ombui JN, Arimi SM, Mc Dermott JJ, Mbugua SK (1995). Quality of raw milk collected and marketed by dairy cooperative societies in kiambu district. Kenya.Bull. Ani. Health Prod. in Africa, pp. 43-44.

Sawa A, Pwczynski D (2002). Somatic Cell Count and milk yield and composition in black and white. Holstien–Friesian cows. Medycna weterynaryjna,. 58: (23-43).

Schallibaum M (2001). Impact of SCC on the quality of fluid milk and cheese. National mastitis council, Inc. 40th Annual Meeting Proceed., 93-100.

Shojaei Z, Yadollahi A (2008). Physico-chemical and microbiological quality of raw, pasteurized and UHT milks in Shops. Asian J. Sci. Res., 1(3): 332-338.

Simona D, Gabriela S, Florentina R, Oana C (2010). Quality control of milk and dairy products. Ovidius Univ. Annals of Chem., 21(1): 91-95. ISSN-1223-7221 ©2010 Ovidius University Press.

Smiddy MA, Martin J, Huppertz T, Kelly AL (2007). Microbial shelf life ofhigh-pressure-homogenised milk. Int. Dairy J., 17: 29-32.

26

The prevalence of *SEN virus* infection in blood donors and chronic hepatitis B and C patients in Chaharmahal Va Bakhtiari province

Payam Ghasemi Dehkordi and Abbas Doosti*

Biotechnology Research Center, Islamic Azad University, Shahrekord Branch, Shahrekord, Iran.

SEN virus (*SEN-V*) is a blood-borne, circular, nonenveloped and single-stranded DNA virus. Phylogenetic analysis demonstrated 9 different genotypes for this virus. *SEN-V* could be related to post-transfusion hepatitis and infections with this virus in blood donors and hepatitis patients differ markedly by geographic region. The purpose of present study was to determine the prevalence of H and D genotypes of *SEN-V* (*SENV-H* and *SENV-D*) infection in blood donors and patients with chronic hepatitis B virus (HBV) and hepatitis C virus (HCV) for the first time in Chaharmahal Va Bakhtiari province located in southwest Iran. *SEN-V* DNA was analyzed in 240 serum samples of the patients with chronic HBV and HCV (172 HBV and 68 HCV) and 60 non-professional blood donors from the blood transfusion organization, hospital and clinical and pathological laboratories in Chaharmahal Va Bakhtiari province. *SEN-V* DNA was amplified by specific primers for *SENV-H* and *SENV-D* genotypes using polymerase chain reaction (PCR) method after extraction of DNA from sera and PCR products were visualized in a 1% agarose gel electrophoresis. *SENV-H* genotype was found to be positive in 54/172 (31.39%), 23/68 (33.82%), and 8/60 (13.33%) and *SENV-D* genotype was detected in 48/172 (27.91%), 27/68 (39.7%), and 6/60 (10%) of patients with chronic HBV, HCV and healthy blood donors, respectively. These results showed that high prevalence of *SEN-V* infection in patients with chronic HBV and HCV compared healthy blood donors in Chaharmahal Va Bakhtiari province using T test statistical analysis (P<0.05). According to these findings examination of serum samples for control and prevention of *SEN-V* infection in hepatitis patients and healthy blood donors seems to be necessary.

Key words: *SEN* virus, hepatitis B virus (HBV), hepatitis C virus (HCV), polymerase chain reaction (PCR), Chaharmahal Va Bakhtiari province.

INTRODUCTION

SEN virus (*SEN-V*) is a member of the Circoviridae family, a group of small, circular DNA virus that includes *TT* virus (*TTV*), *TUS01*, *SANBAN*, and *YONBAN* (Sharifi et al., 2008). This virus is blood-borne and nonenveloped with approximately 3800 nucleotides in length and about 26 nm in size and in these viruses at least three open reading frames (ORF) have been identified (Dai et al.,

2006). The ORF1 with Arg/Lys-rich domains is the largest ORF with hydrophilic characteristic (Mikuni et al., 2002). The role of ORF2 is yet to be determined. The ORF3 translation results in the formation of a protein with a homology amid a DNA topoisomerase I, therefore ORF3 seems to play a significant role in the replication of the virus (Sagir et al., 2004). Interestingly, *SEN-V* have high mutation rate (7.32*10-4 per site per year) and it is more similar to RNA viruses rather than DNA viruses and it could be the ability of persistence of *SEN-V* in the host cells (Umemura et al., 2002).

Viral hepatitis (A to E) is still one of the important agents of acute and chronic liver disease worldwide (Moriondo et al., 2007). *SEN-V* genotype H and D (*SENV-H*, *SENV-D*) can cause post-transfusion hepatitis

---

*Corresponding author. E-mail: biologyshki@yahoo.com.

Abbreviations: HBV, Hepatitis B virus; **HCV,** hepatitis C virus, **PCR,** polymerase chain reaction.

and infections with this virus in blood donors and hepatitis patients differ markedly by geographic region. *SEN-V* is transmitted by blood, as demonstrated by comparing the sequence homology between donors and recipients (Shibata et al., 2001). However, further studies revealed that this virus is distantly related to the *TTV* family (Davidson et al., 1999). To date, phylogenetic analysis of *SEN-V* has demonstrated 9 different genotypes: *SENV-A* to *SENV-H* (Kojima et al., 2003). *SENV-D* and *SENV-H* genotypes are related to transfusion-associated non-A to E hepatitis (Mikuni et al., 2002) and are more prevalent within the population exposed to transfusion (Kao et al., 2002).

SEN-V-B, -A and -E have been less frequently found among blood donors and do not appear to be related to non A to E hepatitis. On the other hand, genotypes D and H have been detected in 1% of blood donors but in more than 50% of non A to E hepatitis cases (Serin et al., 2005). The co-infection with HCV is considered a risk factor for *SEN-V* infection and the exact interaction of this virus with hepatitis C virus (HCV) and hepatitis B virus (HBV) is still unclear (Tahan et al., 2003). Furthermore, the prevalence of *SEN-V* increases in association with *HIV-1* and hepatitis B and C viruses. This strongly supports the hypothesis that *SEN-V* is transmitted via blood (Umemura et al., 2001). Nevertheless, these genotypes have been found at various rates in different populations and the role of *SEN-V* regarding of the pathogenesis of liver disease is not yet known (Mu et al., 2004).

The purpose of present study was to determine the prevalence of *SEN* virus infection in blood donors and patients with chronic HBV and HCV for the first time in Chaharmahal Va Bakhtiari province located in southwest Iran.

MATERIALS AND METHODS

Serum samples and study populations

The serum samples were obtained from 240 patients with chronic HBV and HCV (172 HBV and 68 HCV) and 60 non-professional blood donors from the blood transfusion organization, hospital and clinical and pathological laboratories in Chaharmahal Va Bakhtiari province located in southwest Iran. All serum samples were stored frozen at -20°C until analysis.

Viral DNA extraction

The viral nucleic acid was extracted from 200 μl serums using a QIAamp DNA Blood Mini Kit (Qiagen Ltd, Crawley, UK) according to the manufacturer's recommendation. The extracted genomic DNA was quantified by spectrophotometric measurement at a wavelength of 260 nm according to the method described by Sambrook and Russell (2001).

Detection of *SENV-D* and *SENV-H* genotypes by PCR

The presence of *SENV-H* and *SENV-D* DNA was determined by polymerase chain reaction (PCR) with type-specific primers according to the research articles by Umemura et al. (2001) and Kojima et al. (2003). These primers sequences (D10S: 5'-GTA ACT TTG CGG TCA ACT GCC-3', L2AS: 5'-CCT CGG TT[G/T] [C/G]AA A[G/T]G T[C/T]T GAT AGT-3', C5S: 5'-GGT GCC CCT [A/T]GT [C/T]AG TTG GCG GTT-3' and L2AS: 5'-CCT CGG TT[G/T] [C/G]AA A[G/T]G T[C/T]T GAT AGT-3') were used for *ORF1* gene amplification of *SENV-D* and *SENV-H*, respectively.

PCR was carried out in 25 μl total reaction volumes, each containing 100 ng of template DNA, 0.2 pM of each primer, 2.5 μl of 10X PCR buffer, 1.5 mM MgCl$_2$, 200 mM dNTPs, and 1 unit of *Taq* DNA polymerase (Fermentas, Germany). The amplification reaction consisted of 5 min of pre-denaturing at 94°C, followed by 32 cycles of 1 min denaturation at 94°C, 1 min annealing at 60°C, and 1 min extension 72°C, and then by a final extension at 70°C for 5 min. The samples were amplified in a Gradient Palm Cycler (Corbett Research, Australia). The PCR amplification products (10 μl) were subjected to electrophoresis in a 1% agarose gel in 1X TBE buffer at 80 V for 30 min, stained with Ethidium Bromide, and images were obtained in UVIdoc gel documentation systems (UK).

Statistical analysis

Analysis of data was performed using the SPSS version 17.0 computer software (SPSS, Chicago, IL). Also, the differences between prevalence of *SEN-V* infection in chronic hepatitis patients and healthy blood donors were examined by T test statistical analysis. P values <0.05 were considered significant.

RESULTS

Analysis of PCR products for the presence of *SENV-H* and *SENV-D* DNA on 1% agarose gel revealed 229 base pairs (bp) for *SENV-H* and 222 bp for *SENV-D* (Figure 1). *SENV-H* and *SENV-D* DNA was found to be positive in 54/172 (31.39%), and 48/172 (27.91%), of hepatitis B patients and 23/68 (33.82%), and 27/68 (39.7%) of patients with hepatitis C, respectively. In the control group, *SENV-H* and *SENV-D* was detected in 8/60 (13.33%) and 6/60 (10%) of the healthy blood donors, respectively (Table 1).

DISCUSSION

SEN-V was first reported on July 20, 1999, in the serum of a human immunodeficiency virus type 1 (*HIV*-1)-infected patient possessing hepatitis with unknown etiology in Italy, and was named on the basis of initials of the patient (Yoshida et al., 2002). The prevalence of 5 out of 9 *SEN-V* strains (A, B, C, D, H and E) as well as a consensus sequence as total *SEN-V* have been studied in various donor and patient populations. Among nine genotypes of *SEN-V* (A to I), 5 strains (A, B, H, C, D and E), as well as a consensus sequence and *SENV-H* and *SENV-D* genotypes were extremely associated with non-A to E hepatitis (Schroter et al., 2003). It has been suggested that genotypes *SEN-V* C and H as well as *SEN-V* D and F could be combined due to similarities in ORF 1 (Umemura et al., 2001). Tanaka et al. (2001) found

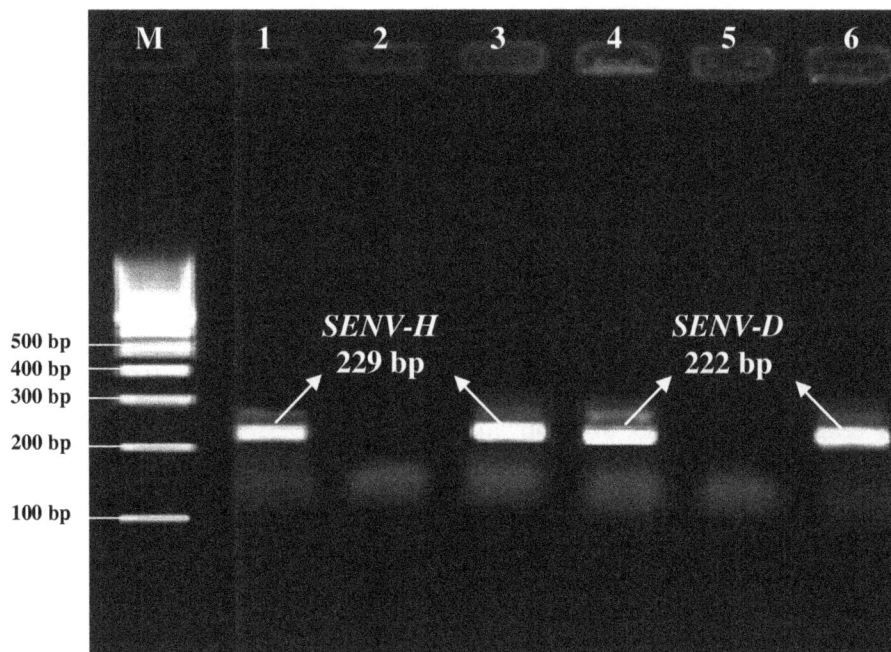

Figure 1. Gel electrophoresis for identification of *SENV-H* and *SENV-D*. Line M: 100 bp molecular weight markers (Fermentas, Germany), lines 1 and 3: 229 bp PCR products of *SENV-H* DNA, line 2: *SENV-H* DNA negative sera sample, lines 4 and 6: 222 bp PCR products of *SENV-D* DNA, and line 5: *SENV-D* DNA negative sera sample.

Table 1. Prevalence of *SEN-V* infection in patients with chronic *HBV* and *HCV* in compared with control groups (healthy blood donors) in southwest Iran (Chaharmahal Va Bakhtiari province).

Patients	Samples	Both *SENV-H/D* positive no. (%)	*SENV-H* positive no. (%)	*SENV-D* positive no. (%)
HBV	172	102 (59.3)	54 (31.39)	48 (27.91)
HCV	68	50 (73.52)	23 (33.82)	27 (39.7)
Controls (healthy blood donors)	60	14 (23.33)	8 (13.33)	6 (10)
Total	300	166 (55.33)	85 (28.33)	81 (27)

that *SEN-V* and *TTV* had similar structure. *SEN-V* has homogeneity of 55% at nucleotide level with *TTV*, but they only have homogenelty of 37% at amino acid level.

The prevalence of *SENV-H* and D in present study was 31.39 and 27.91% in chronic hepatitis B patients and 33.82 and 39.7% in chronic hepatitis C patients, respectively. Furthermore, *SENV-H* and D was detected in 13.33 and 10% of healthy blood donors. Results showed high prevalence of *SEN-V* genotype H and D in patients with chronic HBV and HCV compared blood donors in southwest Iran (Chaharmahal Va Bakhtiari province). The significant differences for prevalence of *SEN-V* infection between the hepatitis patients and healthy blood donors group using T test statistical analysis (P<0.05) were observed in this study. The high prevalence of *SEN-V* infection in hepatitis patients in comparison to blood donors possibly is due to the same risk factors for these viruses such as blood transfusion in

this region.

The distribution of *SENV* strains D and H among hepatitis patients, blood donors and other healthy populations varied in different regions. Also, *SEN-V* was found not to be practical because the prevalence in donors was 13% and the rate in a transfused population exceeded 70% (Umemura et al., 2001).

Umemura et al. (2001) showed the strong association of *SEN-V* with transfusion-associated non–A to E hepatitis and it is similar to the results of the present study. The prevalence of *SEN-V* in healthy blood donors from various geographic areas such as Japan (10 to 22%) (Shibata et al., 2001), Taiwan (15%) (Kao et al., 2002), Thailand (5%) (Tangkijvanich et al., 2003), United States (1.8%) (Umemura et al., 2001), Germany (8 to 17%) (Schroter et al., 2002), and at least 13% in Italy (Pirovano et al., 2002). These findings were same to the results of the present study which confirmed it. Mikuni et al. (2002) found *SEN-V* infection with a high prevalence among blood donors and none had a history of serious illness or blood transfusion. In their study, the less considerable frequency of blood transfusion in subjects with non-B, non-C hepatitis liver disease than in those with chronic HCV-related liver disease, which suggests the existence of non-transfusion related routes. The study of Mu et al. (2000) in China showed that 31% of the blood donors had *SENV-D/H* DNA and it was higher than those in America and Italy (2%), and in Japan and Taiwan (15 to 20%). The prevalence of *SENV-H* among Taiwanese chronic hepatitis C patients with combination therapy of high-dose interferon-alfa and ribavirin was 19.2% (Dai et al., 2004). In the study of Omar et al. (2008) on Egyptian patients with hepatitis C virus related chronic liver disease and patients undergoing hemodialysis, *SEN* virus-D/H DNA was detected in 13.5% of patients with chronic liver disease, 11.1% of patients undergoing hemodialysis, and 7.1% of healthy controls. They showed no significant differences between patients and the control group. The study of Sharifi et al. (2008) on prevalence of *SEN-V* infection in Iranian blood donors showed that 4 (1.5%) of the 260 were infected. *SENV-H* viremia was detected in 47 (18.08 %) of the 260 blood donors and both *SENV-D* and *SENV-H* viremia were detected in 9 (3.4%) of the 260 blood donors. Furthermore, in total *SENV-D* or *SENV-H* viremia was identified in 60 (23.08%) of the 260 blood donors. This results also showed that the high prevalence of *SEN-V* in healthy blood donors with no history of blood transfusion in Iran (Sharifi et al., 2008). The results of their study confirmed the high prevalence of this virus that observed in current research. The study of Karimi-Rastehkenari and Bouzari (2010) in Iran showed that frequency of *SEN-V* strains (*SENV-H* or *SENV-D*) and co-infection (both *SENV-D* and *SENV-H*) viremia was significantly higher among thalassemic patients than healthy individuals.

The results of the current study indicate that PCR amplified with genotype-specific primers could be useful for detection and screening *SENV-H* and D genotype. Furthermore, the findings of the present study in southwest Iran demonstrate that *SENV-D* and *SENV-H* was detected in a high proportion of patients with chronic HBV and HCV compared with healthy blood donors.

In conclusion, the results of present study showed that *SENV-H* and D are more prevalent in hepatitis patients in comparison with blood donors in our region. This virus is parenterally transmitted, and therefore, its spread could be controlled by appropriate screening of hepatitis patients, blood donors and blood products.

ACKNOWLEDGEMENTS

This study was supported by Biotechnology Research Center of Islamic Azad University of Shahrekord Branch in sowthwest of Iran.

REFERENCES

Dai CY, Chuang WL, Chang WY, Chen SC, Lee LP, Lin ZY, Hou NJ, Hsieh MY, Wang LY, Yu ML (2004). The prevalence and clinical characteristics of coinfection of *SENV-H* among Taiwanese chronic hepatitis C patients with combination therapy of high-dose interferon-alfa and ribavirin. Antiviral. Res., 64(1): 47-53.

Davidson F, MacDonald D, Mokili JL, Prescott LE, Graham S, Simmonds P (1999). Early acquisition of *TT* virus (*TTV*) in an area endemic for *TTV* infection. J. Infect. Dis., 179: 1070-1076.

Kao JH, Chen W, Chen PJ, Lai MY, Chen DS (2002). Prevalence and implication of a newly identified infectious agent (*SEN virus*) in Taiwan. J. Infect. Dis., 185: 389-392.

Karimi-Rastehkenari A, Bouzari M (2010). High frequency of *SEN virus* infection in thalassemic patients and healthy blood donors in Iran. Virol. J., 7(1): 1-7.

Kojima H, Kaita KD, Zhang M, Giulivi A, Minuk GY (2003). Genomic analysis of a recently identified virus (*SEN virus*) and genotypes -D and -H by polymerase chain reaction. Antiviral. Res., 60: 27-33.

Mikuni M, Moriyama M, Tanaka N, Abe K, Arakawa Y (2002). *SEN virus* infection does not affect the progression of non-A to -E liver disease. J. Med. Virol., 67(4): 624-629.

Moriondo M, Resti M, Betti L, Indolfi G, Poggi GM, de Martino M, Vierucci A, Azzari C (2007). *SEN virus* co-infection among *HCV*-RNA-positive mothers, risk of transmission to the offspring and outcome of child infection during a 1-year follow-up. J. Viral. Hepat., 14(5): 355-359.

Mu SJ, Du J, Zhan LS, Wang HP, Chen R, Wang QL, Zhao WM (2004). Prevalence of a newly identified *SEN virus* in China. World J. Gastroenterol., 10(16): 2402-2405.

Omar M, El-Din SS, Fam N, Diab M, Shemis M, Raafat M, Seyam M, Hssan M, Badawy A, Akl M, Saber M (2008). *SEN virus* infection in Egyptian patients with chronic hepatitis C and patients undergoing hemodialysis. Medscape J. Med., 10(12): 290.

Pirovano S, Bellinzoni M, Matteelli A, Ballerini C, Albertini A, Imberti L (2002). High prevalence of a variant of *SEN-V* in intravenous drug user *HIV*-infected patients. J. Med. Virol., 68: 18-23.

Sagir A, Kirschberg O, Heintges T, Erhardt A, Haussinger D (2004). *SEN virus* infection. Rev. Med. Virol., 14(3): 141-148.

Sambrook J, Russell DW (2001). Molecular cloning: A laboratory manual. 3rd Edition. Cold Spring Harbor Laboratory Press, Cold Spring Harbor, New York.

Schroter M, Laufs R, Zollner B, Knodler B, Schafer P, Feucht HH (2003). A novel DNA virus (*SEN*) among patients on maintenance hemodialysis: prevalence and clinical importance. J. Clin. Virol., 27: 69-73.

Schroter M, Laufs R, Zollner B, Knodler B, Schafer P, Sterneck M, Fischer L, Feucht H (2002). Prevalence of *SENV-H* viremia among healthy subjects and individuals at risk for parenterally transmitted diseases in Germany. J. Viral. Hepat., 9: 455-459.

Serin MS, Koksal F, Oksuz M, Aslan G, Tezcan S, Yildiz C, Kayar B, Emekdas G (2005). *SEN virus* prevalence among non-B and non-C hepatitis patients with high liver function tests in the south of Turkey. Jpn. J. Infect. Dis., 58(6): 349-352.

Sharifi Z, Mahmoodian-Shooshtari M, Talebian A (2008). The prevalence of *SEN virus* infection in blood donors in Iran. Arch. Iranian Med., 11(4): 423-426.

Shibata M, Wang RY, Yoshiba M, Shih JW, Alter HJ, Mitamura K (2001). The presence of a newly identified infectious agent (*SEN virus*) in patients with liver diseases and in blood donors in Japan. J. Infect. Dis., 184: 400-404.

Tahan V, Ozdogan O, Tozun N (2003). Epidemiology of viral hepatitis in the Mediterranean basin. Rocz. Akad. Med. Bialymst., 48: 7-11.

Tanaka Y, Primi D, Wang RY, Umemura T, Yeo AE, Mizokami M, Alter HJ, Shih JW (2001). Genomic and molecular evolutionary analysis of a newly identified infectious agent (*SEN virus*) and its relationship to the *TT virus* family. J. Infect. Dis., 183: 359-367.

Tangkijvanich P, Theamboonlers A, Sriponthong M, Thong-Ngam D, Kullavanijaya P, Poovorawan Y (2003). *SEN virus* infection in patients with chronic liver disease and hepatocellular carcinoma in Thailand. J. Gastroenterol., 38: 142-148.

Umemura T, Alter HJ, Tanaka E, Orii K, Yeo AET, Shih JWK, Matsumoto

A, Yoshizawa K, Kiyosawa K (2002). *SEN virus*: response to interferon alfa and influence on the severity and treatment response of coexistent hepatitis C. Hepatology, 35(4): 953-959.

Umemura T, Yeo AE, Sottini A, Moratto D, Tanaka Y, Wang RYH, Wai-Kuo Shih J, Donahue P, Primi D, Alter HJ (2001). *SEN virus* infection and its relationship to transfusion-associated hepatitis. Hepatology, 33: 1303-1311.

Yoshida EM (2002). Is there an association between *SEN virus* and liver disease? Reviewing the evidence. Can. Commun. Dis. Rep., 28(7): 53-60.

27

Sodium periodate inhibits the binding efficiency of influenza A virus (H3N2) with mammalian cell lines

M. Paulpandi[1], R. Thangam[2], P. Gunasekaran[2] and S. Kannan[1]*

[1]Proteomics and Molecular Cell Physiology Laboratory, Department of Zoology, School of Life Sciences, Bharathiar University, Coimbatore – 641 046, TN, India.
[2]King Institute of Preventive Medicine and Research, Department of Virology, Chennai-600032, India.

Influenza epidemics cause numerous death and thousands of hospitalization each year. Because of the alarming emergence of resistant to anti-influenza drugs, there is a need to identify new anti-viral therapeutic agents. Viral tropism was stabilized in three mammalian celllines of different origin. The selected celllines are treated with sodium periodate at various concentrations to assess the rate of plaque reduction. Pretreated MDCK cells with sodium periodate at a concentration of 5 mM and 30% of plaque reduction was observed when compared to the untreated group. In A549 and Vero cells the plaque inhibition was found to be 11% and 22% respectively when compared with controls. The ability of influenza A virus (H3N2) binds to cells of canine, human and simian origin are reported here on the basis of cytopathic effect (CPE). H3N2 is more efficiently bound to cells of canine origin and the cytopathic effect was decreased with increasing the evolutionary complexity of the cell lines. The result suggested that dislodging of sialicacid receptors with sodium periodate were inhibiting the binding efficiency of human influenza A virus to mammalian cells.

Key words: Influenza A virus, sialicacid receptor protein, cytopathic effect, sodium periodate.

INTRODUCTION

Influenza A virus is a significant human pathogen that causes annual epidemics in the human population. The only two antigenic subtypes of influenza A virus circulate in human population namely H1N1 and H3N2 (Thompson et al., 2004). Human influenza A virus strain preferentially recognize sialicacid linked via an α-2, 6 glycosidic linkage (2, 6) to the ultimate carbohydrate (Pekosz et al., 2009). The initial step in influenza virus infection is the firm attachment of a virus particle to the target cell surface which is accomplished through the interaction of a glycoprotein found on the viral surface (Hemagglutinin; HA), with cell-surface oligosaccharides containing sialicacids (Eisen et al., 1997). Many viruses recognize specific sugar residues, particularly sulfated or sialylated glycans, as the infection receptors. Avian influenza virus and human influenza virus use different sugar residues as reorganization sites, resulting in different host range of infections. Influenza viruses isolated and propagated

solely in certain mammalian cell lines such as MDCK, Vero, LLC, MK2 and MRC-5 (Katz et al., 1990). It is generally believed that large glycoproteins better exposed on the cell surface are more likely to serve as receptor for the initial virus attachment, whereas subsequent binding to gangliosides could bring the viral and cell membranes into closer proximity and thus, facilitate the membrane fusion which is a mandatory event for the entry of the viral genome into the cell (Matrosovich, 2006).

Influenza viruses are able to replicate in a variety of primary, diploid, and continuous cell cultures (Kilbourne, 1987). Although the susceptibility of most cell lines to influenza virus infection is low, human influenza viruses preferentially attaches to sialic acid (SA) with α-2,6 galactose (α-2,6 Gal) oligosaccharides (Rogers and Paulson, 1983; Carroll and Paulson, 1985) however, the distribution of these receptors on most mammalian cells has not been determined, and their influence on virus attachment and replication is still unclear (Govorkova et al., 1996). Nevertheless some strains of influenza virus A might grow extremely well in A549 cells. (Huang and

*Corresponding author. E-mail: sk_protein@buc.edu.

Turchek, 2000). There are two strategies for blocking the attachment of a virus to the target cell; One is the blocking of the sugar-binding site of HA by peptides13 or Neu5Ac-containing derivatives. (Totani et al., 2003; Tsuchida, et al., 1998; Reuter et al., 1999; Guo, et al., 2002). (Sato et al., 2002) identified HA-binding peptides by using the phage-display system and showed inhibitory activity of the peptides for viral infections. Inhibitory activities of Neu5Ac modified polymers (Totani et al., 2003) dendrimers, and lipids are also reported. These compounds bound to the Neu5Ac-binding site of HA and resulted in the inhibition of HA-Neu5Ac interaction (Tsuchida, et al., 1998; Reuter et al., 1999).

Sodium periodate has the ability to destroy carbohydrate moieties without altering the protein or lipid structures (Stevenson et al., 2004) Lymphocytes that have been transformed by sodium periodate, provide an excellent system for investigating alterations in surface structure. Since the cells have not been coated by foreign protein such as, lectins or antigens, sialicacid form a negatively charged sugar molecules usually found at the ends of oligosaccharides, attached to glycopoteins, glycolipids and proteoglycons. A number of viruses including enveloped and non-enveloped RNA and DNA viruses have been shown to use sialicacids as a component of their cellular receptor (Suzuki et al., 2000). Transduction of Madin-Darby bovine kidney (MDBK) cells pretreated with neuraminidase to remove cell surface sialicacid and with either sodium periodate to remove sialicacid conjugated carbohydrates (Li et al., 2009). The present study was therefore designed to define the interaction of influenza virus (H3N2) with specific population of cells *in vitro* and *in vivo.* Our result provides evidence for sialicacid as a component of influenza virus receptor; further more to find out viral tropism towards various cell lines of different origin.

MATERIALS AND METHODS

Virus

Human Influenza A (H3N2) were obtained from King Institute of Preventive Medicine and Research, Department of Virology, Chennai. It was propagated in MDCK cells as viral stocks.

Cell culture

Continuous MDCK, A549 and Vero cells were grown in minimal essential medium (MEM) contained 10% heat-inactivated fetal bovine serum (FBS) 100 Units/ml penicillin G and 100 µg/ml streptomycin incubated at 37°C and 5% CO_2 for 72 h to get 90% confluency.

Test compound

Sodium periodate were purchased from Himedia chemicals. It was used in different dilutions to assess its cytotoxic concentration (CTC). The compound prepared in ten different dilutions of 0.01 M to 0.1 M concentration for further assay.

Viral sensitivity assay

Confluent (90%) monolayer of MDCK, A549 and Vero cells were grown in six well plates. Ten fold serial dilution of H3N2 were used to infect each of the cell lines in six well plates. CPE was detected by fixation of cell monolayers with 4% formaldehyde in PBS and staining with 0.01% carbomyl fuscin solution.

Determination of effective minimal cytotoxic concentration of sodium periodate

Cytotoxicity of the compound against MDCK, A549 and Vero cells were evaluated in terms of CTC_{50} (50% cytotoxic concentration). MDCK, A549 and Vero cell cultures were exposed to the compound at ten different concentrations 0.01 to 0.1 M. Following 1 h of incubation at 37°C and washing with PBS, After 48 h incubation under the same conditions, the viability of the cells was measured by MTT method. Effective minimal cytotoxic concentration was determined by statistical analyses. The cell viability were assessed by using the formula:

$$\text{Cell viability} = \frac{\text{OD of the treated}}{\text{OD of control}} \times 100$$

Colorimetric MTT assay

Stock MTT (10x), was prepared by dissolving tetrazolium in PBS at pH 7.2 (Phosphate buffer saline) at a concentration of 5 mg/ml and filtered through 0.45 µm of pore size.

The medium of the confluent cells was removed, then 100 µl of 1x MTT was added to each well. Following incubation at 37°C with 5% CO_2 for 2 h, 100 µl of acidic isopropanol was added and mixed to release the colour from the cells. Optical density was measured at 540 nm using ELISA reader (Stat Fax-200) to elevate live cells.

Plaque assay

Confluent MDCK, A549 and Vero cells were grown in six well plate and the cultures were treated subtoxic concentration of sodium periodate 0.01 M. After 1 h incubation 0.5 ml of viral suspension was added to both control and treated plates. Monolayer was inoculated with 0.5 ml of virus dilution, which was adsorbed for 1 h at 36°C. The inoculum was removed and the cells were washed twice with phosphate-buffered saline (pH 7.2) and were covered with 3 ml of an agar medium consisting 100 ml of 0.6% Agarose. After 2 days, a second agar overlay (1.0 ml) containing 1:1000 carbomyl fuscin was added to facilitate plaque counting. The same procedure was followed for A549 and Vero cells and plaque were counted for further analysis. Plaques are counted under microscope and the percentage of plaque reduction calculated by:

$$\% \text{ of Plaque reduction} = \frac{\text{Number of Plaque in treated}}{\text{Number of Plaque in control}} \times 100$$

RESULTS

H3N2 binds efficiently canine origin cells

Human influenza virus A (H3N2) was found to bind with all the selected mammalian cell lines. Among the three, MDCK cell was reported to be more susceptible to H3N2

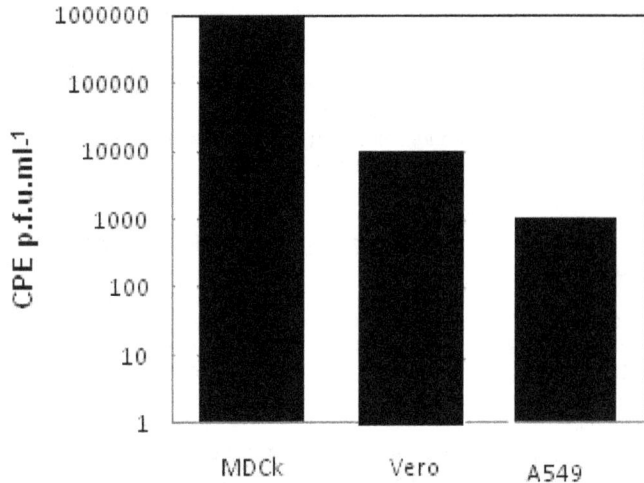

Figure 1. Ten fold serial dilutions of H3N2 were used to infect each of the cell lines in six well plates. CPE was detected by fixation of cell monolayers with 4% formaldehyde in PBS and staining with 0.1% carboyl fuscin. These experiments were repeated four times and one representative set of result is shown.

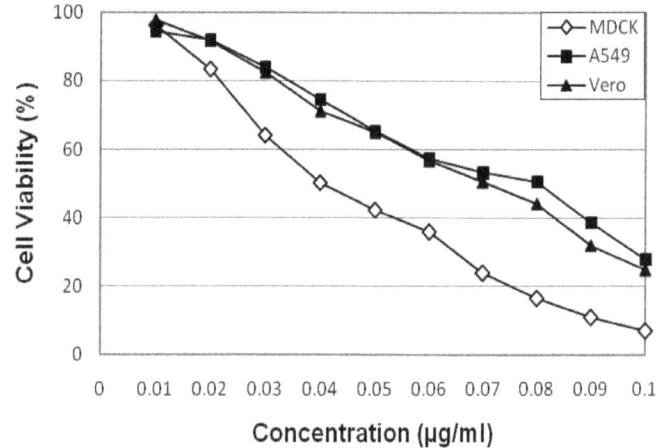

Figure 2. (CTC) effective minimal cytotoxic concentrations of sodium periodate. Optical Density at 540 nm (Y axis) of different dilutions of the compound (X axis) was measured by M.

strain and hence the cytopathic effect was found to be 10 lacks plaque forming units (pfu) per ml of cell culture medium. In the case of Vero and A549 cell lines, the cytopathic effect was expressed as 10,000 and 1000 pfu /ml (Figure 1). This part of result clearly demonstrated that the level of cytopathic effect due to H3N2 viral infection was 100 fold greater than the rest of selected two cells. Thus the cyotopathic effect was well pronounced in canine cells when compared to monkey and human cell lines.

MDCK cells contained numerous sialicacid species on their surface. A549 and Vero cells were observed to minimal availability of these silalicacid epitopes and their cell surface with carbohydrate moiety results negligible rate of cytopathic effect. The H3N2 influenza virus-infected Vero cells shown to have morphological changes in similar to those observed in MDCK cells. It was interesting both MDCK and A549 cells.

Data represented in Figure 2 clearly describing the subtoxic concentration of sodium periodate was deter-mined as 0.1 µg/ml at which the cell viability was recorded. The cytotoxic effect was noticed at very close proximity for Vero and A549 cells, whereas, the MDCK cells observed to easily susceptible to sodium periodate than the rest of the two selected cells.

H3N2 binding / infection requires sialylated carbohydrate moieties

It has been shown that sodium periodate (NaIO4) destroyed carbohydrate moieties by oxidation of vicinal hydroxyl groups of sugars into dialdehydes at acidic pH without altering protein or lipid structures (Stevenson et

al., 2004). To further examine the role of cell surface carbohydrate in H3N2 binding, MDCK, A549 and Vero cells were pretreated with sodium periodate which revealed the different levels of plaque reduction.

Pretreated MDCK cells with sodium periodate at a concentration of 0.005 M 30% inhibited compared to the untreated group. Whereas pretreated A549 and Vero cells with sodium periodate at the same concentration 11% in A549 and Vero cells 22% plaque inhibited com-pared to untreated controls(Figure 3) represents cytotoxic concentration 50 (CTC50) of sodium periodate in selected cell lines. Plaques formed by influenza A virus (H3N2) pre treated with subtoxic concentration of sodium periodate in selected cell lines (Figure. 4).

DISCUSSION AND CONCLUSION

Viruses should penetrate the host cells in order to cause infection. Like most of the enveloped viruses, the influenza virus use receptor binding and fusion as principal route of entry. The HA protein of the virus interact with the host cell sialicacid receptors and enters by receptor mediated endocytosis. Prevention of viral entry is an attractive anti-viral strategy as it can minimize the chance of virus evaluation and subsequent drug resistant ant strain development.

The earliest events in virus infection involve the interaction of virions with cell surface molecules. In this study we examined modulating the cell surface sialcacid receptor carbohydrate moieties with the sodium periodate were inhibiting the efficiency of virion-cell binding. Receptor specificity is an important mechanism governing the susceptibility of cells to virus infection. In the absence of the proper sialic acid receptors, influenza viruses may be unable to bind to the cell surface, thus eliminating the opportunity for productive infection. Although Vero cells

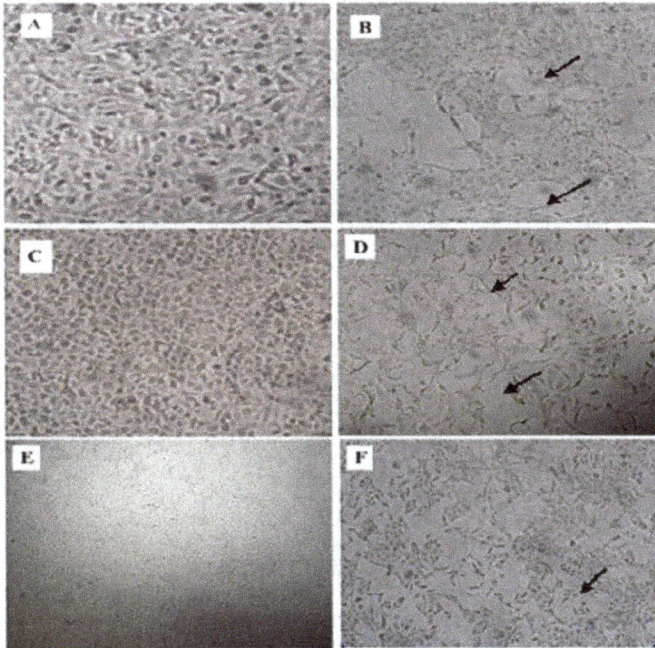

Figure 3. Cytotoxic concentration 50 (CTC $_{50}$) of Sodium periodate in selected cell lines; (A) Untreated control MDCK; (B) 0.04M treated Sodium periodate of MDCK with (CTC$_{50}$); (C) untreated A549; (D)0.08M treated A549 50% cytotoxicity; (E) Control Vero cells; (F) 0.07M exposed CTC$_{50}$ of Vero cells.

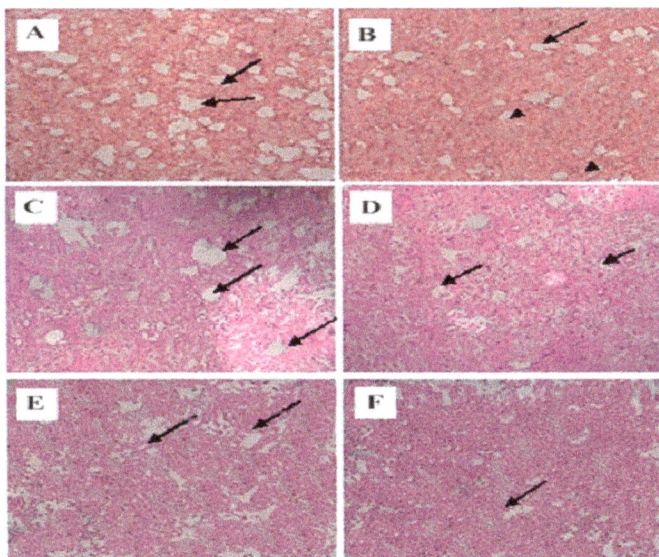

Figure 4. Plaques formed by influenza virus (H3N2) pretreated with subtoxic concentration of sodium periodate in MDCK, A549 and Vero cells respectively. (A) untreated control;(B) 0.01M treated MDCK; (C) untreated A549; (D) 0.01M treated A549; (E) untreated control; (F) 0.01M treated Vero cells.

bore a relatively low level of the NeuAc α2, 6 Gal linkages by comparison with MDCK cells, this relative abundance did not appear to affect their susceptibility to either influenza A

or B viruses.

Govorkova, et al., (1996) finding raises the possibility that linkages other than NeuAc a2, 3 Gal and NeuAc a2, 6 Gal are involved in the attachment of influenza viruses to host cells.

These findings suggested that MDCK cells were very sensitive to sodium periodate it showed more cell death at low concentration of compound. An early study by (Pekosz et al., 2009) showed sialicacid play a vital role in binding of influenza A virus and interaction of HA with saccharides outside the terminal sialicacid. The alteration of the sialicacid and other saccharides can also affect the ability of HA to recognize sialicacid containing carbohydrates (Russel et al., 2006).

Our results also shows that influenza virus binding is dramatically reduced upon pretreatment of selected cells with sodium periodate. Figure 1 demonstrate Neu5Acα 2-6 Gal and Neu5Acα 2-3 Gal containing sialicacid species abundantly bound in MDCK cells (Gambaryan et al., 2005), Hence the H3N2 formed well cytopathic effect even low number of plaque forming units (pfu) of viral suspension. (Pekosz et al., 2009) also attempted to reduce the carbohydrates moieties on the cell surface receptor using O-glyconase and neuraminidase in combination, but the individual had no effect, we found that sodium periodate alone could reduce binding efficiency of H3N2 to its host.

Binging efficiency of H3N2 not only these but also depends upon the carbohydrate moieties present on the cell surface sialicacid receptor. Chu and Whittaker et al., (2004) stated influenza virus entry into susceptible cells appears to be dependent on sialicacid residues attached to N- linked carbohydrates. Therefore there may be a distinction between sialicacid residues that allow for binding of influenza residues that can mediate efficient entry of the virus, other carbohydrate residues besides the terminal sialicacid can contribute significant interacttions with HA that can stabilize and facilitate viral binding (Nicholls et al., 2008).

This experiment shows that entry routes blocks the early viral binding to its receptor and viral fusion. We have proven experimentally that the efficient binding and fusion of H3N2 virus is required carbohydrate moieties present on the cell surface sialicacid receptor. Our study observed the differences in cell specific binding efficiency of H3N2 with selected mammalian cells, increasing evolutionary complexity of the cell lines and to resistant capacity increases against the particular viral strains.

REFERENCES

Carroll SM, Paulson JC (1985). Differential infection of receptor modified host cells by receptor-specific influenza viruses. Virus Res., 3: 165-173.

Chu VC, Whittaker GR (2004). Influenza virus entry and infection require host cell N-linked glycoprotein. Proc. Natl. Acad. Sci., 101: 18153-18158.

Eisen MB, Sabesan S, John J, Skehel S, Wiley DC (1997). Binding of the Influenza A Virus to Cell-Surface Receptors: Structures of Five Hemagglutinin– Sialyloligosaccharide Complexes Determined by

X-Ray Crystallography. Virology, 232: 19-31.

Gambaryan AS, Karasin AI, Tuzikov AB, Chinarev AA, Pazynina GV, Bovin NV, Matrosovich MN, Olsen CW, Kilimov AI (2005). Receptor-binding properties of swine influenza viruses isolated and propagated in MDCK cells. Virus Res., 114: 15-22.

Govorkova EA, Murti G1, Meignier B, Taisne D, Webster RG (1996). African Green Monkey Kidney (Vero) Cells Provide an Alternative Host Cell System for Influenza A and B Viruses. J. Virol., 5519-5524.

Guo CT, Sun XL, Kanie O, Shortridge KF, Suzuki T, Miyamoto D, Hidari KI, Wong CH, Suzuki Y (2002). An O-glycoside of sialic acid derivative that inhibits both hemagglutinin and sialidase activities of influenza viruses. Glycobiology, 12: 183-190.

Huang YT, Turchek BR (2000). Mink Lung Cells and Mixed Mink Lung and A549 Cells for Rapid Detection of Influenza Virus and Other Respiratory Viruses. J. Clin. Microbial., 24: 265-268.

Katz JM, Wang M, Webster RG (1990). Direct sequencing of the HA gene of influenza (H3N2) virus in original clinical samples reveals sequence identity with mammalian cell grown virus. J. Virol., 64: 1808-1811.

Kilbourne ED (1987). Cytopathogenesis and cyto-pathology of influenza virus infection of cells in culture, p. 89–110. In E. D. Kilbourne (ed.), Influenza. Plenum Publishing Corp. New York.

Li X, Bankari DS, Sharma A, Mittal SK (2009). Bovine adenovirus serotypes 3 utilizes sialicacid as a cellular receptor for virus entry. Virology, 392: 162-168.

Matrosovich M, Matrosovich T, Uhlendorff J, Garten W, Klenk HD (2006). Avian-virus-like receptor specificity of the hemagglutinin impedes influenza virus replication in cultures of human airway epithelium. Virology, 361: 384-390

Nicholls JM, Chan RW, Russel RJ, Air GM, Peiris SM (2008). Evolving complexities of influenza virus and its receptors. Trends in Microbiol., 16(4): 149-157.

Pekosz A, Newby C, Bose PS, Lutz A (2009). Sialicacid recognition is a key determinant of influenza A virus tropism in murine trachea epithelial cell cultures. Virology, 386: 61-67.

Reuter JD, Myc A, Hayes MM, Gan Z, Roy R, Qin D, Yin R, Piehler LT, Esfand R, Tomalia DA, Baker JR (1999). Inhibition of viral adhesion and infection by sialic-acid-conjugated dendritic polymers. Bioconjugate Chem., 10: 271-278.

Rogers GN, Paulson JC (1983). Receptor determinants of human and animal influenza virus isolates: differences in receptor specificity of the H3 hemagglutinin based on species of origin. Virology, 127: 361-373.

Russel RJ, Gamblin SJ, Haire LF, Russell RJ, Stevens DJ, Xiao B, Ha Y, Vasisht N, Steinhauer DA, Daniels RS, Elliot A, Wiley DC, Skehel JJ (2006). Avian and human receptor binding by hemagglutinis of influenza A viruses. Glycoconj. J., 23: 85-92.

Sato T, Sumi M, Ogino K, Taki T (2002). Inhibition of influenza virus infection by hemagglutinin-binding peptides. Pept. Sci., 38: 329-330.

Stevenson RA, Huang JA, Studdert MJ, Hartley CA (2004). Sialic acid acts as a receptor for equine rhinitis A virus binding and infection. J. Gen. Virol., 85: 2535-2543.

Suzuki Y, Ito T, Suzuki T, Holland RE, Chambers TM, Kiso M, Ishida H, Kawaoka Y (2000). Sialic acid species as a determinant of the host range of influenza A viruses. J. Virol., 74: 11825-11831.

Thompson WW, Shay DK, Weintraub E, Brammer L, Cox N, Anderson LJ, Fukuda K, Bridges CP (2004). Influenza associated hospitalization in the United States. JAMA, 292: 1333-1340.

Totani K, Kubota T, Kuroda T, Murata T, Hidari KI, Suzuki T, Suzuki Y, Kobayashi K, Ashida H, Yamamoto K, Usui T (2003). Chemoenzymatic synthesis and application of glycopolymers containing multivalent sialyloligosaccharides with a poly- (L-glutamic acid) backbone for inhibition of infection by influenza viruses. Glycobiology, 13: 315-326.

Tsuchida A, Kobayashi K, Matsubara N, Muramatsu T, Suzuki T, Suzuki Y (1998). Simple synthesis of sialyllactose-carrying polystyrene and its binding with influenza virus. Glycoconjugate J., 15: 1047-1054.

28

Immunohistochemical localization and characterization of distinct angiotensin II AT$_1$ and AT$_2$ receptor isoforms in the adrenal glands of the Sprague-Dawley rat and the desert rodent *Meriones crassus*

Al-Qattan, K., Mansour, M. H.* and Al-Naser, M.

Department of Biological Sciences, Faculty of Science, Kuwait University, P. O. Box 5969, Safat 13060, Kuwait.

Employing specific polyclonal anti-AT$_1$ and anti-AT$_2$ antibodies, AT$_1$ and AT$_2$ receptor expression was immunohistochemically demonstrable within adrenal tissues in Sprague-Dawley rats and the desert rodent *Meriones crassus*. Among adrenal cortical zones in rats, AT$_1$ receptor labeling was evident in zona glomerulosa and zona reticularis. In contrast, AT$_1$ receptor labeling was confined to the zona glomerulosa and the deep zona fasiculata in *Meriones crassus*. AT$_1$ receptor labeling was, however, equally observed among ganglion and chromaffin cells constituting the adrenal medulla of both animal models. AT$_2$ receptor labeling was faint in all adrenal regions in rats. However, intensity was high in the deep zona fasiculata, and medullary chromaffin and ganglion cells in *Meriones crassus*. Two-dimensional Western blotting, in the presence or absence of endoglycosidase-F, revealed that structurally distinct spectra of AT$_1$ and AT$_2$ receptor isoforms are expressed in the adrenal tissues of each animal model. These spectra were constituted by molecular isoforms with distinct patterns of charge microheterogeneity unique to each receptor type in each animal model. In both species, heterogeneity of AT$_1$ and AT$_2$ receptor isoforms may be attributed in part to differential post-translational glycosylation mechanisms of the receptor polypeptide backbones, which may be critical in differentially fine-tuning adrenal functions in lab-reared and desert rodents.

Key words: Angiotensin II receptors, adrenal gland, desert rodents, N –linked glycosylation, immunohistochemistry.

INTRODUCTION

The renin-angiotensin system is one of the most potent systems that regulate blood pressure, electrolyte balance, and extracellular fluid volume (Ichihara et al., 2004). The effects of the principal component of this system, angiotensin II (Ang II), are triggered by its interaction with specific receptors in a variety of tissues (Belloni et al., 1998; Johren et al., 2003; Bird and Pattison, 2004). Major advances in understanding the physiological consequences of Ang II-Ang II receptor interactions have been attributed to the development of selective non-peptide Ang II receptor antagonists, which

established the occurrence of at least two major types of Ang II receptors referred to as AT$_1$ and AT$_2$ (Chiu et al., 1994). At a more fundamental level, the structural heterogeneity of these receptor types has been confirmed by cDNA sequence data (Koike et al., 1995), which revealed only 30% homology in the coding regions of both types. Two subtypes of the rodent AT$_1$ receptor (AT$_{1A}$ and AT$_{1B}$) have been identified and sequenced and these have about 96% sequence identity in the coding region (Inagami et al., 1994). The deduced sequence of the rat AT$_1$ receptor shows a protein of 359 amino acids with a molecular weight of 40 kDa, seven putative transmembrane domains, and three potential N-glycosy-lation sites (Murphy et al., 1991). The AT$_2$ receptor is also a seven transmembrane domain receptor, represented by a 363

---

*Corresponding author. E-mail: mheshamshaker@yahoo.com.

amino acid polypeptide with a molecular mass of 41 kDa and five potential N- glycosylation sites located in in the extracellular N-terminal domain (Mukoyama et al., 1993). In a variety of tissues, AT_1 receptors are coupled to ion channels and phosphoinositide breakdown and appear to mediate most of the known physiological effects of Ang II (Capponi, 1996). On the other hand, the coupling of AT_2 receptors has been implicated in modulating guanylate cyclase activity and T-type calcium channels in selective cell types but, as yet, is not fully explored (Dinh et al., 2001; Porrello et al., 2009).

In the adrenal gland, both Ang II receptor types have been localized in different species by a variety of detection methods (Allen et al., 2000). Based on immunohistochemistry (Wang et al., 1998; Giles et al., 1999), in vitro autoradiography (Zhuo et al., 1996; Lehoux et al., 1997) and Northern blot and radioligand binding studies (Gasc et al., 1994; Wang et al., 1998), all mammals studied express both AT_1 and AT_2 receptors in the cortex and medulla, but with marked species-dependent differences observed in the distribution and proportions of both receptors. In the cortex, AT_1 receptors predominate in the zona glomerulosa of all species including rodents, monkeys and humans (Zhuo et al., 1996). Ang II receptors occur in very low or undetectable concentrations in the zona fasciculata and reticularis in most mammals examined except in the human, canine, and bovine adrenals, where a moderate to high density of AT_1 and AT_2 receptors coexists (Ouali et al., 1993; Allen et al., 2000; Bird and Pattison, 2004). In the medulla, moderate levels of AT_1 receptors occur in catecholamine-releasing chromaffin cells of most species examined (Israel et al., 1995; Dinh et al., 2001). In contrast, high levels of AT_2 receptors occur in the medulla of most species, although the density is much lower in humans (Zhuo et al., 1996). The distribution of AT_1 receptors in zona glomerulosa cells of the cortex and chromaffin cells of the medulla in various mammalian adrenal glands is consistent with the known regulatory effects of Ang II on the biosynthesis and release of aldosterone and catecholamines from the adrenal glands (Allen et al., 2000; Jezova et al., 2003). The direct involvement of AT_1 receptor in aldosterone secretion has been particularly strengthened by the observed up-regulation of mRNA and protein levels of this receptor type under conditions of low sodium diet, renovascular hypertension or water deprivation (Giacchetti et al., 1996; Chatelain et al,. 2003). The physiological role of AT_2 receptors in the adrenal gland is, on the other hand, largely unknown, but accumulating evidence suggest that their major role may involve the functional antagonism of the mineralocorticoid-releasing and growth-promoting effects of AT_1 receptors in cortical zona glomerulosa cells (AbdAlla et al., 2001), the stimulating effect in the secretion of endogenous ouabain from cortical cells (Laredo et al., 1997) as well as the synergistic effect with AT_1 receptors in regulating catecholamine synthesis and release by adrenomedullary cells (Mazzocchi et al., 1998; Jezova et al., 2003).

Animals living in arid environments (e.g. the desert rodent Meriones crassus) must deal with major problems of a high ambient temperature, a rapid evaporative water loss, and the scarcity of water. In these animal models, Ang II is expected to play a central widespread role in the regulatory mechanisms for electrolytes and blood volume control. Although the physiological significance of Ang II has been the focus of numerous studies in various animal models, the cellular and molecular bases of Ang II effects on the tissues of desert animals in general, as well as the nature of the receptors involved in particular, are still to be resolved (Al-Qattan et al., 2006). The present study is thus designed to explore, in a comparative context, the tissue distribution pattern and structural characteristics of Ang II AT_1 and AT_2 receptor isoforms expressed in the adrenal gland of the lab-reared Sprague-Dawely (SD) rats and the desert rodent Meriones crassus (M. crassus).

Our investigation is aimed at providing a correlation between the distinctive tissue localization and structural characteristics of the Ang II receptor family, and the differential Ang II- Ang II receptor interactions, putatively operating in lab-reared and desert rodents.

MATERIALS AND METHODS

Animals and reagents

Adult male SD rats (England) weighing 100 g raised at the animal house of the Department of Biological Sciences, Kuwait University were used. Adult males of the desert gerbil M. crassus, Sundevall's jird, weighing 100 g were captured from Kabid area (40 km west of Kuwait City) and used within few days of captivity. Adult female New Zealand rabbits (weighing 2 to 3 kg) were used for immunization. All animals were given standard laboratory chow (170 mmol Na^+/kg) and water ad libitum and kept under standard conditions (23 ± 2°C, 12 h light, 12 h darkness). Animals were treated in strict accordance with the recommendations of the declaration of Helsinki and the guidelines for animal experimentation of Kuwait University, Faculty of Science.

Except where noted, all chemicals were reagent grade and purchased from Sigma Chemical Company (St. Louis, MO, USA). Polyclonal anti-AT_1 receptor antibody (sc-1173, rabbit IgG specific to an epitope mapping within the N-terminal extracellular domain of the human AT_1 polypeptide), polyclonal anti-AT_2 antibody (sc-7421, rabbit IgG specific to an epitope mapping within the N-terminal extracellular domain of the human AT_2 polypeptide) and peroxidase–conjugated goat anti-rabbit IgG antibody were purchased from Santa Cruz Biotechnology, Inc. (Santa Cruz, Ca, USA). Gel electrophoresis reagents, ampholines, peroxidase-conjugated molecular weight and pI standards, and nitrocellulose membranes (0.45 μ) were obtained from BioRad (Richmond, CA, USA). Eight amino acid peptides, corresponding to amino acids 14 to 21 (Ile-Gln-Asp-Asp-Cys-Pro-Lys-Ala) of the first extracellular domain as deduced from the published AT_1 receptor cDNA sequence (Murphy et al., 1991), and corresponding to amino acids 10 to 17 (Thr-Ser-Arg-Asn-Ile-Thr-Ser-Ser) of the first extracellular domain as deduced from the published AT_2 receptor cDNA sequence (Feng et al., 2005), were synthesized manually on the base-labile linker 4-(hydroxymethyl)- benzoyloxymethyl and supplied by The Protein/DNA Technology Center (The Rockefeller University, NY, USA). The octapeptides were conjugated to bovine serum albumin (BSA) and coupled to CNBr-activated Sepharose 4B as described previously (Al-Qattan et al., 2006).

Preparation of adrenal gland sections

The left adrenal gland of anesthetized SD rats and *M. crassus* was excised and placed in 3 ml of Bouin's fixative for 24 to 48 h at room temperature. The tissues were processed for routine paraffin embedding, which included dehydration through a series of ethanol concentrations 50, 70, 90 and 100%, clearing in toluene, embedding in paraffin wax, and finally 3 to 4 μm sections were cut on a rotary microtome. The sections were picked up on clean slides after spreading them in a water bath at 40°C. The slides were air-dried to be used for subsequent staining.

Labeling of adrenal gland sections with anti-AT$_1$ and anti-AT$_2$ receptor antibodies

Tissue sections were examined for AT$_1$ and AT$_2$ receptor distribution by an indirect immunohistochemical labeling technique. Adrenal gland sections were dewaxed in xylene, hydrated with a series of 90, 75 and 60% ethanol and washed with PBS, pH 7.2. Sections were quenched by the addition of 10% normal goat serum and 0.3% hydrogen peroxide in PBS, pH 7.2 for 1 h and then individually labeled for 45 min in a humidified chamber with 300 μl of either the anti-AT$_1$ or anti-AT$_2$ receptor antibody (1:100 dilution, each). After several washes with 200 μl PBS, pH 7.2, the sections were incubated for 45 min with 300 μl of peroxidase-conjugated goat anti-rabbit IgG antibody (diluted to 1:200 in PBS, pH 7.2), followed by a 10 min treatment with 400 μl of 3,3-diaminobenzidine tablets (fast DAB) reconstituted in water. All sections were counter stained with 100 μl haematoxylin (Gill #1) for 1 min and examined by light microscopy for positive labeling of cells expressing AT$_1$ or AT$_2$ receptors. Control sections were identically stained by replacing the specific antibodies with either the anti-AT$_1$ or the anti-AT$_2$ antibody, pre-absorbed with their respective AT$_1$ or AT$_2$ octapeptide/BSA complex-coated CNBr- activated Sepharose 4B beads. Other hydrated sections were stained with haematoxylin for 3 to 4 min and eosin for 2 to 3 min and examined in parallel for comparative purposes by light microscopy. Photomicrographs were taken using Olympus AH-3 automated microscope (Tokyo, Japan), equipped with an Olympus Vanox camera.

Solubilization of adrenal cell-membranes in deoxycholate

Adrenal glands collected from SD rats and *M. crassus* were individually homogenized, solubilized and extracted in 10 mM Tris/HCl, pH 8.0 containing 2 mM phenylmethylsulfonyl fluoride and 2% deoxycholate (Al-Qattan et al., 2006) by an automatic homogenizer followed by sonic disruption and stirring at room temperature for 2 h and three cycles of freezing at -20°C and thawing at room temperature. Solubilized cell membrane lysates were recovered in supernatants following centrifugation of reaction mixtures at 100,000 g for 1 h, and their protein content determined by the method of Lowry et al. (1951) using BSA in the same buffer as a standard.

Endoglycosidase treatments

Adrenal cell-membrane lysates of SD rats and *M. crassus* (120 μg protein) were separately precipitated with 20% trichloroacetic acid and ice-cold acetone for I h at -20°C, washed for 1 h with acetone at -20°C and reconstituted in 50 μl of 100 mM sodium phosphate, pH 6.1, 50 mM EDTA, 1% Nonidet P-40 containing 200 mU of Endo-F (Endo-ß-N-acetylglucosaminidase F, from *Flavobacterium meningosepticum*, 600 U/mg, Sigma Chem. Comp. , St. Louis, MO). Samples were incubated for 18 h at 37°C, precipitated with equal volume of 20% trichloroacetic acid, washed with cold acetone and

dried under nitrogen gas before analysis by polyacrylamide gel electrophoresis. Control samples were similarly treated but in the absence of Endo-F.

Two-dimensional sodium dodecylsulfate-polyacrylamide gel electrophoresis

Aliquots of solubilized cell-membrane lysates (120 μg protein) collected from adrenal tissues of SD rats and *M. crassus*, and either untreated or treated with Endo-F, were individually resolved by two-dimensional (2-D) sodium dodecylsulfate-polyacrylamide gel electrophoresis (SDS-PAGE). Protein samples were reconstituted for 4 h at 37°C in first dimension sample buffer (9.5 M urea, 2% NP-40, 0.4% ampholines "pH 3 to 10" and 1.6% ampholines "pH 5 to 7") and analyzed essentially as described by O'Farrell (1975). First dimension isoelectric focusing tube gels were focused at 750 V (constant voltage) for 3.5 h, equilibrated for 15 min in equilibration buffer (62.5 mM Tris/HCl, pH 6.8, 2.3% SDS and 10% glycerol) and resolved in the second dimension with 12% slab SDS-PAGE using a Bio Rad Mini-Protein II 2-D cell according to Laemmli (1970). Slab gels were equilibrated in electrophoretic transfer buffer (25 mM Tris, 192 mM glycine, pH 8.3 containing 20% methanol) for 15 min for subsequent analyses by Western blotting.

Western blotting

Protein samples in adrenal cell-membrane lysates resolved by 2-D SDS-PAGE were electrophoretically transblotted to 0.45 μ nitrocellulose membranes at 100 V (constant voltage) for 1.5 h at 4°C in electrophoretic transfer buffer, pH 8.3 using a Bio Rad Mini Trans-Blot electrophoretic transfer cell. Nitrocellulose membranes were washed 3 times with 200 mM PBS, pH 7.2 containing 0.05% Tween 40 (PBS – Tween buffer), each for 15 min with constant agitation and nonspecific binding sites blocked by incubation for 1 h in blocking buffer (3% BSA in PBS - Tween buffer, pH 7.2). Membranes were then washed thrice in PBS – Tween buffer, pH 7.2 and subsequently probed by either anti-AT$_1$ or anti-AT$_2$ antibody (each diluted to 1:500 in PBS – Tween buffer, pH 7.2, respectively) by incubation for 2 h at room temperature and then overnight at 4°C with constant agitation. Control blots were prepared by substituting the specific antibodies with either the anti-AT$_1$ or the anti-AT$_2$ antibody, pre-absorbed with their respective AT$_1$ or AT$_2$ octapeptide/BSA complex-coated CNBr-activated Sepharose 4B beads. Following the incubation time, membranes were washed thrice in PBS – Tween buffer, pH 7.2, treated for 1 h at room temperature with peroxidase-conjugated goat anti-rabbit IgG antibody (diluted to 1:1000 in PBS – Tween buffer, pH 7.2) and the reactions visualized by treatments with DAB tablets reconstituted in distilled water. Membranes were allowed to air-dry, photographed and relative molecular weights and pIs estimated using peroxidase-conjugated Bio Rad broad range molecular weight and pI standards, which were analyzed under identical conditions in parallel to the protein samples.

RESULTS

Localization of the AT$_1$ and AT$_2$ receptors in adrenal tissues of SD rat and *M. crassus*

The expression and zonal distribution of the AT$_1$ and AT$_2$ receptors was immunohistochemically investigated and compared in both SD rats and *M. crassus*.

Of the different adrenal areas, light microscopy revealed

that both AT_1 and AT_2 receptor expression was primarily associated with cortical zones, and to a relatively much lesser extent in the medulla, in both animal models. Nonetheless, a uniformed labeling of the two receptor types was only evident within the deep region of the cortical zona fasiculata and the medulla of *M. crassus* (Figure 1). Among adrenal cortical zones in SD rat, intense labeling with the anti-AT_1 receptor antibody was selectively evident in both the zona glomerulosa and the zona reticularis and was also observed among scattered cells within the zona fasiculata, which were otherwise uniformly marked by their less pronounced labeling pattern (Figures 2A and C). Labeling with the anti-AT_2 antibody was selectively prominent in confined areas outlining the capillary sinusoids separating the cell cords of the zona fasiculata, but was notably of faded intensity among all cellular constituents of the three cortical zones (Figures 3A and C). In contrast, labeling of *M. crassus* adrenal with either anti-AT_1 or anti-AT_2 antibodies resulted in a similar pattern, in which apparent labeling was observed within the deep region of the zona fasiculata (Figures 1B and D). Within this area, the labeling appeared to be more intense compared to the less intense labeling observed within the zona glomerulosa, and the much faded staining of the outer region of the zona fasiculata or the zona reticularis, using either antibodies (Figures 2B and D; 3B and D).

Similarly, distinct labeling patterns of the SD rat adrenal medulla were observed with the anti-AT_1 and the anti-AT_2 antibodies. As shown in Figure 2E, labeling of high intensity with the anti-AT_1 was selectively confined to ganglion cells as well as the cell-surface of few chromaffin cells, but was of faded intensity in both cell types with the anti-AT_2 antibody (Figure 3E). Conversely, intense to moderate labeling was equally observed with both antibodies among ganglion cells and the cytoplasm of several, but not all, chromaffin cells constituting the adrenal medulla of *M. crassus* (Figures 2F and 3F). It is noteworthy that none of the specific cortical and medullary labeling patterns were observed with adrenal sections, treated with aliquots of the anti-AT_1 antibody pre-absorbed with an AT_1 octapeptide/BSA complex (Figures 2G and H) or with anti-AT_2 antibody pre-absorbed with an AT_2 octapeptide/BSA complex (Figures 3G and H), of either animal models.

Structural characterization of the AT_1 receptor expressed in SD rat and *M. crassus* adrenal tissues

The anti-AT_1 antibody was utilized (at a dilution 1:500) in probing whole adrenal solubilized proteins (120 µg), which were either untreated or treated with Endo-F and resolved by 2-D Western blotting. As judged by 2-D Western blots conducted in the absence of Endo-F treatments, the reactivity of the anti-AT_1 antibody was selectively targeted towards a major 74.1 kDa component,

in addition to another minor 69.2 kDa component, in both SD rat (Figure 4A) and *M. crassus* (Figure 4B) adrenal lysates. However, each of these components was constituted by a number of equal-sized molecular isoforms that express distinct patterns of charge microheterogeneity in each animal model. In SD rat, the 74.1 kDa component was constituted by two major acidic charge variants with pIs of 6.2 and 6.3 and three less-prominent isoforms with pIs of 5.9, 6.1 and 6.6 (Figure 4A), whereas its counterpart in *M. crassus* resolved into a molecular cluster exhibiting both size and charge microheterogeneity (Figure 4B). This cluster included six 74.1 KDa isoforms with distinct pIs of 5.0, 5.1, 5.2, 5.3, 5.5 and 5.7 in addition to three minor 76.0 kDa charge variants with pIs of 5.7, 6.1 and 6.2. Similarly, the 69.2 kDa component was focused as a single fuzzy spot with a pI of 6.3 in SD rat (Figure 4A), whereas its counterpart in *M. crassus* resolved into a cluster of size and charge variants of 69.2 to 72.3 kDa in the pI range of 5.4 to 5.6 (Figure 4B).

The putative association of the 74.1 and 69.2 kDa components with oligosaccharides was investigated by testing their susceptibility to Endo-F treatments and analyses 2-D Western blotting. Treatments of the SD rat and *M. crassus* adrenal lysates with 200 mU of Endo-F reduced the extensive molecular weight and acidic charge microheterogeneity expressed by the untreated 74.1 kDa and 69.2 kDa components into homogeneous spots, with an apparent shift in molecular weight as well as charge towards relatively more basic pIs, in both animal models (Figures 4C and D). Given the known specificity of Endo-F in cleaving linkages in the core of complex- and high-mannose-type N-linked glycans (Elder and Alexander, 1982), the shift towards a lower molecular and basic pI was consistent with the removal of complex-type glycan units carrying variable acidic moieties. In SD rat, the deglycosylated targets of the anti-AT_1 antibody focused as four 40 to 44 kDa components, each with an identical pI of 7.1 to 7.2 (Figure 4C). Interestingly, the deglycosylated targets of the anti-AT_1 antibody exhibited in *M. crassus* a much simpler pattern and focused as a major 41 kDa and a minor 42 kDa components, each with an identical pI of 7.0 (Figure 4D). It is noteworthy that none of the glycosylated or deglycosylated components were observed in Western blots, of either animal models, analyzed under identical conditions but probed by aliquots of the anti-AT_1 antibody, pre-absorbed with AT_1 octapeptide/BSA complex-coated CNBr-activated sepharose 4B beads.

Structural characterization of the AT_2 receptor expressed in SD rat and *M. crassus* adrenal tissues

The anti-AT_2 antibody was utilized (at a dilution 1:500) in probing whole adrenal solubilized proteins (120 µg), which were either untreated or treated with Endo-F and

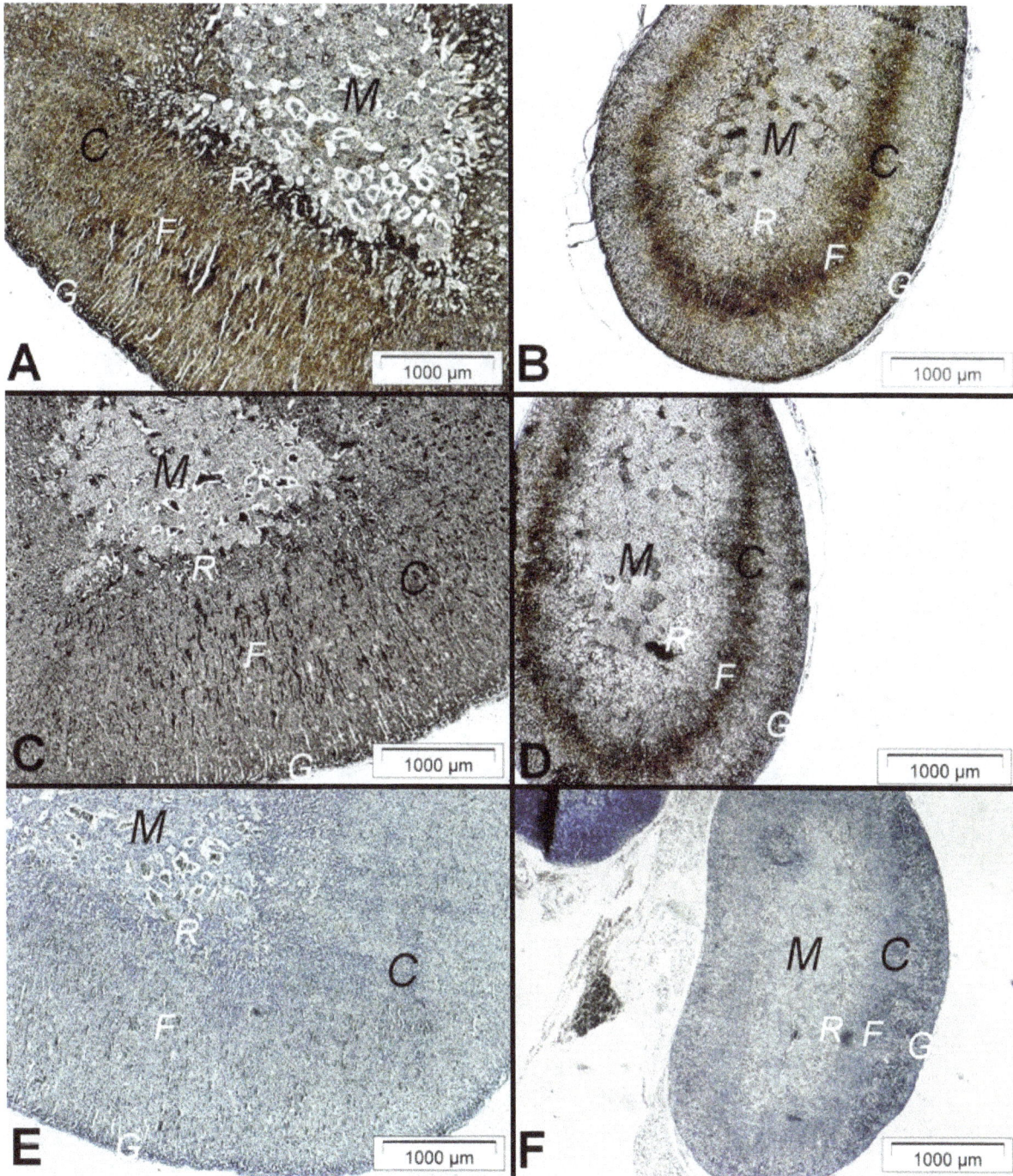

Figure 1. Photomicrograph of SD rat (A, C, E) and *M. crassus* (B, D, F) adrenal glands. Immunohistochemical distribution patterns of AT_1 (A, B) and AT_2 (C, D) receptors. (C) cortex, (G) zona glomerulosa, (F) zona fasciculata, (R) zona reticularis, (M) medulla. No specific labeling with either the anti-AT_1 or the anti-AT_2 antibodies pre-absorbed with their respective AT_1 or AT_2 octapeptide/BSA complex-coated CNBr-activated Sepharose 4B beads (E, F). X40, Bar = 1000 μm.

resolved by 2-D Western blotting. Figure 5 shows that, in the absence of Endo-F, the reactivity of the polyclonal anti-AT_2 receptor antibody were selectively targeted towards a 71.3 kDa component, in both SD rat and *M. crassus*. This component was constituted by five equal-sized acidic isoforms exhibiting charge microheterogeneity in

the pI range of 5.2 to 5.6 (Figure 5A), whereas its counterpart in *M. crassus* resolved into a molecular cluster exhibiting both size and charge microheterogeneity (Figure 5B). This cluster included five 70.2 kDa acidic isoforms in the pI range of 5.0 to 5.4, in addition to 71.3, 72.1 and 77.4 kDa components, each being focused as a fuzzy

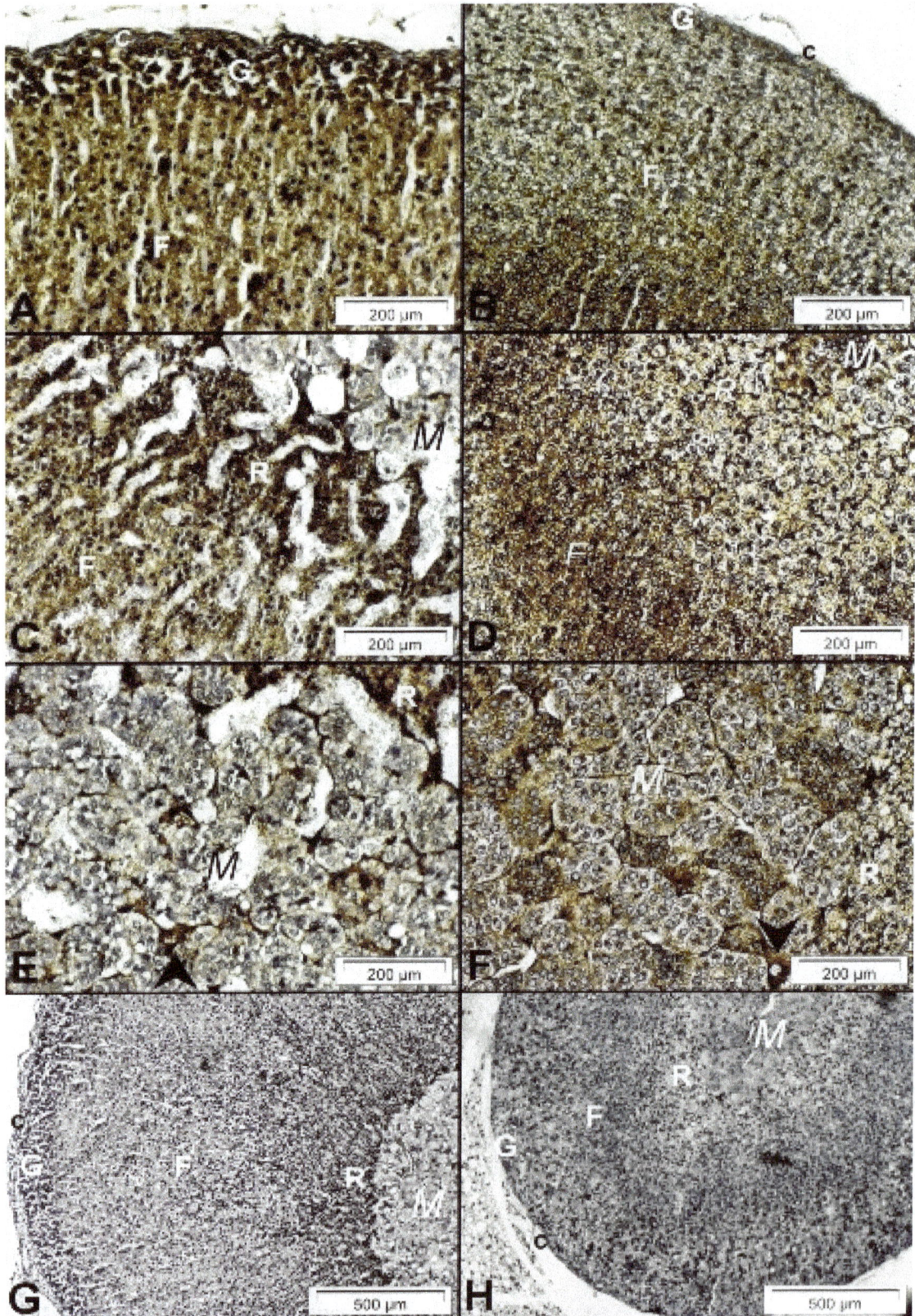

Figure 2. Immunohistochemical localization of the AT_1 receptor in the adrenal tissues of SD rat (A, C, E; X200, G; X100) and *M. crassus* (B, D, F; X200, H; X100). (*c*) capsule, (C) cortex, (G) zona glomerulosa, (F) zona fasciculata, (R) zona reticularis, (M) medulla, (arrowhead) ganglion cells. No specific labeling was observed in the adrenal of SD rat (G) or *M. crassus* (H) labeled with the anti-AT_1 antibody pre-absorbed with an AT_1 octapeptide/BSA complex. A-F: Bar = 200 μm, G and H: Bar = 500 μm.

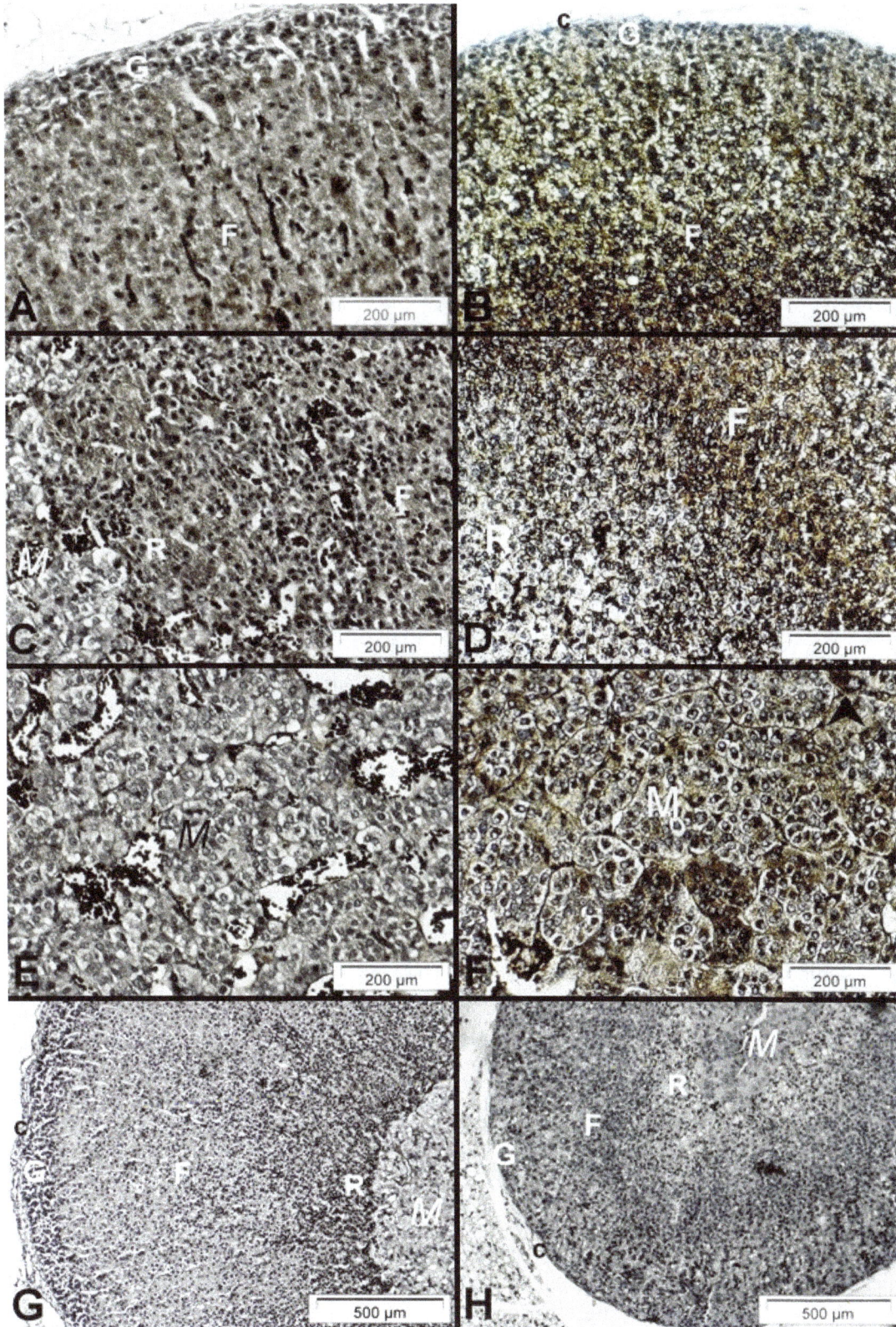

Figure 3. Immunohistochemical localization of the AT$_2$ receptor in the adrenal tissues of SD rat (A, C, E; X200, G; X100) and *M. crassus* (B, D, F; X200, H; X100). (*c*) capsule, (C) cortex, (G) zona glomerulosa, (F) zona fasciculata, (R) zona reticularis, (M) medulla, (arrowhead) ganglion cells. Note the nonspecific labeling of erythrocytes located within cortical and medullary sinusoids in SD rat adrenal (C, E). No specific labeling was observed in the adrenal of SD rat (E) or *M. crassus* (F) labeled with the anti- AT$_2$ antibody preabsorbed with an AT$_2$ octapeptide/BSA complex. A-F: Bar = 200 μm, G and H: Bar = 500 μm.

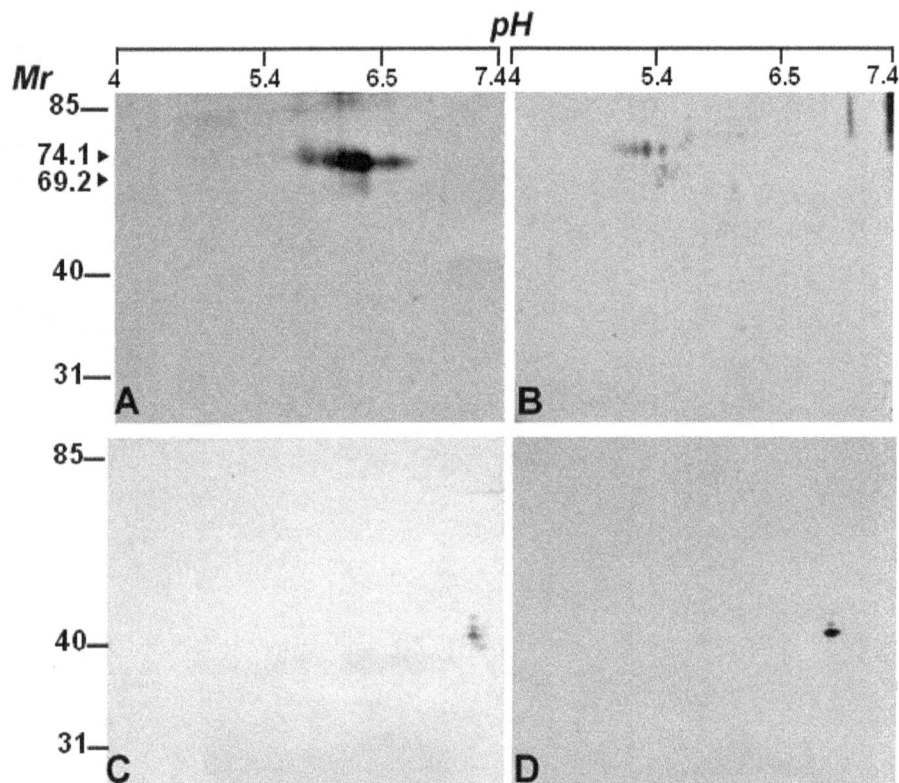

Figure 4. Western blots of two-dimensional SDS-PAGE analysis of AT_1 receptor isoforms expressed in SD rat (A and C) and *M. crassus* (B and D) adrenal glands. (A and B) untreated, (C and D) Endo-F-treated. Positions of Bio Rad 2-D SDS-PAGE isoelectric points and molecular weight (Mr X 10^{-3}) standards are indicated. Also shown are estimated molecular weights of the glycosylated AT_1 receptor isoforms.

spot with a pI of about 5.4. In addition, a 66.8 kDa component, which focused as multiple spots within the pI ranges of 5.2 to 5.6, and a 58.2 kDa component, which also focused as multiple spots within the pI ranges of 5.0 to 5.5, were observed in SD rat and *M. crassus*, respectively. Following treatments with Endo-F, the microheterogeneous pattern exhibited by the 71.3 kDa, in addition to either the 66.8 kDa or the 58.2 kDa components, disappeared and was replaced by two 42 kDa charge variants with pIs of 6.7 and 7.0 in SD rat (Figure 5C) and two 41 kDa spots with pIs of 6.8 and 7.2 in *M. crassus* (Figure 5D), which apparently represent the completely deglycosylated AT_2 receptor forms in both animal models. In both SD rat and *M. crassus*, none of the glycosylated or deglycosylated components were observed in Western blots analyzed under identical conditions but probed by the anti-AT_2 antibody pre-absorbed with AT_2 octapeptide/BSA complex-coated CNBr- activated Sepharose 4B beads.

DISCUSSION

Employing specific polyclonal anti-AT_1 and anti-AT_2 antibodies,

selective zonal expression patterns of the AT_1 and AT_2receptor types were immunohistochemically evident in the adrenal gland of both SD rats and *M. crassus*. These specific labeling patterns were not detectable utilizing either antibodies, pre-absorbed with their respective synthetic AT_1 and/or AT_2 receptor octapeptides, thus confirming the specificity of both antibodies in binding their respective Ang II receptor type. Within the adreno-cortical region of both animal models, the relatively more intense AT_1 receptor labeling among zona glomerulosa cells was paralleled by the faded labeling of the AT_2 receptor type. This imbalanced expression of both receptor types within this zone is in direct agreement with previous reports on different mammalian species, including humans (Zhuo et al., 1996), employing immunohistochemical (Giles et al., 1999) or radioligand binding and autoradiography (Lehoux et al., 1997) to detect the expressed polypeptide, or *in situ* hybridization and PCR techniques detecting intra-cytoplasmic mRNA (Gasc et al., 1994; Wang et al., 1998) of both receptors. In all incidences, irrespective of the species tested or variations in detection methods, the zona glomerulosa seemed to be the common structural denominator for high density AT_1 receptor expression,

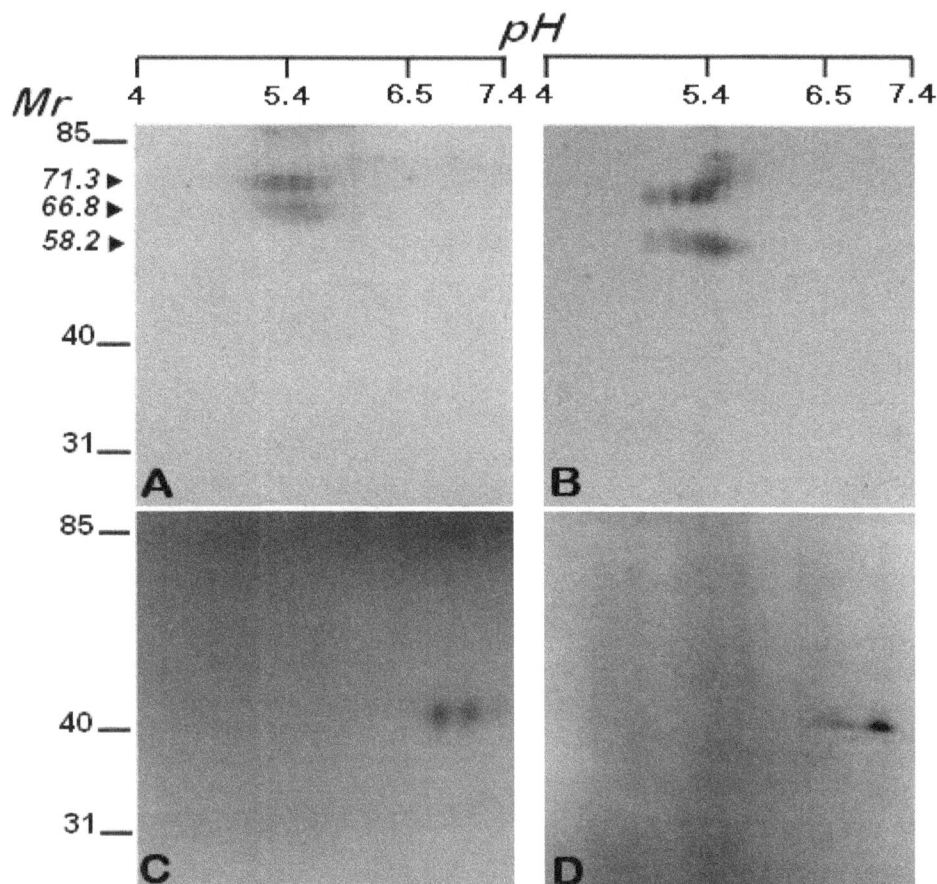

Figure 5. Western blots of two-dimensional SDS-PAGE analysis of AT_2 receptor isoforms expressed in SD rat (A and C) and *M. crassus* (B and D) adrenal glands. (A and B) untreated, (C and D) Endo-F-treated. Positions of Bio Rad 2-D SDS-PAGE isoelectric points and molecular weight ($Mr \times 10^{-3}$) standards are indicated. Also shown are estimated molecular weights of the glycosylated AT_2 receptor isoforms.

and may thus be confirmed as the primary adrenal site for the regulatory role of Ang II in the synthesis and release of aldosterone (Gupta et al., 1995; Kakiki et al., 1997; Belloni et al., 1998) in all mammals, including desert rodents.

As suggested earlier, the putative significance of the AT_2 receptor, expressed with low density in the zona glomerulosa, in either counterbalancing the AT_1–mediated aldosterone-releasing and/or mitogenic func-tions (Bottari et al., 1992; Nakajima et al., 1995; Tanabe et al., 1998; AbdAlla et al., 2001), or in targeting the Ang II-mediated secretion of endogenous ouabain (Laredo et al., 1997; Shah et al., 1998), may also not be excluded.

As revealed in the present investigation, a clear inter-species difference in AT_1 and AT_2 receptor expression was observed among cells within the zona fasiculata/zona reticularis. In these deep cortical zones, the low to moderate expression of the two receptor types in SD rats was contrasted by their intense co-expression in the deep region of the zona fasiculata in *M. crassus*. As reported in previous studies (Zhuo et al., 1996; Allen et al., 2000;

Dinh et al., 2001), the observed low expression of both receptors in the zona fasiculata/zona reticularis of SD rats is typical of rodents, as well as most mammals studied, where Ang II signaling through either receptor types is apparently not directly implicated in regulating corticosterone synthesis or release (Lehoux et al., 1997; Tanabe et al., 1999; Aguilar et al., 2004). However, apparently these cortical zones are targets of inter-species discrepancies, since as observed in *M. crassus*, and contrary to most mammals, cells of both zones express high density of both AT_1 and AT_2 receptors also in canine, bovine as well as human adrenal glands (Ouali et al., 1993; Zhuo et al., 1996; Allen et al., 2000). Interestingly, the level of expression of both receptor types, within these zones, seems to be selectively prone to various modulatory signals in lab-reared rats stimulated by different physiological conditions. As revealed by immune-flourescence, autoradiography and *in situ* hybridization, an up-regulatory shift in both AT_1 and AT_2 receptor expression within the zona fasiculata/zona reticularis was observed in rats kept under dietary sodium restriction

(Lehoux et al., 1997; Wagner et al., 1998). However, within these cortical zones, this up-regulatory shift was selective for the AT_1 receptor, with no alterations in the level of expression of the AT_2 receptor, in rats subjected to sodium loading (Sun et al., 1996), stimulated levels of endogenous Ang II (Wagner et al., 1998) or water deprivation (Chatelain et al., 2003). On the other hand, in aldosterone-treated rats, the highest adrenal AT_1 receptor expression was shifted from the zona glomerulosa to the zona fasciculata, and was paralleled by a significant reduction in the level of expression of the AT_2 receptor in all regions of the gland (Wang et al., 1998). Collectively, within deep adrenal cortical zones, the exact functional significance of this species-dependent or experimentally-induced variability in levels of the expressed Ang II receptor type, is still to be elucidated. Nonetheless, given the suggested implication of an Ang II-dependent phospholipase D-mediated cortisol secretion in bovine zona fasiculata cells (Rabano et al., 2004), its is tempting to speculate that the observed high level of AT_1 and AT_2 receptors, co-expressed in the zona fasiculata/zona reticularis in M. crassus, may represent a significant adaptive variant in selectively shaping Ang II-mediated adrenal functions in rodents coping with a stressful arid environment.

In accordance with previous reports (Israel et al., 1995; Zhuo et al., 1996; Dinh et al., 2001), the expression of both AT_1 and AT_2 receptors was demonstrable among selective chromaffin and ganglion cell bodies in the adrenal medulla of both SD rat and M. crassus. However, in SD rat, the labeling of both receptor types was confined to a relatively few number of chroma-ffin/ganglion cells, and was marked by a high intensity labeling of the AT_1 receptor compared to the faded labeling of the AT_2 receptor type. In M. crassus, labeling of both receptors was equally of a moderate-high intensity and involved more frequent, but not all, chromaffin/ganglion cells. Although the predominant biological effects of Ang II within the adrenal medulla is believed to be mediated primarily via the AT_1 receptor type (Wong et al., 1990; Armando et al., 2001), early autoradiographic (Israel et al., 1995; Allen et al., 2000) and immunoflourescence (Jezova et al., 2003) studies have indicated that the AT_1 receptor represents only 5 to 10% of the total number of Ang II receptors and that, with the exception of humans (Zhuo et al., 1996), AT_2 receptors predominate in the medulla of all mammals, including rats.

Compared to these reports, the discrepancy in the ratio of the expressed AT_1/AT_2 receptors in the medulla, as observed in SD rat in the present immunohistochemical investigation, may be attributed to variations in the sensitivity and specificity of the detection methods and/or the utilized antibodies. Nonetheless, both receptor types have been suggested to act synergistically, yet through independent pathways, in regulating tyrosine hydroxylase transcription, and subsequently controlling basal and stress-induced adrenomedullary catecholamine synthesis and release (Jesova et al., 2003; Armando et al., 2004).

Indeed, the distinct pattern of AT_1/AT_2 receptor expression observed in M. crassus, as well as humans (Cavadasa et al., 2003), still suggest that this ratio may be governed by species-dependent mechanisms, which may prove significant in balancing the proposed AT_1/AT_2 receptor cross-talk to regulate secretagogue activities of the adrenal medulla under different physiological/environmental conditions.

A novel observation of the present investigation was stemmed from the structural assessment of the expressed AT_1 and AT_2 receptors by 2-D Western blotting. Employing the anti-AT_1 receptor antibody in this assay, in the presence and absence of Endo-F, revealed that the reactivity of the antibody was selectively targeted towards a major 74.1 kDa fully-glycosylated component, in addition to another minor 69.2 kDa partially-glycosylated component, in both SD rat and M. crassus. Under similar conditions, the reactivity of the polyclonal N-19 anti-AT_2 receptor antibody was selectively targeted towards a fully-glycosylated 71.3 kDa component, in addition to another partially-glycosylated 66.8 kDa component in SD rat or a 58.2 kDa component in M. crassus. However, each of these components was apparently constituted by a number of molecular isoforms that express distinct patterns of charge and/or size micro heterogeneity unique to each receptor type and/or animal model. The estimated range of molecular weight for the constituent AT_1 and AT_2 receptor isoforms in both species was comparable to the predominant 60 to 78 KDa Ang II receptors previously detected in the adrenal gland (Belloni et al., 1998), as well as other tissues in different species (Marsigliante et al., 1996; Servant et al., 1996; Giles et al., 1999; Al-Qattan et al., 2006), thus confirming the specificity of the antibodies used in this study in binding their respective Ang II receptor type. Given that the anti-AT_1 receptor antibody used is not subtype-specific, the expected co-detection of the structurally homologous AT_{1A} and AT_{1B} receptor subtypes (Sandberg et al., 1992; Inagami et al., 1994) may account, at least in part, for the micro heterogeneous pattern observed in SD rat AT_1 receptors. The putative occurrence of a subtype heterogeneity for the AT_2 receptor as reported in early reports (Reagan et al., 1996) has been recently ruled-out (Feng et al., 2005), and thus may not contribute to the micro heterogeneous pattern observed in SD rat AT_2 receptors. Nonetheless, the reported inter-species conservation of sequences within N-terminal extra-cellular domains of either AT_1 or AT_2 receptor types, as revealed in various mammalian AT_1 and AT_2 receptor cDNAs (Inagami et al., 1994; Feng and Douglas, 2000), may provide the basis for the detection of cross-reacting homologues to AT_1 and AT_2 polypeptides in M. crassus.

As revealed by Endo-F treatments, and given its known specificity in cleaving linkages in the core of N-linked glycans (Elder and Alexander, 1982), it seems plausible that the differential addition of high-mannose type or complex type N-linked glycan units, of different sizes and compositions, is a major source of the intra- and inter-species structural variability of the expressed AT_1 and AT_2 receptor isoforms. This assumption is supported by the observed collapsed micro heterogeneity, as well as the obvious shift towards lower molecular weight, following the enzymatic deglycosylation of the mature, fully-expressed AT_1 and AT_2 receptors in both animal models. Also supportive of this assumption is the documented presence of three (Murphy et al., 1991) and five (Servant et al., 1996), highly-conserved N-linked glycosylation sites along the AT_1 and AT_2 receptor sequences, respectively. Interestingly, the deglycosylated AT_1 receptor forms in SD rat were represented by a pattern of multiple components in the molecular weight range of 40 to 44, which are likely representing allelic variants corresponding to the structurally-related subtypes known to constitute the AT_1 receptor family (Dinh et al., 2001). In contrast, AT_1 subtype heterogeneity was apparently limited in *M. crassus*, since the deglycosylated receptor pattern was represented by only a major 41 kDa and a minor 42 kDa components. In both animal models, the deglycosylated AT_2 receptor pattern was represented by two 41 to 42 kDa components, which may reflect the lack of subtype heterogeneity of the AT_2 receptor gene, as previously suggested by Feng et al. (2005). It is noteworthy that the predominant molecular weight range of 41 to 42 kDa estimated for the deglycosylated receptors was in direct agreement with the predicted 41 to 42 kDa of the protein back-bones of cDNA sequences of both AT_1 and AT_2 receptor types, in various mammals (Murphy et al., 1991; Mukoyama et al., 1993; Feng et al., 2005).

The observed expression of structurally-distinct AT_1 and AT_2 receptor isoforms, dictated genetically as well as by differential degrees of glycosylation, clearly supports early biochemical and pharmacological studies indicating the extensive heterogeneity of Ang II receptors in different tissues and/or species (Chiu et al., 1994; Zhuo et al., 1996; Dinh et al., 2001). Apparently, differential tissue-specific and/or species-specific post-translational regulatory mechanisms would be decisive in the expression of differentially glycosylated AT_1 and AT_2 receptor isoforms within the adrenal tissue of SD rat, and may account for the structurally unique spectra of AT_1 and AT_2 receptor isoforms observed in *M. crassus*. The potential for the differential glycosylation state to alter the optimal expression of selective isoforms of either receptor types (Servant et al., 1996; Jayadev et al., 1999) may have important implications in regulating the various adaptive physiological responses mediated by Ang II-receptor interactions in key tissues. In particular, the variable pattern of expression of distinct AT_1 and AT_2 receptor isoforms in the adrenal may add another level of complexity to mechanisms controlling the synthesis and release of aldosterone, cortisol and catecholamines in different species. In this regard, the present observations are indicative of a significant role of AT_1 and AT_2 receptor isoform spectra in differentially fine-tuning adrenal functions in water/electrolyte homeostasis and stress responses in lab-reared and desert rodents.

ACKNOWLEDGEMENT

This work was supported by Research Project # SL 05/08 funded by Kuwait University.

REFERENCES

AbdAlla S, Lother H, Abdel-tawab AM, Quitterer U (2001). The angiotensin II AT_2 receptor is an AT_1 receptor antagonist. J. Biol. Chem., 276: 39721-39726.

Aguilar F, Lo M, Claustrat B, Saez JM, Sassard J, Li JY (2004). Hypersensitivity of the adrenal cortex to trophic and secretory effects of angiotensin II in Lyon genetically-hypertensive rats. Hypertension, 43(1): 87-93.

Allen AM, Zhuo J, Mendelsohn FAO (2000). Localization and function of angiotensin AT_1 receptors. Am. J. Hypertens., 13: 31S-38S.

Al-Qattan K, Al-Akhawand S, Mansour MH (2006). Immunohistochemical localization of distinct angiotensin II AT_1 receptor isoforms in the kidneys of the Sprague-Dawley rat and the desert rodent *Meriones crassus*. Anat. Histol. Embryol., 35: 130-138.

Armando I, Carranza A, Nishimura Y, Hoe KL, Barontini M, Terron JA, Falcon-Neri A, Ito T, Jourio AV, Saavedra JM (2001). Peripheral administration of and angiotensin II AT_1 receptor antagonist decreases the hypothalamic-pituitary-adrenal response to isolation stress. Endocrinol., 142: 3880-3889.

Armando I, Jezova M, Bregonzio C, Baiardi G, Saavedra JM (2004). Angiotensin II AT1 and AT2 receptor types regulate basal and stress-induced adrenomedullary catecholamine production through transcriptional regulation of tyrosine hydroxylase. Ann. N. Y. Acad. Sci., 1018: 302-309.

Belloni AS, Andreis PG, Macchi V, Gottardo G, Malendowicz LK, Nussdorfer GG (1998). Distribution and functional significance of angiotensin-II AT1- and AT2-receptor subtypes in the rat adrenal gland. Endocr. Res., 24: 1-15.

Bird IM, Pattison JC (2004). Expression of AT1-R in marmoset whole adrenal glands and adrenocortical cells in culture. Endocr. Res., 30: 753-757.

Bottari SP, King IN, Reichilin S, Dahlstroem I, Lydon N, de Gasparo M (1992). The angiotensin AT2 receptor stimulates protein tyrosine phosphatase activity and mediates inhibition of particulate guanylate cyclase. Biochem. Biophys. Res. Commun., 183: 206-211.

Capponi AM (1996). Distribution and signal transduction of angiotensin II AT_1 and AT_2 receptors. Blood Pressure, 2: 41-46.

Cavadasa C, Granda D, Mosimannb F, Cotrimc MD, Ribeirod CF, Brunnera HR, Grouzmanna E (2003). Angiotensin II mediates catecholamine and neuropeptide Y secretion in human adrenal chromaffin cells through the AT1 receptor. Reg. Pept., 111: 61-65.

Chatelain D, Montel V, Dickes-Coopman A, Chatelain A, Deloof S (2003). Trophic and steroidogenic effects of water deprivation on the adrenal gland of the adult female rat. Reg. Pept., 110: 249-255.

Chiu AT, Smith RD, Timmermans PBMWM (1994). Defining angiotensin receptor subtypes. In "Angiotensin receptors" Ed by JM Saavedra, PBMWM Timmermans, Plenum Press, New York, pp. 49-63.

Dinh DT, Frauman AG, Johnston CI, Fabiani ME (2001). Angiotensin receptors: distribution, signaling and function. Clin. Sci., 100: 481-492.

Elder JH, Alexander S (1982). Endo-β-N-acetylglucosaminidase F: Endoglycosidase from Flavobacterium meningosepticum that cleaves high mannose and complex glycoproteins. Proc. Natl. Acad. Sci., USA, 79: 4540-4549.

Feng YH, Douglas JG (2000). Angiotensin receptors: An overview. In "Angiotensin II receptor antagonists" Ed by M Epstein, H Brunner, Hanley and Belfus, Philadelphia, pp. 29-48.

Feng YH, Zhou L, Sun Y, Douglas JG (2005). Functional diversity of AT_2 receptor orthologues in closely related species. Kidney Int., 67(5): 1731-1738.

Gasc JM, Shanmugam S, Sibony M, Corvol P (1994). Tissue-specific expression of type 1 angiotensin II receptor subtypes. An in situ hybridization study. Hypertension, 24: 531-537.

Giacchetti G, Opocher G, Sarzani R, Rappelli A, Mantero F (1996). Angiotensin II and the adrenal. Clin. Exp. Pharmacol. Physiol., 3: S119-S124.

Giles EM, Fernley RT, Nakamura Y, Moeller I, Aldred GP, Ferraro T, Penschow JD, McKinley MJ, Oldfield BJ (1999). Characterization of a specific antibody to the rat angiotensin II AT_1 receptor. J. Histochem. Cytochem., 47: 507-515.

Gupta P, Franco-Saenz R, Mulrow PJ (1995). Locally generated angiotensin II in the adrenal gland regulates basal, corticotropin-, and potassium-stimulated aldosterone secretion. Hypertension, 25: 443-448.

Ichihara A, Kobori H, Nishiyama A, Navar G (2004). Renal renin-angiotensin system. Contrib. Nephrol., 143: 117-130.

Inagami T, Guo DF, Kitami Y (1994). Molecular biology of angiotensin receptors: an overview. J. Hypertens., 12(10): S83-S94.

Israel A, Stromberg C, Tsutsumi K, GarridoMdel R, Torres M, Saavedra JM (1995). Angiotensin II receptor subtypes and phosphoinositide hydrolysis in rat adrenal medulla. Brain Res. Bull., 38: 441-446.

Jayadev S, Smith RD, Jagadeev G, Baukal AJ, Hunyady LS, Catt KJ (1999). N-linked glycosylation is required for optimal AT_{1a} angiotensin receptor expression in Cos-7 cells. Endocrinol. 140(5): 2010-2017.

Jezova M, Armando I, Bregonzi C, Yu Z-X, Qian S, Ferrans VJ, Imboden H, Saavedra JM (2003). Angiotensin II AT_1 and AT_2 receptors contribute to maintain basal adrenomedullary norepinephrine synthesis and tyrosine hydroxylase transcription. Endocrinology, 144(5): 2092-2101.

Johren O, Golsch C, Dendorfer A, Qadri F, Hauser W, Dominiak P (2003). Differential expression of AT1 receptors in the pituitary and adrenal gland of SHR and WKY. Hypertension, 41: 984-990.

Kakiki M, Morohashi K, Nomura M, Omura T, Horie T (1997). Regulation of aldosterone synthase cytochrome P450 (CYP11B2) and 11 beta-hydroxylase cytochrome P450 (CYP11B1) expression in rat adrenal zona glomerulosa cells by low sodium diet and angiotensin II receptor antagonists. Biol. Pharm. Bull., 20(9): 962-968.

Koike G, Winer ES, Horiuchi H, Browm DM, Szpirer C, Dzau VJ, Jacob HJ (1995). Cloning, characterization and genetic mapping of the rat type 2 angiotensin II receptor gene. Hypertension, 26: 998-1002.

Laemmli UK (1970). Cleavage of structural protein during the assembly of the head of bacteriophage T4. Nature, 227: 680-685.

Laredo J, Shah JR, Lu ZR, Hamilton BP, Hamlyn JM (1997). Angiotensin II stimulates secretion of endogenous ouabain from bovine adrenocortical cells via angiotensin type 2 receptors. Hypertension, 29: 401-407.

Lehoux J-G, Bird IM, Brier N, Martel D, Ducharme L (1997). Influence of dietary sodium restriction on angiotensin II receptors in rat adrenals. Endocrinology, 138: 5238-5245.

Lowry OH, Rosebrough NJ, Farr AL, Randall RJ (1951). Protein measurement with the folin phenol reagent. J. Biol. Chem., 193: 265-275.

Marsigliante S, Muscella A, Vilella S, Nicolardi G, Ingrosso L, Ciardo V, Zonno V, Vinson GP, Ho MM, Storelli C (1996). A monoclonal antibody to mammalian angiotensin II AT1 receptor recognizes one of the angiotensin II receptor isoforms expressed by the eel (*Anguilla anguilla*). J. Mol. Endocrinol., 16(1): 45-56.

Mazzocchi G, Gottardo G, Macchi V, Malendowicz LK, Nussdorfer GG (1998). The AT2 receptor-mediated stimulation of adrenal catecholamine release may potentiate the AT1 receptor-mediated aldosterone secretagogue action of angiotensin-II in rats. Endocr. Res., 24(1): 17-28.

Mukoyama M, Nakajima M, Horiuchi M, Sasamura H, Pratt RE, Dzau VJ (1993). Expression cloning of type 2 angiotensin II receptor reveals a unique class of seven-transmembrane receptors. J. Biol. Chem., 268: 24539-24542.

Murphy TJ, Alexander RW, Griendling KK, Runge MS, Beinstein KE (1991). Isolation of a cDNA encoding the vascular type 1 angiotensin II receptor. Nature, 351: 233-236.

Nakajima M, Hutchinson HG, Fujinaga M, Hayashida W, Morishita R, Zhang L, Horiuchi M, Pratt RE, Dzau VJ (1995). The angiotensin II type 2 (AT_2) receptor antagonizes the growth effects of the AT1 receptor: gain-of-function study using gene transfer. Proc. Natl. Acad. Sci., USA, 92: 10663-10667.

O'Farrell PH (1975). High resolution two-dimensional electrophoresis of proteins. J. Biol. Chem., 250: 4007-4021.

Ouali R, Leberthon MC, Saez JM (1993). Identification and characterization of angiotensin-II receptor subtypes in cultured bovine and human adrenal fasciculata cells and PC 12 W cells. Endocrinology, 133: 2776-2772.

Porrello ER, Delbridge LM, Thomas WG (2009). The angiotensin II type 2 (AT2) receptor: an enigmatic seven transmembrane receptor. Front. Biosci., 14: 958-972.

Rábano M, Peña A, Brizuela L, Macarulla JM, Gómez-Muñoz A, Trueba M (2004). Angiotensin II-stimulated cortisol secretion is mediated by phospholipase D. Mol. Cell Endocrinol., 222(1-2): 9-20.

Reagan LP, Yee DK, He PF, Fluharty SJ (1996). Heterogeneity of angiotensin type 2 (AT2) receptors. Adv. Exp. Med. Bio., 396: 199-208.

Sandberg K, Ji H, Clark AJ, Shapira H, Catt KJ (1992). Cloning and expression of a novel angiotensin II receptor subtype. J. Biol. Chem., 267(14): 9455-9458.

Servant G, Dudley DT, Escher E, Guillimette G (1996). Analysis of the role of N-glycosylation in cell-surface expression and binding properties of angiotensin II type-2 receptor of rat pheochromocytoma cells. Biochem. J., 313: 297-304.

Shah JR, Laredo J, Hamilton BP, Hamlyn JM (1998). Different signaling pathways mediate stimulated secretions of endogenous ouabain and aldosterone from bovine adrenocortical cells. Hypertension, 31: 463-468.

Sun JZ, Han B, Liu DQ, Ding JF, Chang C (1996). Effects of high salt-loading on the regulation of angiotensin II receptor mRNA expression. Sheng. Li. Xue. Bao., 48(4): 361-367.

Tanabe A, Naruse M, Arai K, Naruse K, Yoshimoto T, Seki T, Imaki T, Miyazaki H, Zeng ZP, Demura R, Demura H (1998). Gene expression and roles of angiotensin II type 1 and type 2 receptors in human adrenals. Horm. Metab. Res., 30(8): 490-495.

Tanabe A, Naruse M, Naruse K, Yoshimoto T, Tanaka M, Mishina N, Imaki T, Sugaya K, Miyazaki H, Demura H (1999). Angiotensin II receptor subtype in human adrenal glands. Nippon. Rinsho, 57(5): 1042-1048.

Wagner C, Kurtz A (1998). Positive feedback regulation of angiotensin II-AT1B receptor gene expression in rat adrenal glands. Pflugers. Arch., 436(3): 323-328.

Wang DH, Qiu J, Hu Z (1998). Differential regulation of angiotensin II receptor subtypes in the adrenal gland: Role of aldosterone. Hypertension, 32: 65-70.

Wong PC, Hart SD, Zaspel AM, Chiu AT, Ardecky RJ, Smith RD, Timmermans PB (1990). Functional studies of nonpeptide angiotensin II receptor subtype-specific ligands: DuP 753 (AII-1) and PD123177 (AII-2). J. Pharmacol. Exp. Ther., 255: 584-592.

Zhuo J, MacGregor DP, Mendelsohn FAO (1996). Comparative distribution of angiotensin II receptor subtypes in mammalian adrenal glands. In "Adrenal glands, vascular system and hypertension" Ed by GP Vinson, DC Anderson, Ltd, Bristol, UK, J. Endocrinol., pp. 53-68.

Permissions

All chapters in this book were first published in JCAB, by Academic Journals; hereby published with permission under the Creative Commons Attribution License or equivalent. Every chapter published in this book has been scrutinized by our experts. Their significance has been extensively debated. The topics covered herein carry significant findings which will fuel the growth of the discipline. They may even be implemented as practical applications or may be referred to as a beginning point for another development.

The contributors of this book come from diverse backgrounds, making this book a truly international effort. This book will bring forth new frontiers with its revolutionizing research information and detailed analysis of the nascent developments around the world.

We would like to thank all the contributing authors for lending their expertise to make the book truly unique. They have played a crucial role in the development of this book. Without their invaluable contributions this book wouldn't have been possible. They have made vital efforts to compile up to date information on the varied aspects of this subject to make this book a valuable addition to the collection of many professionals and students.

This book was conceptualized with the vision of imparting up-to-date information and advanced data in this field. To ensure the same, a matchless editorial board was set up. Every individual on the board went through rigorous rounds of assessment to prove their worth. After which they invested a large part of their time researching and compiling the most relevant data for our readers.

The editorial board has been involved in producing this book since its inception. They have spent rigorous hours researching and exploring the diverse topics which have resulted in the successful publishing of this book. They have passed on their knowledge of decades through this book. To expedite this challenging task, the publisher supported the team at every step. A small team of assistant editors was also appointed to further simplify the editing procedure and attain best results for the readers.

Apart from the editorial board, the designing team has also invested a significant amount of their time in understanding the subject and creating the most relevant covers. They scrutinized every image to scout for the most suitable representation of the subject and create an appropriate cover for the book.

The publishing team has been an ardent support to the editorial, designing and production team. Their endless efforts to recruit the best for this project, has resulted in the accomplishment of this book. They are a veteran in the field of academics and their pool of knowledge is as vast as their experience in printing. Their expertise and guidance has proved useful at every step. Their uncompromising quality standards have made this book an exceptional effort. Their encouragement from time to time has been an inspiration for everyone.

The publisher and the editorial board hope that this book will prove to be a valuable piece of knowledge for researchers, students, practitioners and scholars across the globe.

List of Contributors

L. O. Odokuma
University of Port-Harcourt, Port-Harcourt, Rivers State, Nigeria

E. Akponah
Delta State University, Abraka, Delta State, Nigeria

Arash Esfandiari
Deparment of Anatomical Sciences, School of Veterinary Medicine. Islamic Azad University, Kazerun Branch, Kazerun, Iran
School of Veterinary Medicine, Islamic Azad University, Kazerun Branch, Iran

Reza Dehghani
Deparment of Anatomical Sciences, School of Veterinary Medicine. Islamic Azad University, Kazerun Branch, Kazerun, Iran
School of Veterinary Medicine, Islamic Azad University, Kazerun Branch, Iran

I. S. Akande
Department of Biochemistry, Faculty of Basic Medical Sciences, College of Medicine, University of Lagos, P. M. B. 12003, Idi Araba, Lagos, Nigeria

A. A. Oseni
Department of Biochemistry, Faculty of Basic Medical Sciences, College of Medicine, University of Lagos, P. M. B. 12003, Idi Araba, Lagos, Nigeria

O. A. Biobaku
Department of Biochemistry, Faculty of Basic Medical Sciences, College of Medicine, University of Lagos, P. M. B. 12003, Idi Araba, Lagos, Nigeria

Huthail Najib
Department of Animal and Fish Production, College of Agricultural Sciences and Food, King Faisal University, Al-Hofuf 31982, Saudi Arabia

Yousef M. Al-Yousef
Department of Animal and Fish Production, College of Agricultural Sciences and Food, King Faisal University, Al-Hofuf 31982, Saudi Arabia

B. Mahmoudi
Islamic Azad University, Meshkinshahr Branch, Meshkinshahr, Ardabil, Iran

M. Sh. Babayev
Department of genetic, Faculty of Biology, Baku State University, Baku, Azerbijan

F. Hayeri Khiavi
Islamic Azad University, Meshkinshahr Branch, Meshkinshahr, Ardabil, Iran

A. Pourhosein
Islamic Azad University, Meshkinshahr Branch, Meshkinshahr, Ardabil, Iran

M. Daliri
Department of Animal Science, Genetic Engineering and Biotechnology Institute, Tehran, Iran

A. O. Eweka
Department of Anatomy, School of Basic Medical Sciences, College of Medical Sciences, University of Benin, Benin City, Edo State, Nigeria

A. B. Eweka
School of Nursing, University of Benin Teaching Hospital, Benin City, Edo State, Nigeria

Fatemeh Abedini
Institute of Bioscience, Faculty of Medicine, University of Putra Malaysia, 43400 UPM Serdang, Selangor Darul Ehsan, Malaysia

Maznah Ismail
Institute of Bioscience, Faculty of Medicine, University of Putra Malaysia, 43400 UPM Serdang, Selangor Darul Ehsan, Malaysia
Faculty of Medicine, University of Putra Malaysia, 43400 UPM Serdang, Selangor Darul Ehsan, Malaysia

Hossein Hosseinkhani
School of Biomedical Engineering, National Yang Ming University, Taipei 112, Taiwan

Tengku azmi
Institute of Bioscience, Faculty of Medicine, University of Putra Malaysia, 43400 UPM Serdang, Selangor Darul Ehsan, Malaysia
Faculty of Veterinary, University of Putra Malaysia, 43400 UPM Serdang, Selangor Darul Ehsan, Malaysia

AbdolrahmanOmarb
Institute of Bioscience, Faculty of Medicine, University of Putra Malaysia, 43400 UPM Serdang, Selangor Darul Ehsan, Malaysia
Faculty of Veterinary, University of Putra Malaysia, 43400 UPM Serdang, Selangor Darul Ehsan, Malaysia

Chong PeiPei
Faculty of Medicine, University of Putra Malaysia, 43400 UPM Serdang, Selangor Darul Ehsan, Malaysia

Norsharina Ismail
Institute of Bioscience, Faculty of Medicine, University of Putra Malaysia, 43400 UPM Serdang, Selangor Darul Ehsan, Malaysia

Ira-Yudovin Farber
Department of Medicinal Chemistry and Natural Products, School of Pharmacy, the Hebrew University-Hadassah Medical School, Jerusalem, Israel

Abraham J. Domb
Department of Medicinal Chemistry and Natural Products, School of Pharmacy, the Hebrew University-Hadassah Medical School, Jerusalem, Israel

Pham Van Phuc
Laboratory of Stem cell Research and Application, University of Science, VNU-HCM, Vietnam

Tran Thi Thanh Khuong
Laboratory of Stem cell Research and Application, University of Science, VNU-HCM, Vietnam

Le Van Dong
Military Medical University, Ha Noi, Vietnam

Truong Dinh Kiet
University of Medicine - Pharmacy, HCM city, Vietnam

Tran Tung Giang
The University of New South Wales, Sydney, Australia

Phan Kim Ngoc
Laboratory of Stem cell Research and Application, University of Science, VNU-HCM, Vietnam

A. Ashayerizadeh
Department of Animal Science, Ramin Agricultural and Natural Resources University, Ahvaz, Iran

N. Dabiri
Department of Animal Science, Ramin Agricultural and Natural Resources University, Ahvaz, Iran
Faculty of Agricultural, Animal Science Department, Islamic Azad University, Karaj Branch, Karaj, Iran

K. H. Mirzadeh
Department of Animal Science, Ramin Agricultural and Natural Resources University, Ahvaz, Iran

M. R. Ghorbani
Department of Animal Science, Ramin Agricultural and Natural Resources University, Ahvaz, Iran

R. Salamatdoust Nobar
Department of Animal Science, Islamic Azad University, Shabestar Branch, Shabestar, Iran

A. Gorbani
1Department of Animal Science, Islamic Azad University, Shabestar Branch, Shabestar, Iran

K. Nazeradl
Department of Animal Science, Islamic Azad University, Shabestar Branch, Shabestar, Iran

A. Ayazi
Agriculture and Natural Resources Animal Science Department, East Azerbaijan Research Center, Tabriz, Iran

A. Hamidiyan
Agriculture and Natural Resources Animal Science Department, East Azerbaijan Research Center, Tabriz, Iran

A. Fani
Agriculture and Natural Resources Animal Science Department, East Azerbaijan Research Center, Tabriz, Iran

H. Aghdam Shahryar
Department of Animal Science, Islamic Azad University, Shabestar Branch, Shabestar, Iran

J. Giyasi ghaleh kandi
Department of Animal Science, Islamic Azad University, Shabestar Branch, Shabestar, Iran

V. Ebrahim Zadeh Attari
Department of Biochemistry and Nutrition, Tabriz University of Medical Science, Tabriz, Iran

E. A. Dwomoh
Cocoa Research Institute of Ghana, P. O. Box 8, New Tafo-Akim, Ghana

S. K. Ahadzie
Cocoa Research Institute of Ghana, P. O. Box 8, New Tafo-Akim, Ghana

G. A. Somuah
Cocoa Research Institute of Ghana, P. O. Box 8, New Tafo-Akim, Ghana

A. D. Amenga
Jaman South District Directorate, Ministry of Food and Agriculture, Jaman South, Ghana

Djoko Winarso
Faculty of Veterinary Medicine of Brawijaya University, Indonesia

Budi Purwo
Extension Agriculture College, Magelang, Indonesia

Y. Rina Kusuma
Extension Agriculture College, Magelang, Indonesia

O. M. Ogundele
Trinitron Biotech LTD, Science and Technology Complex, Sheda, Abuja, Nigeria

B. U. Enuaibe
University of Ilorin, Department of Anatomy, Ilorin. Nigeria

J. Igwe
National Hospital, Department of Histopathology, Central Area, Abuja, Nigeria

A. A. Olu-Bolaji
Department of Epidemiology and Community Health, University of Ilorin, Nigeria

E. A. Caxton-Martins
University of Ilorin, Department of Anatomy, Ilorin. Nigeria

May M. EL Rehima
Ministry of Agriculture and Animal Resources, Khartoum State, Sudan

Atif E. Abdelgadir
Department of Preventive Medicine and Veterinary Public Health, Faculty of Veterinary Medicine, University of Khartoum, Sudan

Khitmat H. ELMalik
Department of Preventive Medicine and Veterinary Public Health, Faculty of Veterinary Medicine, University of Khartoum, Sudan

A. W. Alhassan
Department of Human Physiology, Faculty of Medicine, Ahmadu Bello University Zaria, Kaduna State, Nigeria

A. Y. Adenkola
Department of Physiology and Pharmacology, College of Veterinary Medicine, University of Agriculture, Makurdi, Benue State, Nigeria

A. Yusuf
Department of Human Physiology, Faculty of Medicine, Ahmadu Bello University Zaria, Kaduna State, Nigeria

Z. M. Bauchi
Department of Human Anatomy, Faculty of Medicine, Ahmadu Bello University, Zaria, Kaduna State, Nigeria

M. I. Saleh
Department of Human Physiology, Faculty of Medicine, Ahmadu Bello University Zaria, Kaduna State, Nigeria

V. I. Ochigbo
Department of Physiology and Pharmacology, College of Veterinary Medicine, University of Agriculture, Makurdi, Benue State, Nigeria

Bharucha Bhavna
Division of Avian Biology, Department of Zoology, Faculty of Science, M. S. University of Baroda, Vadodara, Gujarat, India

Padate Geeta
Division of Avian Biology, Department of Zoology, Faculty of Science, M. S. University of Baroda, Vadodara, Gujarat, India

O. M. Ogundele
Science and Technology Complex, Trinitron Biotech LTD, P. M. B. 186, Garki, Abuja, Nigeria

E. A. Caxton-Martins
Department of Anatomy, University of Ilorin, Ilorin, Nigeria

O. K. Ghazal
Unilorin Stem Cell Research Laboratory, Ilorin, Nigeria

O. R. Jimoh
Department of Anatomy, University of Ilorin, Ilorin, Nigeria

Safieldin A. Mohammed
Federal Ministry of Animal Resources and Fisheries, P. O. Box 293, Khartoum, Sudan

Gadir E. Atif Abdel
Faculty of Veterinary Medicine, University of Khartoum, P. O. Box 32, Khartoum North, Sudan

Elmalik H. Khitma
Faculty of Veterinary Medicine, University of Khartoum, P. O. Box 32, Khartoum North, Sudan

I. A. Umar
Department of Biochemistry, Ahmadu Bello University, Zaria, Kaduna State, Nigeria

M. A. Ibrahim
Department of Biochemistry, Ahmadu Bello University, Zaria, Kaduna State, Nigeria

N. A. Fari
Department of Biochemistry, Ahmadu Bello University, Zaria, Kaduna State, Nigeria

S. Isah
Department of Biochemistry, Ahmadu Bello University, Zaria, Kaduna State, Nigeria

D. A. Balogun
Department of Biochemistry, Ahmadu Bello University, Zaria, Kaduna State, Nigeria

Awa Ndiaye
Faculty of Science and Technology, University Cheikh Anta Diop, P. O. Box 5005 Dakar, Senegal
Horticultural Development Centre CDH/ ISRA Camberene Dakar Senegal, Box 3120, Senegal

Mbacké Sembène
Faculty of Science and Technology, University Cheikh Anta Diop, P. O. Box 5005 Dakar, Senegal

Fathi Mokhtar
Department of Animal Science, Islamic Azad University, Shabestar Branch, Iran

Nazer adl Kambiz
Department of Animal Science, Islamic Azad University, Shabestar Branch, Iran

Ebrahim Nezhad Yahya
Department of Animal Science, Islamic Azad University, Shabestar Branch, Iran

Aghdam Shahryar Habib
Department of Animal Science, Islamic Azad University, Shabestar Branch, Iran

Daneshyar Mohsen
Department of Animal Science, Urmia University, Iran

Tanha Taimor
Department of Animal Science, Islamic Azad University, Shabestar Branch, Iran

Mukaram Shikara
Biotechnology Division, Department of Applied Sciences, University of Technology, Baghdad, Iraq

Sanjoy Kumar Kabiraj
Reproductive Biotechnology Laboratory, Department of Animal Breeding and Genetics, Bangladesh Agricultural University, Mymensingh-2202, Bangladesh

S. A. Masudul Hoque
Department of Animal Breeding and Genetics, Bangabandhu Sheikh Mujibur Rahman Agricultural University, Gazipur-1706, Bangladesh

M. A. M. Yahia Khandoker
Department of Animal Breeding and Genetics, Bangladesh Agricultural University, Mymensingh-2202, Bangladesh

Syed Sakhawat Husain
Department of Animal Breeding and Genetics, Bangladesh Agricultural University, Mymensingh-2202, Bangladesh Patuakhali Science and Technology University, Dumki, Patuakhali-8602, Bangladesh

AdilM. A. Salman
Department of Veterinary Preventive Medicine and Veterinary Public Health, College of Veterinary Science, University of Bahr El Ghazal, Sudan

Hind A. Elnasri
Department of Biochemistry, College of Veterinary Science, University of Bahr El Ghazal, Sudan

Payam Ghasemi Dehkordi
Biotechnology Research Center, Islamic Azad University, Shahrekord Branch, Shahrekord, Iran

Abbas Doosti
Biotechnology Research Center, Islamic Azad University, Shahrekord Branch, Shahrekord, Iran

M. Paulpandi
Proteomics and Molecular Cell Physiology Laboratory, Department of Zoology, School of Life Sciences, Bharathiar University, Coimbatore – 641 046, TN, India

R. Thangam
King Institute of Preventive Medicine and Research, Department of Virology, Chennai-600032, India

P. Gunasekaran
King Institute of Preventive Medicine and Research, Department of Virology, Chennai-600032, India

S. Kannan
Proteomics and Molecular Cell Physiology Laboratory, Department of Zoology, School of Life Sciences, Bharathiar University, Coimbatore – 641 046, TN, India

K. Al-Qattan
Department of Biological Sciences, Faculty of Science, Kuwait University, P. O. Box 5969, Safat 13060, Kuwait

M. H. Mansour
Department of Biological Sciences, Faculty of Science, Kuwait University, P. O. Box 5969, Safat 13060, Kuwait

M. Al-Naser
Department of Biological Sciences, Faculty of Science, Kuwait University, P. O. Box 5969, Safat 13060, Kuwait